HEAT PUMPS AND THEIR CONTRIBUTION TO ENERGY CONSERVATION

NATO ADVANCED STUDY INSTITUTES SERIES

Proceedings of the Advanced Study Institute Programme, which aims at the dissemination of advanced knowledge and the formation of contacts among scientists from different countries.

The series is published by an international board of publishers in conjunction with NATO Scientific Affairs Division

A	Life Sciences	Plenum Publishing Corporation
B	Physics	London and New York
C	Mathematical and Physical Sciences	D. Reidel Publishing Company Dordrecht and Boston
D	Behavioural and Social Sciences	Sijthoff International Publishing Company Leyden, The Neth. and Reading, Mass., USA
E	Applied Sciences	Noordhoff International Publishing Leyden, The Neth. and Reading, Mass., USA

Series E: Applied Science - No. 15

HEAT PUMPS AND THEIR CONTRIBUTION TO ENERGY CONSERVATION

edited by

E. CAMATINI
professor of advanced technology
Department of Mechanical Engineering
Polytechnic Institute of Milan, Italy

and

T. KESTER
doctor of physical sciences
Scientific Affairs Division
NATO, Brussels

NOORDHOFF - LEYDEN - 1976

Proceedings of the NATO Advanced Study Institute on
Heat Pumps and Their Contribution to Energy Conservation,
Les Arcs, France, June 16-27, 1975.

ISBN 978-94-011-7573-9   ISBN 978-94-011-7571-5 (eBook)
DOI 10.1007/978-94-011-7571-5

PREFACE

This book is the direct result of a NATO Advanced Study Institute on "Heat Pumps" held at Les Arcs, France, June 1975, under the directorship of Professor E. Camatini, Politecnico di Milano, Italy.

Leading world experts in the field were invited to attend this Institute and to lecture on the heat pump and its potential applications in the broader context of energy conservation. The lectures focussed on today's energy problems, fundamental principles of thermodynamics and their application to the heat pump, potential heat sources for heat pumps, alternate heat pump drives, heat transport and special applications in some areas where considerable savings could be obtained.

When the Institute was planned, publication of the lectures in the form of a book was not envisaged. It was felt, however, during a round-table discussion with the lecturers that there was a definite need for publication of the lectures presented, since a comprehensive scientific/technical book on the heat pump seemingly did not exist. The resulting volume might not provide the reader with a full picture of the heat pump in the sense that the technical design of individual components and existing large installations of heat pumps are not addressed in detail.

It is hoped, however, that this publication which puts more emphasis on basic scientific aspects, design criteria and systems considerations including economical aspects meets its goal, namely to provide a set of lectures which are representative of the state-of-the-art and to draw the attention of many - scientists, technologists, planners and architects alike - to the heat pump and its potential to contribute to a better management of energy.

The late decision to publish the proceedings of this Advanced Study Institute, the timeliness of the meeting and the consequent need to prepare the texts of the lectures speedily put a heavy burden on many:

The editors wish to thank the lecturers for their excellent collaboration. We are grateful to the Head of the NATO Conference Service, Mr. J. R. E. Dobson, the Head of the Typing Section, Mrs. E. Adamson, and the Head of the Graphics Service, Mr. L. Blinn, who greatly facilitated our task in providing the necessary assistance. Many of the illustrations were redrawn in the Graphic Service and all contributions were retyped on camera-ready sheets by Miss Linda Minshull, who did this difficult work competently and with enthusiasm. Particular thanks are due to Miss Judy van Stekelenburg of the Scientific Affairs Division, NATO, for her dedication and diligence in the co-ordination of these various tasks.

The authorities of the Politecnico di Milano, in particular its Rector, Prof. L. Dadda, co-operated in the preparation of the meeting. We are thankful to Dr. M. di Lullo of the Scientific Affairs Division, NATO, Dr. E. Camerini and Prof. E. Sacchi of the Politecnico di Milano, for their advice concerning the outline of the scientific programme presented at the Advanced Study Institute.

E. CAMATINI
T. KESTER

TABLE OF CONTENTS

VIII

OPENING REMARKS

T. Kester

Advanced Study Institutes Programme,
NATO Scientific Affairs Division

Professor Camatini, Ladies and Gentlemen,

On behalf of the Assistant Secretary General for Scientific
and Environmental Affairs and the NATO Science Committee,
I should like to welcome you to this Advanced Study Institute.
This meeting is organized by Professor Camatini of the Technical
University of Milan under the NATO ASI Programme, the aim of
which is to contribute to the dissemination of advanced knowledge
in almost all fields of science and to create favourable conditions
for scientists who can help solve problems of our society in a
peaceful and constructive way.

In most of the industrially-developed countries, the early
sixties marked a period during which a shift took place in the use
of primary energy resources. Gradually more emphasis was put
on petroleum which, at that time, was promised to be available
in the future in abundance and at low prices.

As a result of the low prices practised during the past years
and the absence of a long-term strategy in the area of energy and
energy supplies, dependence on imported energy was allowed to
grow drastically. At the same time, energy losses were taken
into consideration only in those cases where their prevention was
economically feasible. Long-term supply was not considered and
the efforts to develop alternative sources of energy were therefore inadequate.

The correlation between gross national product, welfare and energy consumption of a society is well-established and applies to developed and less well-developed countries. It was for this reason that for many years in the past, proud announcements were made that the energy requirements or production in a particular country had doubled over a period of ten years or so.

The new policy of the oil-producing countries has forced the industrialised countries to take a more realistic attitude in formulating their energy policies. It has now been realized that it is of vital importance to reduce the growth rates for consumption of primary energy while, at the same time, trying to maintain the standard of living or even improving the quality of life. It is now recognized that if current practices in the use of energy were to continue, they would result not only in an early depletion of fossil fuels but also in further degradation of the environment.

Education of the public might help to reduce what might be qualified as waste of energy. The alleviation of taxation on energy-saving investments might have an immediate and long-lasting effect but would not be sufficient to solve the problem of future energy supply. Therefore, promotion of scientific research in these fields, new energy resources and energy conservation, which were badly neglected so far, might be the only way to contribute to the solution of the energy problem. New energy resources, i.e. the fusion reactor, the breeding reactor, geothermal and solar energy, are at present either undeveloped or not applicable on a large scale. For the time being, therefore, energy conservation seems to be the only appropriate and immediate means to reduce the unacceptable growth rates in consumption of primary energy resources.

The aim should be to reduce the high percentage of primary energy use for domestic purposes including "space heating" which in the European Communities amounts to about one third of the energy demand.

The "heat pump", which is the subject of this meeting can be used to recover heat losses once all possible steps have been taken to minimize them. In this context, one should also bear in mind that the cleanest form of energy is the energy we save!

You have come here to learn about and to discuss, one aspect of the wide field of energy conservation. You will be concentrating on the heat pump, the physical principle of which has been known for a long time, whereas its practical use on a larger scale has been discouraged to a certain extent, by the availability of cheap fuel.

Depending on the temperature levels of the sources, electrically driven heat pumps have an efficiency of a few hundred per cent, if the heat output is related to the electrical energy used. The efficiency remains in the order of a hundred per cent when the thermal energy produced is related to the primary energy required to generate and transport the electrical energy to drive the heat pump. This suggests that alternate pump drives should be considered and I am pleased to note that some lectures of this ASI are devoted to this problem.

This Advanced Study Institute has been included in our Programme - amongst the 45 other meetings scheduled this year - because we felt that the subject is both timely and promising. The scientific and technical cross-breeding which this meeting will no doubt originate, will be reflected in a publication we plan to issue as an authorative text book. Thus, we are confident that, thanks to your contributions and their dissemination, the development and use of the heat pump will be strongly boosted.

I hope you will find this meeting beneficial and enjoyable. May I extend to you my very best wishes for the success of the meeting and your future activities.

E. Camatini

INTRODUCTION

Changes in rational goals, public attitudes and private life-styles may reduce the rate of growth of our energy consumption, but those who believe we can reduce the total energy consumption fail to take into account three things:

(1) We are going to have a significant increase in population over the next few decades, even if we are successful in our control effort.

(2) The basic physical needs - and hence basic energy demands - of that population will be enormous because we are in the midst of a social revolution that will inevitably raise the standard of living for the world's underprivileged peoples.

(3) Vast amounts of energy - needed by energy-intensive industries - hold the key to saving, not destroying, the environment as we grow to meet the human demands ahead.

The basis for this last claim is that, properly used, energy can create materials that substitute the massive consumption of "natural" materials: that with new technologies - and intelligent, far-sighted planning - it can do so with less impact on less land; and that it can be used to conserve vast quantities of natural resources while allowing us to return to nature a minimum of waste in its most acceptable form. Much of this has to do, of course, with recycling. In connection with the above, the use of low-temperature heat sources and heat pumps at differentiated temperature levels, and possibility of large-scale applications, is well worth considering as a potential contribution to facing the energy crisis and demand.

A heat pump, like a refrigeration unit, consists basically of a compressor, an evaporator, a condenser and the necessary controls. The compressor is the key component in a heat pump system, its main function being to pump a refrigerant vapour from a relatively low suction pressure to a higher head pressure. The evaporator is placed in the "heat source area", for instance, the outside air. The refrigerant liquid, in boiling and evaporating, draws latent heat from this source. The vapour is "pumped" by the compressor to the condenser, where, again becoming a liquid, it releases heat to the "heat receiving area", e.g. the air inside the building. The energy requirement of the heat pump is, therefore, that required to operate the compressor. It will depend on the pressure difference between the evaporator and condenser, and thus on the difference in temperature and the thermodynamic properties of the refrigerant selected. The coefficient of performance (COP) of a heat pump is the ratio of the heat energy obtained to the electrical energy required to run the compressor. The heat energy obtained is equal to the temperature difference between inlet and outlet temperatures of the condenser cooling fluid (e.g. air), multiplied by its mass flow and specific heat. The electrical energy requirement is obtained by placing a wattmeter on the electrical supply line. Heat pumps are usually classified according to the heat source mentioned first, followed by the heat receiver used. The most commonly used are: air/air, water/air, ground/air, air/water, water/ water, ground/water.

Recognition of the shortcoming of the heat pumps of the early 1960s has resulted in redesign - especially the compressor. Significantly, the availability of a new generation of heat pumps comes at a time when electric rates are rising, and wise use of electricity is now a nationwide goal in most industrialised countries. If then the heat pump is, in large measure, the answer to today's problems, why is it not an overnight national success? Why does there seem to be such a lack of recognition? A brief history is required.

Manufacturers began to produce heat pumps in small quantities in the early 1950s. In those early days it was an immediate success, based on its concept and its promise of efficiency. History can be simplified by stating that the heat pump was produced before it was sufficiently developed; it was oversold, and failed miserably from a reliability standpoint. This memory is vivid for many in the utility and heating and air-conditioning business. In fact, it is so vivid that a review of the technical progress made is necessary to appreciate how much is new in today's heat pump designs. This review can also stimulate an open debate on the future development of the heat pump as far as its application. Actually, at present, there is a conflict in attitudes towards this development. A report recently published by Du Pont de Nemours International in Geneva, on "Market Prospects for the Heat Pump in Europe", gives a summary of arguments for and against the heat pump gathered by interviewing representatives of 25 different companies in 9 European countries.

1. ARGUMENTS AGAINST THE HEAT PUMP

1.1 General arguments

Arguments of a general nature against the future development of the heat pump are:

- Resources of natural gas are plentiful and cheap. (There is thus no need for the heat pump.)

- Market has lack of knowledge of heat pumps.

- Conventional heating system manufacturers are working against any competitive alternatives.

- Installers and contractors are not prepared to deal with heat pumps.

- Maintenance of heat pumps requires specialized personnel.

- Inertia of equipment manufacturers related to the requirements for investment needed to develop and introduce a new product to the market.

- Widespread use of heat pumps will require an increased demand for electricity which the Electricity Boards are not ready to satisfy either by production or distribution.

- The price increase of oil will automatically cause an increase in electricity prices and thus decrease the economic advantages of the heat pump.

- Heat pumps will only be economical in the case of large installations.

-   Initial investment and reduced space requirements
    for the equipment are the most important factors
    to be considered, not long term economy.

-   Investment reduction, when heat pumps are
    considered, can only be achieved when mass-
    produced units are available. This is not possible
    because of different climatic conditions, available
    heat sources, and considerable diversity in local
    construction regulations in Europe.

1. 2 Types of Heat Pumps

The last of the above arguments brings us to the problem
of the choice of type of heat pump.

Temperature differences in Europe make optimization of
the installation difficult when summer cooling is also required.
Only air heating is possible if summer cooling is also required
and the European market is not in favour of air heating. Because
of the above, and possibly other local factors, the heat pump can
only be used in occasional special cases. If the heat pump is
built with heating as its first objective, traditional heating methods
are more reliable especially for bigger installations. If the heat
pump is considered only as a means to recover waste heat, other
cheaper methods can be used.

TECHNICAL ASPECTS OF THE USE OF DIFFERENT TYPES OF HEAT PUMPS

Most heat pumps existing on the European market were
designed for cooling under different climatic conditions than those
usually found in Europe. The result is lack of heating capacity
during cold weather, overloading of the compressor, and risk
of failure. In general, the working conditions of the heat pump
compressor are tougher than in other applications, and
relatively delicate reciprocating compressors can be used
because of their comparatively flat efficiency curves, which make
them effective over a wide range of conditions. There may be
oil miscibility problems when starting up under cold conditions.
Generally more controls are required. The above factors all
call for more maintenance. Particular subjects related to the
different heat sources are:

(a) Air heat source - only workable under temperate climatic
    conditions. Ice build-up requires de-icing, which in turn
    means energy consumption, extra controls and thermal

shock when heating cycle is reversed. Depending on climate, air heat source requires additional electrical heaters during cold periods.

(b) Water heat source - large quantity water supply is necessary. Local regulations usually tend to force economy upon users. Water is sometimes locally expensive. It is not always possible for the heat pump to work as a cooler in summer.

(c) Ground heat source - Heat transfer coefficient is subject to excessive variation. Big investment is required for a large-sized evaporator. During the summer the system cannot work as a cooler. There are also problems in burying and servicing the evaporator.

2. ARGUMENTS FOR THE HEAT PUMP

2.1 General Arguments

Arguments of a general nature for the future development of the heat pump are:

(1) A definite increase in interest and demand for heat pumps has occurred since the start of the energy price crisis.

(2) A definite demand for more efficient systems has been generated.

Heat pumps give a definite energy saving as compared to all-electrical heating. They also compare favourably with other conventional heating methods, taking into account their losses, and the COP of the heat pump.

Government and Electricity Boards' promotion campaigns encourage the public to think in terms of more intelligent use of energy. Diversification of energy sources and future use of nuclear power plants imply a wider use of electricity, and heat pumps will help us to use it more rationally.

The inevitable result of the above factors will be to create a wider knowledge about the heat pump.

The newer generation of architects is more conscious of the necessity for energy saving, and is thus more open to solutions which were considered too sophisticated by the older generation.

2.2 Types of Installation

The secret of an efficient heat pump of any type lies in the correct dimensioning of heat exchanger surfaces. Provided this is done, statements that more maintenance is necessary for heat pumps of any dimension (due to tougher working conditions), are largely unfounded. The necessity for increased maintenance, due to a larger number of controls, has little foundation, as controls required are no more sophisticated than with normal heating and air conditioning systems. When only heat recovery type pumps are considered, maintenance becomes a minor problem as working conditions are more constant. This gives the further advantage that the more sturdy centrifugal compressors can be used. Load variation problems can be overcome by the capacity variation potential of the heat pump, using tiers of reciprocating compressors automatically started up as required. When an air heat source is considered, ice build-up is not a problem as this only occurs for limited periods. It can also be reduced by improving the evaporator design. Ice can, in any case, be eliminated by reversing the cycle and using strip heaters for short periods. The economy is still superior to any other type of "all-electric" heating. Even in northern Europe where heating alone can account for up to 40% of a nation's total energy requirement, heating ability is the major reason for selecting one system rather than another. It has been found that, even here, peak heating is required for only 5% of the total annual heating season, where an air source heat pump is in use. Therefore investment in supplementary heating means (usually electrical heaters) should be minimal. There is no ice build-up when the heat pump takes up heat from expelled waste ventilation air.

Examining the arguments listed above for and against the heat pump, the following points become clear.

GENERAL ARGUMENTS

The fact that natural sources of energy, such as North Sea Gas, are cheap and plentiful, might be true of specific areas of Europe, but certainly not everywhere. It should, in any case, not eliminate the need for energy saving. The general opinion of the market seems to have taken this into account, and this is confirmed by the definitely increased interest in heat pumps. Despite this trend, it seems that one of the reasons the heat pump is not better known on the market may be because it represents something new or out of the ordinary which inspires caution among the more conservative. This can sometimes account for the heat pump being accused of complicated and frequent maintenance. However the European nations' growing concern regarding over-dependence on a few energy sources, is leading them to look elsewhere. As almost all the economically feasible hydro-electric possibilities in Europe are already exhausted, countries look more and more towards nuclear power. This will lead to a wider availability of electricity, and wasteful uses will be a thing of the past. Total electrical heating is and will remain expensive and the heat pump appears, in this respect, to be an excellent future candidate for energy saving. This viewpoint is generally shared by the Electricity Councils, and marketing work by these organizations will most certainly help to increase the use of heat pumps by making them better known. So much for the heat pump as a heating medium. But this type of installation can also be considered as a method to rationalize air conditioning installations in general. Some hostility has been shown towards air conditioning as an unnecessary consumer of energy. In many cases however air conditioning is a technical must. It was admitted by several equipment manufacturers, that many existing installations are over-dimensioned and could be used more rationally. The heat pump is a means of reaching this objective by offering auxiliary heating possibilities in climates where cooling is the primary need; and cooling possibilities in climates where heating is the main objective. The fact that the heat pump can be used as a recuperator of waste heat is, without doubt, a point which will contribute to its future development, as energy becomes more expensive. The interest in heat pumps is increasing as the average outside temperature is decreasing. This is easily explained, as heating, and not cooling, is of primary concern in northern and mid-European countries. The oil crisis made clear that no fuel means no heat, even at a higher price. It is therefore, first of all, an alternative source which is wanted. The heat pump, although using electricity, appears to be an economical solution. As the heat pump concept becomes better known, its use is likely to become more universal, since the market should see a change from the large-size tailor-made heat recovery unit to mass-produced standard air: air unit for small houses. In this field of application a comparison study with traditional gas-furnace heating has recently been made by Westinghouse Electric Corporation: the result of this conclusively favours the heat pump.

TECHNICAL ARGUMENTS

One of the first arguments quoted against the heat pump is the difficulty of optimizing such an installation because of climatic temperature differences.

No agreement with this view is possible. The heat pump must be built to perform its primary function - cooling with additional heating in warm climates, and heating with additional cooling in cold ones. In order to obtain this result, the heat exchanger surfaces must be properly dimensioned. Only then will the working conditions of the compressor be suitable and the failure rate decrease. It is generally admitted that much progress has been made in overcoming the difficulties encountered with the heat pumps of 20 years ago. It is true that load conditions are variable, and this renders the use of centrifugal compressors difficult, except in cases of heat recovery from constant sources. However, it is possible to minimize and overcome the problem of capacity variation requirements by installing a tier of small reciprocating compressors. Also it is possible to minimize ice build-up and de-icing time by an appropriate evaporator construction when air is chosen as the heat source (an air-to-air heat pump is entirely practicable in northern European climates). Concerning the need of inversion for de-icing, this does not involve any complications. Once again it is rather a matter of choosing the right components. Today, the inversion valve itself is no longer a problem, but could be if the other system components are not properly sized, as they may not be able, in the long run, to endure the thermal shock caused by cycle inversion. With the exception of heat recovery heat pumps which are extremely versatile, we tend to agree with people who say that sources other than air are difficult to use. This is not because they create any technical problems as such, but because they are not always available. This is especially the case with water, if one is not fortunate enough to live in the vicinity of a lake or a river or have cooling water in large quantities at one's disposal. Ground heat is of course nearly always available, but the heat transfer coefficient variations due to the nature of the soil and degree of humidity make it difficult to standardize such a system. Furthermore, the uncertainty related to the determination of the heat transfer coefficient calls for a security factor, and consequently oversized evaporators. This in turn means a big investment which is increased by the need to bury the evaporator. But for those who are ready to make this investment, the performance of a ground-to-air or ground-to-water system is undoubtedly excellent because of the small variations of the heat-source temperature which simplifies calculation of the installation requirements.

CONCLUSION

Let us draw some conclusion now. Mainly due to the drive to find alternative energy sources and since heating is a major energy consumer (which is particularly the case in northern

Europe), we believe that the situation has finally reached a point where the heat pump will come back onto the market on a large scale and stay there, primarily as a heating medium and not as an accessory to cooling. To eliminate major air pollution problems in residential areas, only very clean fuel oils are permitted in domestic oil furnaces. These oils are, of course, now in short supply and are becoming steadily more expensive. Two possible solutions are presented: a high efficiency central furnace with heat distribution to neighbouring houses, or a power plant distributing electricity for heating as well as for other purposes. The first method is limited due to the heat losses which would be incurred over anything but the shortest distances; a substantial investment would be required to set up such efficient heat distribution systems; no side benefits other than heating would be obtained. The second method is an improvement from a distribution point of view, but direct heating by means of electricity is wasteful. We are thus confronted with the parallel demands of energy conservation and environmental control, both of which seem to favour central distribution, from an energy conservation and a human point of view. The heat pump can provide a practical solution, by recovering energy from natural or waste-heat sources and converting it into viable domestic heating at only a third of the energy requirements of competitive electric heating. It would be in harmony with today's drive towards diversification of energy resources, and in particular the rational use of nuclear-source electricity production. The growing availability of large-size tailor-made heat recovery units and mass production of standard air-to-air heat pumps could contribute substantially to its growing use in this market. Although most technological problems have been overcome and big improvements made since the first heat pumps were produced, some studies are still to be made especially in the control field where more refining can lead to more economical systems. New compressors can already be found on the market, made especially for heat pumps. Quieter, more compact and lighter, they are economical to run also from a maintenance point of view. The use of proper refrigerant will also contribute to this last point by reducing maintenance and allowing more capacity for smaller installation working with low temperature heat sources. The conversion to a more general use of the heat pump may not happen over night. It will most probably take from 3 to 5 years, at least. This will also allow time for the electricity suppliers to make the required installations, enabling them to satisfy the growing demand resulting from their planned information campaigns. Finally, it is fine to obtain heat rationally, but even better to keep it. The demand for more and improved insulation, and a thorough study of building techniques in order to put insulation material to best use, is a factor which will almost certainly become of prime importance in the future.

The papers presented at the Advanced Study Institute held at Les Arcs by the invited lecturers gave a valuable contribution to focus all the arguments for and against the heat pumps mentioned before, and answer some of the related questions.

ACKNOWLEDGEMENT

This introduction to the volume was written largely using the material and the data supplied in "Market Prospects Heat Pump in Europe", January-April 1974, by Paul Riisager, Freon Products Department, Du Pont de Nemours Int. S.A., Geneva, Switzerland.

THE PRESENT URGENT NEED FOR A BETTER MANAGEMENT OF ENERGY RESOURCES

Arnaldo M. Angelini,

Ente Nazionale per l'Energia Elettrica, Rome, Italy.

1. FOREWORD

The Advanced Study Institute sponsored by the NATO Scientific Affairs Division is devoting this session to a subject of major interest: "Heat Pumps. Their Contribution to Energy Conservation".

As it appears from the programme for the two-week seminar, many reports will deal with the manifold aspects of this theme which, while covering one specific subject, is surprisingly extensive.

The discussion of the variety of applications, developments and technical and economic prospects of heat pumps will be preceded, as called for by the programme, by an introductory report of a general nature, embracing the whole energy sector.

I feel that this was an appropriate decision, since it makes it possible to view in the broader perspective of energy-resource management the role that heat pumps may play in energy conservation.

I was asked to deliver this introductory report; I have welcomed this request, also because heat pumps are one subject I have been closely studying for years and their interesting technological and economic prospects have often induced me to advocate their development, even in years past when the "heat pump" topic had not yet assumed the importance it did after the energy crisis.

Given the general nature of this talk, which deals with the entire energy sector, I shall confine myself to brief references to heat pumps in relation to those aspects of the energy sector where their application appears to be of potential interest; these matters, however, will be discussed in all of the following papers.

The good management of energy resources has been pursued since the beginning of the growth of industrial society. Even in the less recent past, the choices in the energy field have been dictated by a number of criteria - foremost among them, for instance - cost, availability and supply security.

Recently, the relative importance of some of these criteria has undergone a major evolution, prompted first by the increasing importance rightly attributed to the conservation of the environment and then by the oil crises, which has clearly demonstrated the great vulnerability of the energy systems of many industrialized countries.

Today the urgency of a better management, according to new or revised criteria, is strongly felt; essentially, the cost criterion has come to be associated with that of compatibility with the environment and that of a higher degree of source supply security.

A mere listing of these criteria, in its inevitable sketchiness, does not make it possible to illustrate the spirit that pervades them today.

In effect, compatibility with the environment will have to be interpreted in the broader sense, as the choice, among the various possible alternatives, of that which ensures, overall, the best compromise between technical and economic requirements and interactions with the environment.

Supply security may reflect itself, among other things, in the greater emphasis to be placed on the development of indigenous sources, further to an evaluation quantifying the benefit of a higher degree of energy independence.

The cost criterion, finally, will have to be related to an enlarged scope, taking into account, among other things:

- all costs that actually have to be borne, and in particular those connected with environment conservation and interactions or supply security, even when they do not constitute direct burdens on project operators but rather on the community.

- long time spans, affording a guarantee of the soundness of economic choices even beyond the short term, and this both in a prospect looking beyond the needs of the moment and in view of the long time required by many actions in the field of energy, - and

- the irreplaceability of certain energy forms and sources for given applications, such as for instance hydrocarbons in petrochemistry.

These brief remarks, which do not cover the full range of the aims that are now guiding and dictating the energy choices, do give some idea of the complexity of the overall objective of a better management.

In practice, the better management is achieved through:

(a) the rationalization of the use of energy sources, in relation mainly to:

- the availability of primary sources;

- the efficient development of deposits;

- the use in each individual application of the energy source or form that will, with generally acceptable results, involve a smaller total energy consumption, and

- the existence and the development of technically reliable and economically acceptable means of conversion.

(b) energy conservation, through:

- an improvement in the overall utilization efficiency;

- a curtailment of more or less superfluous consumptions and the elimination of wastage;

- the adoption, among the various alternatives sometimes available to achieve certain objectives, of that which, with a generally acceptable result, will require a smaller amount of useful energy (one typical example is the use of telecommunications instead of travelling to attend meetings), and

Table I   Average per capita consumption of primary energy
sources and electricity in some economic areas.

Economic Area	Average per-capita consumption of primary sources			Average per-capita consumption of electricity		
	(tons oil equivalent)		1971/ 1951	(kWh)		1971/ 1951
	1951	1971	Ratio	1951	1971	Ratio
Western Europe	1.60	2.72	1.70	916	3,400	3.72
Africa	0.15	0.24	1.60	81	263	3.24
North America (1)	5.56	7.74	1.39	2,940	8,450	2.88
Latin America	0.33	0.67	2.03	168	612	3.64
Asia:						
- Southern Countries (2)	0.07	0.18	2.56	20	132	6.60
- Japan	0.61	2.29	3.75	562	3,620	6.44
World	0.80	1.35	1.69	418	1,420	3.40

(1) Mexico included in Latin America

(2) Arabia, India, Indonesia and nearby areas.

-   a widespread action of information and education of
    the general public as concerns the need to save energy.

The problem of a better management takes on particular
importance and urgency if we bear in mind that the "quality of
life" and economic and social progress are strictly linked with
the level of energy requirements; in particular, the close connection between per capita energy consumption and per capita
income is universally recognized.

Some indications in this connection may be had from the
Table I, showing the evolution of per capita energy consumptions
in the recent past in a number of economic areas.

It appears evident from the first three columns that in areas
with higher standards of living the per capita energy consumptions
are higher, while they fall in the poorer areas. Thus back in
1951, in North America, 5.6 tons of oil equivalent were consumed
for each citizen in industrial, commercial, household and transport activities; this level has not been reached in Western
Europe, which still is a privileged area. In large areas of Asia,
Africa and Latin America, per capita energy consumptions are
still quite low (in the order of a few hundred kilograms of oil
equivalent a year).

Wide disparities therefore do exist, which have not decreased
with time to the extent that would have been desirable.

In conclusion, the data of the table show that it is reasonable
to expect that, on a world scale, energy requirements, which are
a necessary condition for social and economic progress, will continue to increase in the near future under the pressure of a number
of factors, foremost among which is the achievement of a satisfactory quality of life in increasingly large areas of our earth;
this tendency may only be weakened and to a rather limited
extent, by energy conservation measures.

Now some comments about the last three columns, which
show per capita electricity consumptions.

Since electricity is a particularly valuable form of energy,
such figures show that the quantitative increase in energy consumptions was associated with a significant trend towards a
better quality; for instance, through the world, in 20 years' time
the average per capita primary-source consumptions increased
by about 1.7 times, while those of electricity increased twice
as much, or 3.4 times.

20

Table II   World fossil fuel reserves and production in 1974

Primary source	Reserves	Production in 1974	Years of reserve duration (1)
Coal and lignite (2)	$725 \ 10^9$ tons	$2.4 \ 10^9$ tons	300
Crude oil	$98 \ 10^9$ tons	$2.9 \ 10^9$ tons	34
Natural gas	$72,400 \ 10^9 \ m^3$	$1,300 \ 10^9 \ m^3$	57

(1) At the 1974 production level

(2) Only deposits not less than 30 centimetres thick and no more than 1,200 or 500 meters deep for coal and for lignite respectively.

2.   AVAILABILITY OF AND ACCESS TO PRIMARY SOURCES

The basic elements of the energy problem are the size and distribution of primary-source reserves.

Table II gives an overall picture of the size of fossil fuel reserves, as compared with the latest production levels.

It can be readily seen how plentiful solid fuel reserves are: only those proved in deposits at least one foot thick and located at reasonable depths would last, at the current production rate, a few hundred years (300); regrettably, a large increase in solid fuel production is often opposed by considerable difficulties, such as for instance those of labour recruitment and interactions with the environment, for which no simple and fast solutions are to be seen.

The amounts of the reserves of liquid and gaseous fuels are lower and at 1974 production rates they would last a few decades (34 years for oil and 57 for natural gas); however, if we consider the quick adjustment of these reserves to the increase in requirements that has consistently taken place in the past, through the discovery of new deposits, they do as a whole appear satisfactory.

This table, for the very reason that it shows a world picture, could prove misleading and suggest that an adequate supply of fossil fuels is to be readily had.

As it is well known, the mismatch between geographic location of reserves and consumptions does instead create major difficulties and wide margins of risk.

Solid fuels, which are awkward and often costly to transport, are better suited to massive utilization on site. On the other hand, the quantities available on the international market appear very limited as compared to the overall energy requirements.

Also the access to remote natural gas deposits is made harder by transport problems.

Oil and its products, instead, are easily transportable and this characteristic, combined with what up till recently was the low cost of these products, gave rise to their substantial increase in consumption.

However, serious doubts exist today for oil supply security and its consequences on price levels.

Table III   World uranium resources in some materials.

Material	U contents (grams/ton)	U resources (10^6 tons U_3O_8)	Presumed extraction cost ($/lb of U_3O_8)
Seawater	$3.3 \cdot 10^{-3}$	4,500	100-200?
Granites	2-10	1,000-2,000	200
Uranium-bearing granites	>10	100	50-200
Schists	20-500	20	30-100
Assured and estimated additional resources at less than $ 15/lb	2,000-10,000	3.6	<15

WORLD OIL AND URANIUM RESERVES [*]
AS OF JANUARY 1973

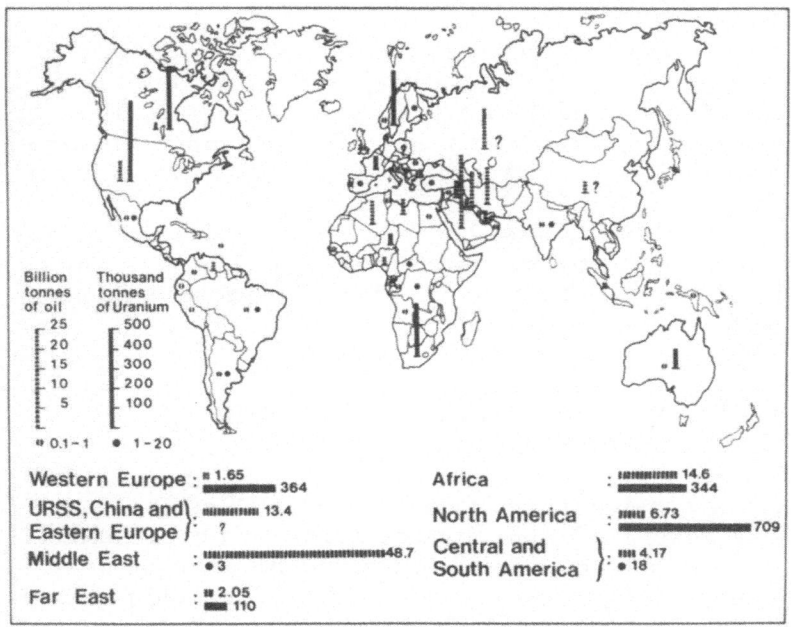

[*] Uranium reserves with extraction cost less than 15 $/lb U_3O_8

Fig. 1.

The fig. 1, showing the distribution of petroleum and uranium reserves, points out one of the reasons for the low security of oil supplies: petroleum reserves are concentrated in some preferential areas, and in particular in those of the Middle East and North Africa. Unlike oil, uranium, which appears distributed in a complementary manner, is extensively present in industrialized nations, with stable political systems, open economies and a large volume of international trade.

The size of uranium resources is shown in the Table III, which indicates assured and estimated resources of uranium workable at reasonable cost, as well as the amounts of this element contained in a number of materials. Uranium, indeed, is widespread in the earth's crust and seawaters and the table shows that its availability, at high but not prohibitive cost levels, is very large. Given the modest level of uranium production, a comparison between resources and present production (19,700 tons

in 1972) has little significance, while it may be more indicative
to mention that the resources at reasonable cost should prove
sufficient to feed the great expansion anticipated for nuclear
production through the end of the century.

Based on this and other considerations, uranium now seems
to afford a high degree of supply security, to the point that in
energy-balance analysis and forecasts it is sometimes equated
with a domestic source, when imported. This assumption is also
warranted by the easiness of transport and storage of this source.

It should be mentioned, however, that very recently measures
have been proposed in certain uranium producing countries which,
if put into effect, could restrict the freedom of the international
uranium market.

3.   THE PRESENT PICTURE AND FUTURE PROSPECTS OF
     ENERGY REQUIREMENTS IN THE EUROPEAN AREA

The following considerations relate mainly to Western
Europe, both because it is of our direct interest and because in
this area the energy problems - due to the volume of requirements and the disparity between energy consumptions and primary
source resources - have assumed a very special importance.
Moreover, the high population density, combined with advanced
industrialization, creates severe constraints in relation to the
conservation of the environment and, as regards the large industrial plants, to land management.

The major components of the present energy picture are
shown in quantitative terms in the chart in figure 2, which portrays a simplified energy balance for the Community of the Nine
and for 1972.

The graph, which develops from the primary sources to
useful energy, highlights:

-    the practical importance and industrial significance of the
     various energy forms and sources;

-    the fact that all primary sources are converted into elec-
     tricity, while fossil fuels only are used in significant
     quantities also for other uses;

-    the fact that the nuclear source becomes available to users
     only through electricity;

Table IV Western Europe - Source production, imports and
requirements in 1972.

Primary source	Production	Net imports	Domestic Requirements [1]
	10^6 tons oil equivalent (2)		
Fossil fuels:			
- coal and lignite	230	27	257
- crude oil	20	664	684
- natural gas	113	5	118
Hydro and geo-thermal (2,400 kcal/kWh)[2]	78	-	78
Nuclear (2,400 kcal/kWh)[2]	16[3]	-	16
Total Western Europe	457	696	1, 153
- European Community	378	554	932

(1) Bunkers not included
(2) 1 ton oil equivalent = 10 million kcal
(3) Assimilated to an indigenous source when imported in view of the
high security of its supply.

26

SIMPLIFIED ENERGY BALANCE

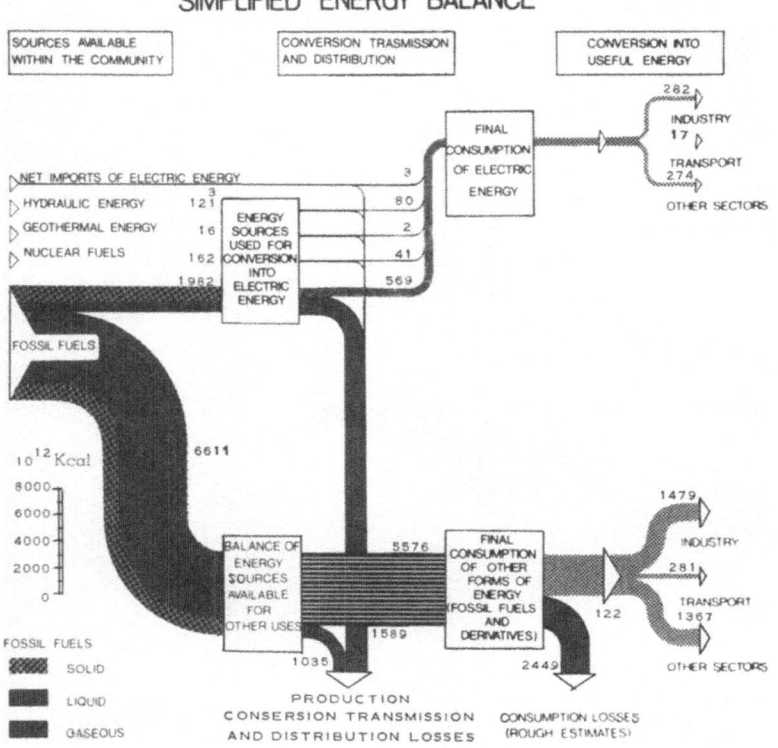

Fig. 2.

- the absolute predominance of liquid fossil fuels (oil and oil products) and

- the amount of the losses in the energy cycle and their breakdown between the two essential stages of production-distribution and of conversion into useful energy.

   It should be noted that the Community's energy requirements account for over 80% of the total requirements of Western Europe, and therefore it can be assumed that the graph does satisfactorily illustrate also the entire European market-economy area.

   As it appears from Table IV, Western Europe meets its primary-source requirements, which in 1972 amounted to more than 1,150 tons of oil equivalent, with a sizeable domestic pro-

WESTERN EUROPE · 1972

DEGREE OF INDEPENDENCE IN THE REQUIREMENTS
OF PRIMARY ENERGY SOURCES AND OIL

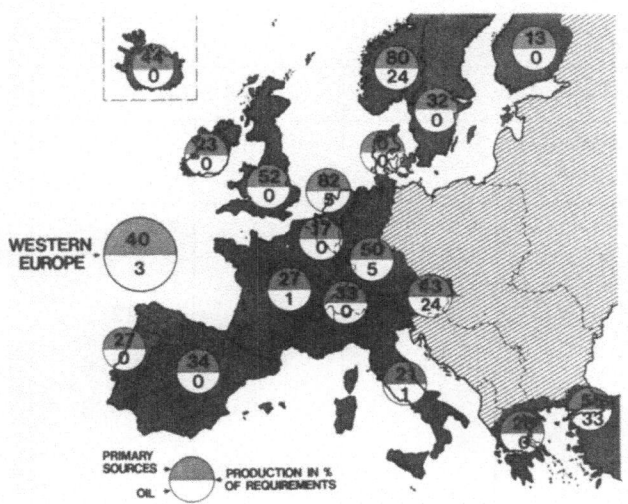

Fig. 3.

duction (over 450 million tons) and, to a greater extent, by
large imports (totalling close to 700 million tons); the table
supplies the figures of domestic production, net imports and
total requirements by sources, and highlights among other
things:

- the nearly total import dependence for oil requirements:
  requirements exceed 680 million tons, as against 20 million
  tons domestically produced;

- the good self-sufficiency in solid fossil fuels and natural
  gas; the domestic production of coal and lignite - which,
  however, is concentrated in the United Kingdom and Western
  Germany - amounts to 230 million tons equivalent, while
  that of natural gas, thanks mainly to the Dutch deposits,
  exceeds 110 million tons oil equivalent, and

- the appreciable, although not major, contribution of the
  hydraulic source: together with the geothermal and nuclear
  it amounts to something less than the equivalent of
  100 million tons of oil.

Table V   Western Europe - Domestic requirements of primary
sources in 1952 and 1972[1].

Primary source	1952		1972	
	10^6 tons oil eq.	%	10^6 tons oil eq.	%
Fossil fuels:				
- coal and lignite	371	79.1	257	22.3
- crude oil [2]	65	13.8	684	59.3
- natural gas	2	0.5	118	10.2
Hydro and geo-thermal(2,400 kcal/kWh/	31	6.6	78	6.8
Nuclear (2,400 kcal/kWh/	-	-	16	1.4
Total	469	100.0	1,153	100.0

(1) All sources converted to tons oil equivalent(one ton equal to 10 million kcal).
(2) Bunkers not included (13 10^6 tons in 1952 and 47 10^6 tons in 1972)

The table provides an average picture, while major differences exist among the various countries. They are depicted by fig. 3, which shows in percentages the degree of independence for all primary sources, and for oil. While European countries as a whole meet four tenths of their requirements with indigenous sources, certain countries like Norway and Holland enjoy a degree of independence of better than 80%, while for others, including Italy, this degree is around 20 to 25 per cent.

The figure also shows the situation of almost total oil deficit which, with few exceptions, affects almost all countries.

The European oil picture, in addition to being characterised by a high degree of import dependence (97% of the oil requirements), is one of substantial dependence on a geographic producing area: the Middle East and North Africa. This appears from the fig. 4, which shows for 1972 the oil import flows in million tons, by areas of origin. The figure also shows, for each consumer country, the percentages of imports from the Middle East and North Africa on total imports: the Arab countries supply 82% of the total European oil requirements; for certain countries, including Italy, imports from this area account for nine tenths of total oil imports.

The picture offered by the preceding tables and figures clearly illustrates some of the reasons of Europe's current vulnerability as concerns oil supplies and prices.

By comparison, I shall mention that other industrialized countries are in a definitely more favourable position. Apart from the USSR, an energy and oil exporting country, also for the United States the degree of dependence is more limited: in 1972 the oil demand amounted to 827 million tons and imports 242 million tons, or 29% of oil consumptions.

Finally, it may be of interest to recall that Western Europe's vulnerability to energy shortage and prices is a recent fact and depends essentially on political reasons and on the great expansion in energy requirements. The figures in Table V, which shows energy consumption by sources in 1952 and 1972, require no comments. I only wish to call attention to oil's share in total requirements, which in 20 years' time has risen from 14% to 59%; I add that the overall energy independence has dropped from 85% in 1952 to 40% in 1972; in the same period, the degree of oil independence declined from 9% to 3%.

Table VI Year 1985 - Estimates of contribution to energy
demand by sources in Western Europe.
(Source: OECD: Energy prospects to 1985 - Paris 1974)

Primary source	1972 (reference)	Estimates for 1985	
		before the crisis	after the crisis[1]
	10^6 tons oil equivalent[2] (percent change:'85/'72)		
Fossil fuels:			
- coal and lignite	257	215 (-16%)	293 (+14%)
- crude oil[3]	731	1,441 (+97%)	966 (+32%)
- natural gas	117	303 (+159%)	435 (+271%)
Hydro and geo-thermal(2,400 kcal/kWh)	78	107 (+37%)	107 (+37%)
Nuclear (2,400 kcal/kWh)	16	235 (+1,370%)	273 (+1,600%)
Total	1,199	2,301 (+92%)	2,074 (+73%)

(1) Crude oil at 9 $ (1972)/bbl

(2) 1 ton oil eq. = 10 million kcal

(3) Bunkers included

OIL IMPORTS IN WESTERN EUROPE IN 1972
(MILLION TONS)

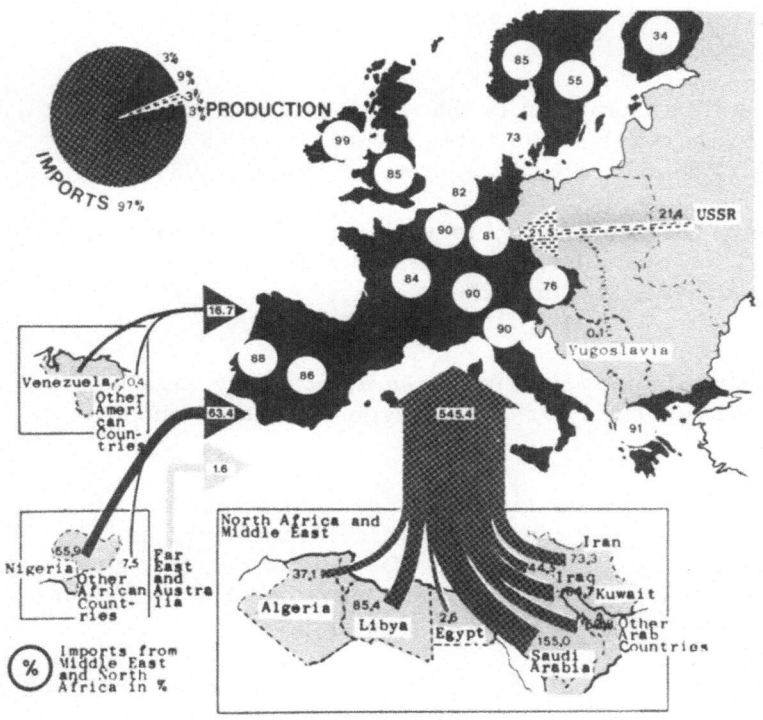

Fig. 4.

And now some mention of the size of the European area's energy requirements in the medium term.

Determining the rates of probable expansion of energy requirements in the various economic areas is a complex exercise in forecasting analysis. It is not within the scope of my talk to go into this subject, a detailed discussion of which would require more time than I am allowed.

I shall confine myself to commenting briefly on the figures in Table VI, taken from a rather recent OECD study. It shows the estimated energy requirements in Western Europe by 1985,

referred to the conditions that were regarded as likely prior to
the crisis, and those that today are expected to prevail. The
assumption underlying these estimates is that requirements will
be determined solely by the mechanisms triggered by price rises
under market conditions, irrespective of specific actions of a
new economic policy.

Another basic assumption is that the gross national product
will continue to increase at the rates forecast before the crisis
(5.1% between 1972 and 1980; 5.2% between 1980 and 1985).

By virtue of the price machinery alone, the estimates in the
table show that the overall energy demand will increase by 73%
between 1972 and 1985 (vs. the 92% estimated at pre-crisis price
levels).

The crude oil demand will increase by 32% only, again over
1972, while the pre-crisis forecasts suggested that it should
almost double.

A large-scale recourse will be had to the nuclear source,
whose contribution will increase by 15 times and more, and by
1985 will amount to over 10%; the OECD experts feel that an
even larger recourse to this source will be limited by the industrial capacity to build nuclear power plants.

The contribution of coal and lignite, which had been declining
for decades, may pick up and increase albeit moderately (+14%).
In Europe, finally, a major role will be assumed by natural gas:
by 1985 its consumption may be 4 times as large as in 1972
(three times in the pre-crisis estimates).

This picture, as I have said, reflects in some sense the
automatic adjustment of the energy market to the pressure of the
new oil prices. Against these values will therefore be confronted
the objectives of a better management which, on the one hand,
will concern primary sources and, on the other, will include
actions of energy consumption guidance and rationalization beyond
and outside those set in motion by the traditional market mechanisms.

4.   THE BETTER MANAGEMENT OF ENERGY SOURCES

Let us now see concretely what the main lines are for
possible actions and how far they can prove justified in terms of
economics, environment and supply security.

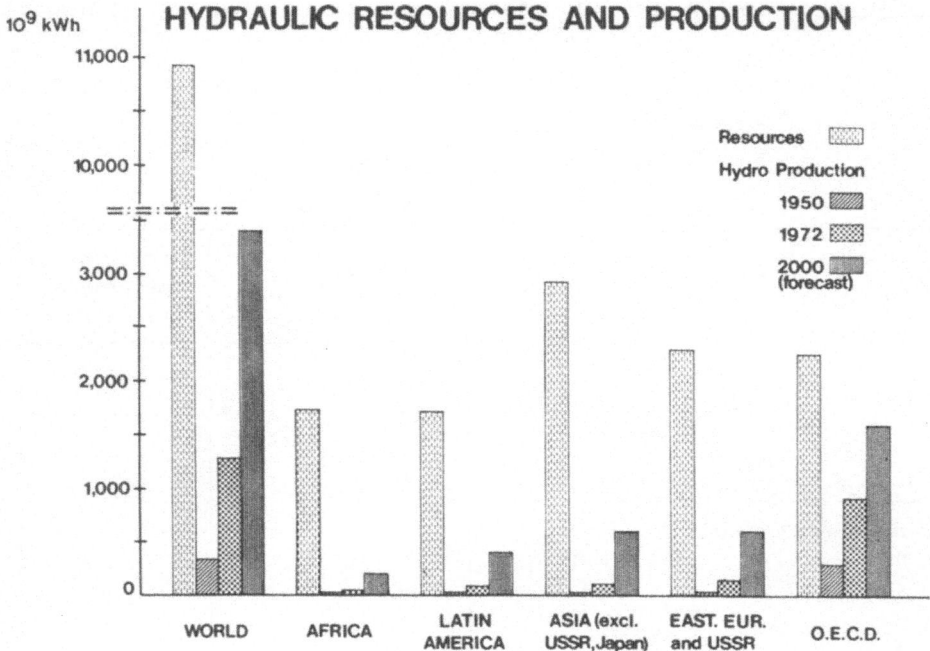

10^9 kWh   **HYDRAULIC RESOURCES AND PRODUCTION**

Resources
Hydro Production
1950
1972
2000 (forecast)

WORLD   AFRICA   LATIN AMERICA   ASIA (excl. USSR, Japan)   EAST. EUR. and USSR   O.E.C.D.

Fig. 5.

The hydraulic source is well compatible with the environment and, as an indigenous and self-renewing source, offers very substantial advantages in supply security and balance of payment terms. It is used industrially for electric power generation, and the picture of the availability of this source and of its past and predictable utilization is shown, on a world scale, in fig. 5. This shows for various geo-economic areas the technically usable resources, totalling about 11,000 billion kWh/year, of which some years ago 4,000 billion kWh/year was regarded as economically usable.

These figures are not high if compared to the total amount of electricity generation (6,000 billion kWh in 1973).

In any event, today the desirability of a fuller utilization of hydraulic resources - although, as shown in the figure, they are located largely in low-industrialization areas - appears unquestionable and was proposed by myself at the last World Energy Conference. On that occasion I recommended certain actions,

Table VII   Hydro-electric resources and production in Europe.

Geo-economic areas	Technically exploitable resources [1]	Hydroelectric production in 1973	Production in percent of resources
	10^9 kWh/year		%
- Western Europe	660	344	52. 1
-- European Community	192	108	56. 2
-- Scandinavian Countries	267	135	50. 6
- Eastern Europe [2]	138	33	23. 9
Total Europe [2]	798	377	47. 2

(1) Exploited and unexploited

(2) USSR not included

albeit quite complex, such as the construction in "symbiosis" of big hydro-electric power plants and factories with a high specific energy consumption (aluminium production, uranium enrichment, etc.) in remote locations.

Actions towards a fuller utilization of hydraulic resources are possible also in the industrialized countries, and in particular in Europe, even though most of the competitive resources in these areas have already been harnessed. Table VII shows those that are in practice the extreme limits for the utilization of these resources and indicates that more than half of them, i.e. almost all those competitive before the crisis, are utilized and deliver annually over 350 billion kWh.

The main lines of these actions are:

- the planning of new projects and the revival of abandoned projects in the light of the new prices and of the general situation in the energy sector;

- the study and implementation of the restructuring of resources already in use towards a higher production, a larger installed capacity and a higher regulation possibility;

- a fuller automation of the existing smaller facilities, which otherwise would have to be deactivated for economic reasons, and

- wherever orographic conditions are favourable, a greater recourse to pumped-storage plants, which with their contribution to electric power regulation, do make a contribution, although an indirect one, to a better management of energy resources, by reducing the capacity required in conventional or nuclear thermal power plants and increasing their utilization, and offer opportunities for a better management of the hydro-electric energy stored in reservoirs.

The possible actions in the area of fossil fuels vary depending on whether they are concerned with solid fuels, crude oil or natural gas.

As for the former, we have noted their steady decline, in spite of their extensive use, where conditions were favourable, for the generation of electric power.

From 1952 through 1972, coal and lignite production in Western Europe has declined by 31% and the pre-crisis OECD

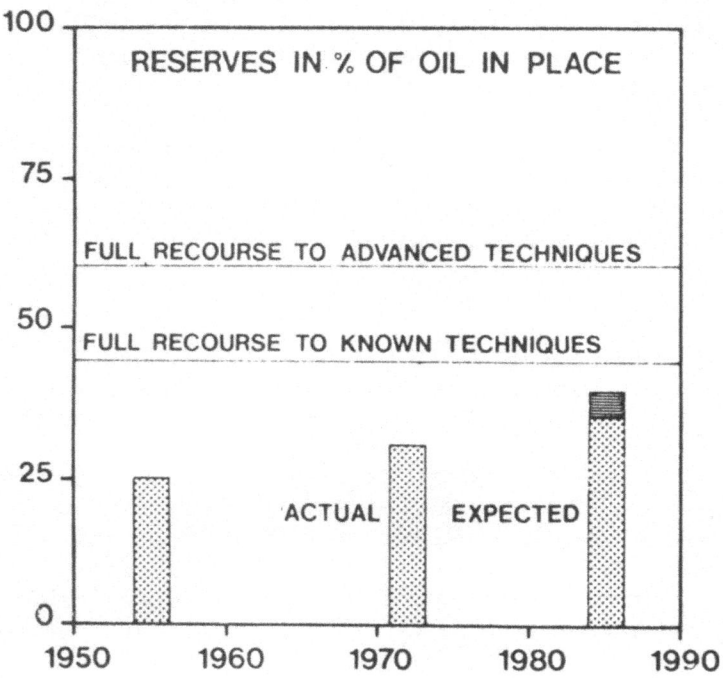

WORLDWIDE AVERAGE RECOVERY FACTOR FOR OIL

Fig. 6.

estimates assumed a further downturn. The dual problem with coal is in the first place that of increasing production from deposits which, even using the latest technologies, is certainly not one of the most socially acceptable activities - particularly for underground deposits - and, in the second place that of new outlets. In this respect no simple solutions are in sight, and the gasification of coal in place, which could solve them both, is a distant prospect.

Of course, in those countries where coal is an indigenous source, the desirability of working deposits even at the present rates and if possible at higher than the present ones, should be carefully considered in a long-term prospect.

As regards oil and natural gas, the actions will concern in the first place the intensification of prospecting for new fields, in particular in the politically more stable areas. Moreover, there is much room for improvement in the working of deposits, as shown by the figure 6, indicating the percentage of oil actually recovered from a deposit: the world average has increased from 25% in the Fifties to the present 30-odd per cent; by the mid-Eighties, extrapolating the present trend, it might be as high as 35% to 37%.

It is believed likely that the energy situation that has arisen from the Kippur war will speed up the recourse to the advanced recovery technologies developed in laboratories or tested on sample fields, so that by 1985 the average resource recovery could be as high as 40%. The figure also shows the maximum recovery rates regarded as possible with the present technologies (45%) and with the more advanced ones (60%).

Natural gas, finally, may benefit from the spreading of the pipeline networks and from the adoption of ships equipped to carry it in the liquid state; it is believed logical that these actions too will be accelerated.

It is from nuclear energy, however, that the greatest benefits are expected in terms of economy, supply security and environment conservation. Today nuclear energy becomes available only through electric power, and was already competitive with fossil fuels before the oil crisis.

Its competitiveness is now largely established; moreover, from the viewpoint of interactions with the environment, the experts feel that nuclear power plants constitute a definite improvement on conventional thermal plants; finally, the supply security is high and nuclear power generation as a substitute for conventional thermal production does involve major balance of payments benefits.

The role that nuclear energy may play in electric power generation in industrialised countries will be a major one and the size of the current programmes suggests that the nuclear source will account for a major portion of the electricity market already in the early Eighties. It will also serve as a major stimulant to a greater "penetration" of electric power. At present the portion of primary energy sources used for conversion into electricity is about 25%, and it should gradually rise until in the year 2000, according to many experts, it will reach or even exceed 50%. Besides the improvement in the competitiveness of electric power that should result from the gradually rising share of nuclear origin, the greater "penetration" will be connected with the in-

Table VIII   Electricity generation development target in the European Community.

| Year | Electricity generation | | | Percent primary source requirements for electricity generation |
| | total | from nuclear sources | | |
	10^9 kWh	10^9 kWh	% of total	
1970	855	41	5	23
1980	1,640	350	21	29
1985	2,400	1,100	45	35
1990	3,650	2,500	68	45
2000	7,000	(5,600)	~ 80	~ 60

creasing preference of the users for this form of energy, due to its features of "flexibility" and "cleanliness" in the use stage and to the general tendency towards a greater reliability, simplification and automation of industrial processes. Moreover, the wider use of electric heating - with respect to which heat pumps could play a basic role - and the advent of electric cars could open up new large markets for electric power.

It appears clear from these considerations that one of the most promising options for a better management of primary sources concerns that group of actions that can promote the penetration of electric power and the use of the nuclear source. This contention is reflected in the European Community's targets for the development of electric power generation shown in Table VIII.

The figures in the table represent the total result of all planned actions, and therefore also including conservation ones; what matters to stress here is:

- the speeding up of the "penetration" of electric power that should rise to 45% by 1990 and then exceed 50% in the early Nineties, and

- the incidence of nuclear power generation on total electricity generation, which will rise from 5% in 1970 to more than one fifth by 1980, to nearly one half by 1985 and then become largely predominant in the Nineties.

A nuclear programme of this magnitude will of course require a firm will from all parties concerned - on the political, administrative and technical levels - and a widespread acceptance by public opinion which should be obtainable through correct and comprehensive information; the difficulties experienced in many countries in the siting of nuclear plants, largely stemming from an alarmist and biased information, do however seriously threaten the possibility of reaching these objectives in the short-medium term.

New reactor types may play an essential role towards a better utilization of the uranium source; I shall mention only sodium breeder reactors, on which research and development efforts are being spent - because of their excellent degree of utilization of uranium s energy content (from 50 to 70 times as much as the reactors now in industrial use). Their introduction, which could take place around 1990 will move ahead in the centuries the life of the reasonable-cost natural uranium resources, now estimated at some decades with the use of the present reactors.

Finally, I shall touch briefly on the subject of the so-called "new" sources and on their predictable role.

Solar energy may find interesting applications in many countries for the heating of water for home uses, for at least partial space heating and, in fewer cases, for the desalting of sea and brackish water. These applications should receive considerable impetus over the next few years; possible actions should be directed both to research and development work on solar collectors suitable for residential uses and on the residential market itself, which due to the nature of the decision-making process that characterises it, lends itself but poorly to the adoption of these applications. In effect, the constructional details of a home are rarely specified by the end user and the party that decides (usually the builder) is particularly sensitive to cost-cutting considerations and sceptical towards innovations aimed mainly at housekeeping problems.

The use of the solar source for the generation of electric power appears very problematic, due both to the discontinuity in time and low density of solar energy and to the very high costs now predictable per kWh generated.

So far the use of the geothermal source has been rather limited and almost exclusively for the generation of electric power at competitive conditions; there are various regions of the world where it appears possible to make more use of endogenous steam from relatively shallow depths. It will be a good thing if exploration work in this sector will be considerably intensified.

It is impossible to estimate the amount of the new geothermal resources that may be discovered, but it appears reasonable to assume that they will not account for more than a rather limited percentage of the electric power generating requirements.

The prospects of the use of geothermal energy could be rather more interesting if competitive methods were developed for the utilization of the heat from the deeper geological strata.

As for the energy from the fusion of light nuclei, it is a well-known fact that it would supply practically unlimited amounts of energy; however, according to the latest estimates, the experts believe that not until the end of the century it will be possible to demonstrate with a prototype of semi-industrial characteristics the possibility of generating electric power from nuclear fusion; hence, fusion energy will necessarily play only a very limited role at least until the year 2000.

5.   BETTER MANAGEMENT THROUGH ENERGY
     CONSERVATION

Energy conservation is understood, in its broadest sense,
as the best and most efficient use of energy, properly taking into
account benefits and possible disadvantages: it would be illogical
to save energy in circumstances that involve a disproportionate
sacrifice of other resources - economic, social or environmental.

From energy conservation measures we expect a contribution
to the solution of a number of energy problems and an improvement in the compatibility of the energy system with the natural
environment.

A general analysis of the possible conservation measures and
of the times involved is presented below, dividing the sectors of
activity into three categories: transport, industry and utilities
and others (household, commercial, etc.).

The chart in figure 7, illustrates for the transport sector
the present situation (right), the main conservation measures
(left) and the estimated possible medium-term savings.

The chart shows the great predominance of car consumptions
(graph to the upper right) and the very sizeable difference in
specific consumption between transport modes (centre and lower
right graphs).

The possible conservation measures are listed on the left
side and are aimed at the following main objectives:

-    meeting the transport demand with energetically more
     efficient systems, some examples being public transport
     for passengers and a greater use of freight trains for long
     hauls;

-    curbing the transport demand:

     (a)   for passengers, through a rational development of the
           urban environments and greater use of communications,
           when acceptable in lieu of travelling, and

     (b)   for freight, through a rationalization of intermodal
           transfers and the use of information systems to co-
           ordinate most effectively freight transport, handling
           and distribution.

ENERGY CONSERVATION

TRANSPORT

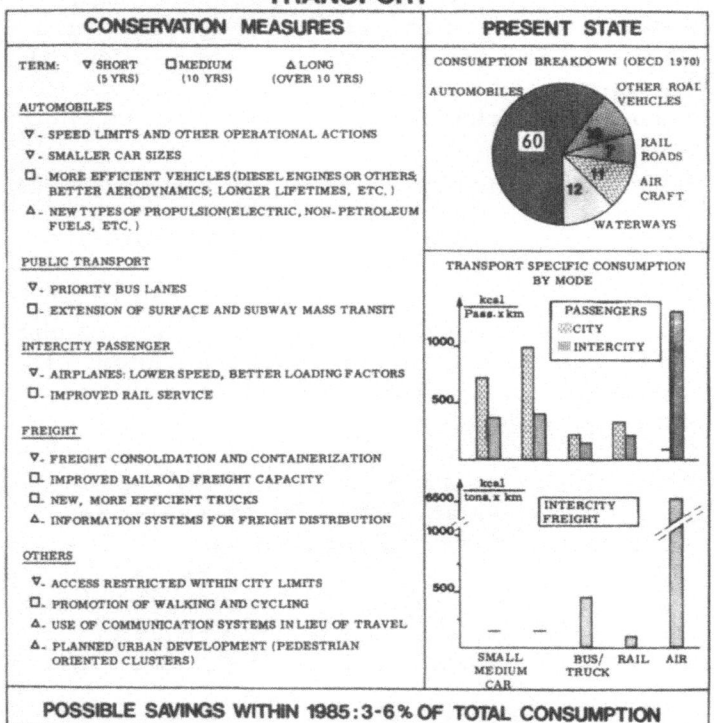

Fig. 7.

In the whole, by 1985 it should be possible to achieve a saving in the order of 3 to 6 per cent in terms of total primary-source requirements; it is important to note, that part of this saving is attributable to price rises which, at least in part, stimulate the taking of several of these actions. Thus, for instance, the energy saving resulting from lower vehicle speeds will result from two almost indistinguishable components: self-restraint on the part of the user, who is interested in saving money through a lower consumption, and speed limits imposed by the authorities.

The chart figure 8 portrays the sector of industries and utilities and is arranged in the same way as that discussed above.

In these sectors, and especially in big industries and in

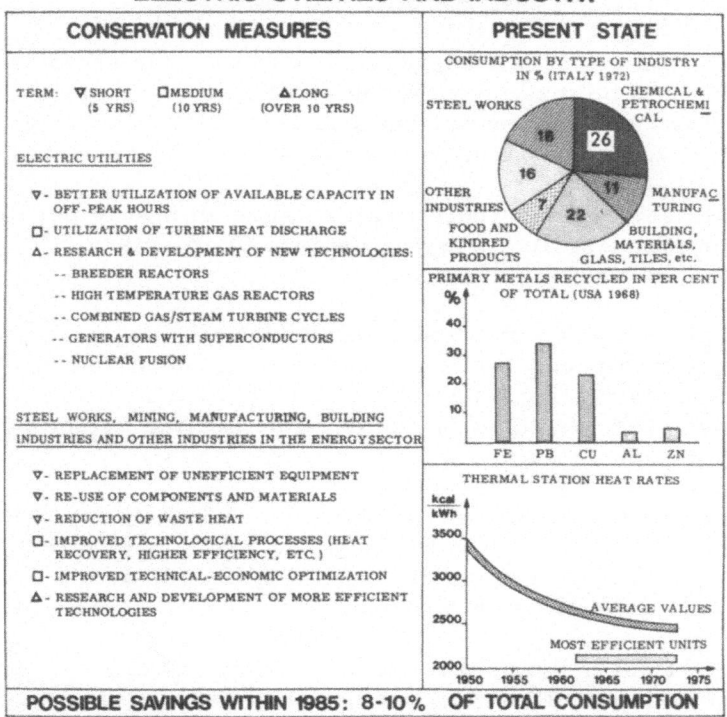

Fig. 8.

electric utilities, the use of energy is already rational. In particular, note should be taken of the consistent improvement in the heat rate of thermo-electric power plants and the narrowness of the remaining margins; in the case of the most modern units, the minimum heat rate possible in practice has been reached in the Sixties.

Margins for savings are expected mainly in the following areas:

- improving the efficiency of the equipment and systems used in industrial processes;

- re-using materials and components;

- developing and adopting new more efficient technologies, and

- utilising waste heat.

By 1985, in terms of overall energy consumption, the sector of end industries could be making a 5% contribution to savings; that of energy industries a somewhat small contribution, counting on electric utilities for a contribution in order of 2 to 3 per cent.

The industrial sector as a whole could achieve an 8 to 10 per cent saving on the foreseeable aggregate demand for primary sources.

The picture of the residential and commercial sector, in figure 9, similar to the preceding ones, shows first (right) the predominance of space heating (upper chart), for which a comparison is made (centre chart) of some of the heating systems now available.

It should be noted that heat pumps, about which I shall say something more later on, are already showing a rather sizeable overall energy advantage, but certainly not a major one, over heating with oil (ca. 15%) and with electricity (ca. 20%).

The lower right chart illustrates the variability of the efficiency of certain types of lamps: sodium ones are now more efficient than incandescent ones, with an efficiency in excess of 50%; the overall gains, however, are small because of the limited incidence of lighting consumptions.

A review of the conservation measures listed on the left side suggests that some savings are possible in this sector even in the very short term. In the medium term, i. e. by 1985, it is estimated that a widespread adoption of the measures listed could result in a 3 to 6 per cent saving on total requirements.

The result of energy conservation measures is presented in summary form in figure 10. By 1985, through conservation, end-user sectors as a whole could cut their demand by 11 to 17 per cent below the amount estimated before the crisis; this percentage saving would rise to 15-20% through the contribution achievable within the area of energy industries.

It should be noted that this is an overall result of the conservation actions in which converge, as mentioned above, the total effect of price rises and that deriving from possible

ENERGY CONSERVATION

RESIDENTIAL / COMMERCIAL AND OTHERS

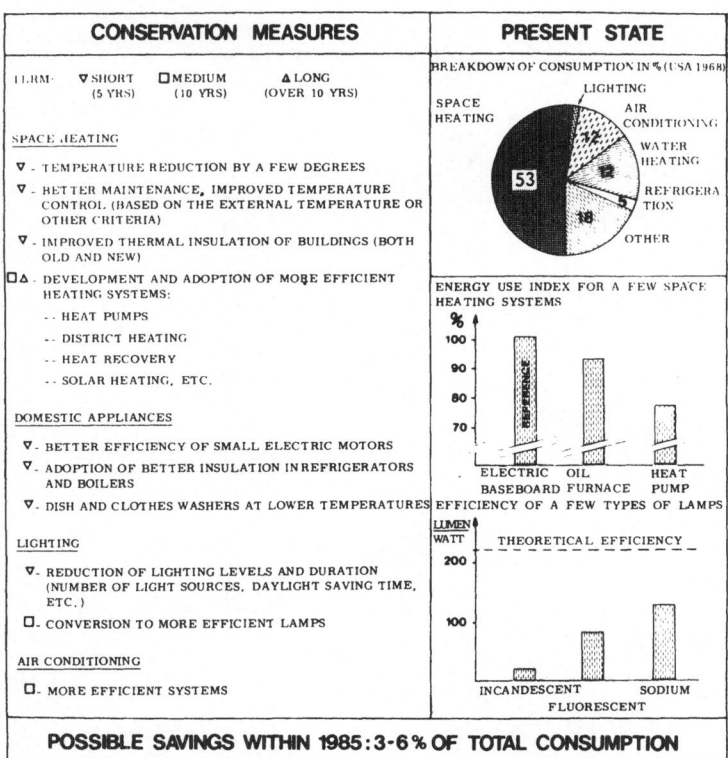

CONSERVATION MEASURES	PRESENT STATE

TERM: ▽ SHORT □ MEDIUM ▲ LONG
(5 YRS) (10 YRS) (OVER 10 YRS)

SPACE HEATING

▽ - TEMPERATURE REDUCTION BY A FEW DEGREES

▽ - BETTER MAINTENANCE, IMPROVED TEMPERATURE
CONTROL (BASED ON THE EXTERNAL TEMPERATURE OR
OTHER CRITERIA)

▽ - IMPROVED THERMAL INSULATION OF BUILDINGS (BOTH
OLD AND NEW)

□△ - DEVELOPMENT AND ADOPTION OF MORE EFFICIENT
HEATING SYSTEMS:

-- HEAT PUMPS

-- DISTRICT HEATING

-- HEAT RECOVERY

-- SOLAR HEATING, ETC.

DOMESTIC APPLIANCES

▽ - BETTER EFFICIENCY OF SMALL ELECTRIC MOTORS

▽ - ADOPTION OF BETTER INSULATION IN REFRIGERATORS
AND BOILERS

▽ - DISH AND CLOTHES WASHERS AT LOWER TEMPERATURES

LIGHTING

▽ - REDUCTION OF LIGHTING LEVELS AND DURATION
(NUMBER OF LIGHT SOURCES, DAYLIGHT SAVING TIME,
ETC.)

□ - CONVERSION TO MORE EFFICIENT LAMPS

AIR CONDITIONING

□ - MORE EFFICIENT SYSTEMS

BREAKDOWN OF CONSUMPTION IN % (USA 1968)

SPACE HEATING — LIGHTING — AIR CONDITIONING — WATER HEATING — REFRIGERATION — OTHER

53 12 18 5

ENERGY USE INDEX FOR A FEW SPACE HEATING SYSTEMS

% 100 90 80 70

ELECTRIC BASEBOARD — OIL FURNACE — HEAT PUMP

EFFICIENCY OF A FEW TYPES OF LAMPS

LUMEN/WATT THEORETICAL EFFICIENCY

200 100

INCANDESCENT — FLUORESCENT — SODIUM

POSSIBLE SAVINGS WITHIN 1985 : 3-6 % OF TOTAL CONSUMPTION

Fig. 9.

energy-policy measures. The latter could yield sizeable savings, as instanced by regulations on residential heat insulation and various forms of encouragement of public transport.

From this brief analysis we infer that a substantial portion of the energy savings concerns oil and its products; they can largely be regarded as reductions in the imports of this raw material.

Here lies one of the greatest benefits of conservation. Another major advantage is the benefit that will accrue to the environment: conservation will bring about a renovation in plants, processes and equipment, individually possessing a greater environmental compatibility. Real benefits to the environment

46

ENERGY CONSERVATION
SHORT AND MEDIUM TERMS

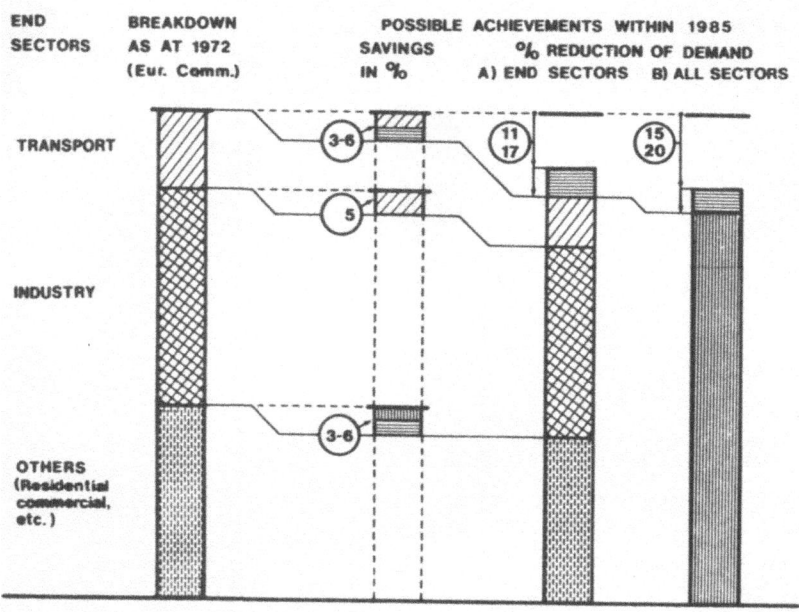

Fig. 10.

will also result from better housekeeping, such as a more careful maintenance of space heating systems.

It should not be forgotten, however, that Operation Conservation will have a cost, and the investments required will have to be co-ordinated with the others, in a logical priority scale.

To conclude these comments on conservation, I wish to take up again briefly the subject of heat pumps, which is potentially one of the most effective tools in a productive conservation programme.

Heat pumps are highly flexible and they can use a wide range of energy sources and different types of motive power to transfer heat from a cold source to a higher temperature level. Today heat pumps have become established mainly in the United States in the field of space heating and air-conditioning; heat pumps in

Table IX   Coefficient of performance of heat pumps for
            central air conditioning and heating.
            (Rating conditions: 7°C outdoors, 21°C indoors)

Heat pump cycles and temperatures	Coeff. of performance (Heating)	Ratios	
a. Carnot cycle source 7°C; sink 21°C	14	-	-
b. Carnot cycle evaporator: -3°C; condenser: 47°C	6.4	b/a	0.46
c. Rankine cycle evaporator: -3°C; condenser: 47°C;	5.0	c/b	0.78
d. Commercial heat pump[^]	2.2 ÷ 3.0	d/c d/b d/a	0.44 - 0.60 0.34 - 0.47 0.16 ÷ 0.21

(^) Losses and aux. consumption accounted for

their commercial versions use electricity for motive power and
employ as sources indoor and outdoor air.

The table IX supplies some elements for an evaluation of
heat pumps from the energy standpoint.

It shows the coefficient of performance, meaning the ratio of
heat delivered indoors to required work, in the heating mode; for
the testing conditions in effect in the United States, case (a) gives
the maximum theoretical coefficient of performance, which is
quite high[14]. If we take into account the practical need for using
an evaporator and a condenser, even in the assumption of an ideal
fluid, the coefficient decreases considerably (6.4); further reductions then occur with the real fluid (case (c)) and with conversion
losses and auxiliary consumptions required in actual operation
(case (d)); the systems now commercially available have a coefficient of performance of 2.2 to 3.0. It should also be recalled
that such data hold true for outdoor temperatures of 7°C. At
lower temperatures the coefficient decreases and this effect,
combined with the existing pumps' capacity decline as outdoor
temperatures drop, makes it difficult to dimension heat pumps:
reconciling the requirements of heating with those of summer
conditioning is a complex job, which calls for a careful choice
of the type of heat pumps to suit local conditions.

In the design and dimensioning process, other very difficult
problems have arisen in the past, and in the United States a pre-
mature marketing, began in the Fifties before a product with
satisfactory performances would be developed, has resulted in a
widespread lack of confidence in this system and in a reduced
sales growth; today, thanks to the major technological progress
achieved, the US market prospects appear favourable.

Heat pumps for use in space year-round conditioning appear
to be a technologically sophisticated system, and it is competitive
as it replaces both the heating furnace and the central conditioning
unit. With the present pumps, use for heating purposes only can
hardly be justified in view of their cost; this possibility should be
investigated towards an ad hoc solution along new design lines,
with a view of obtaining, with the help of present technology,
better technical and above all economic performance; this would
also make it possible to adapt heat pumps to the custom prevailing
in Europe, where heat distribution into home spaces is carried
out mainly by water and not by air as it is done in the United States.

I would also like to mention that, for heat pumps, certain
research lines appear promising even in the short term. It will

suffice to mention but one aspect relating to the coefficient of performance which, averaged over one heating season in a city like Milan, Italy, is now about 2, while the experts feel that a coefficient of 3 or 4 is possible quite soon and with a relatively limited effort, combined with careful design.

In the very short term, finally, it seems logical to assume that the present heat pumps may find advantageous applications also in Europe in those cases where year-round conditioning is in any way required, such as for instance in many office buildings.

Allow me to conclude this brief discussion and my talk by expressing a personal hope in relation to the future developments of heat pumps. In my opinion, it is basically important that the European industry operating in the field of space-heating and air-conditioning qualifies through an effort proportional to the magnitude of the potential applications and the difficulty of the task.

A heat pump is a complex system, and under the present conditions it is necessary to avoid such errors as were committed in the past in the United States. I also believe it desirable that the participation of the industry concerned should take place not only through actual production, in such cases where the application of heat pumps is now possible, but also through a major research and development effort.

Finally, in the job of establishing heat pumps, the industry will rely on an effective co-operation from power utilities. Indeed, the latter may come to play a major role under several aspects, mainly in relation with user information, especially important in the very early market development stages, and setting goals in applied research.

ENERGY ECONOMICS AND EXERGY -
COMPARISON OF DIFFERENT HEATING SYSTEMS BASED ON
THE THEORY OF EXERGY

L. Borel

Lausanne Federal Polytechnic, Thermodynamics
Institute.

1. INTRODUCTION

In the present context of the energy crises, the actual
handling of energy (production, operation, distribution and consumption) has become a vital problem, not only on the scale of
the individual, but also on the scale of the company, the region
and the whole country, to say nothing of the world.

In the past, this problem was in most cases preceded by
financial and political considerations and it should be pointed out
that the purely energetic aspect was relegated to the background.
One can even say that the thermodynamic aspect was largely
ignored and often happily rejected. But times have changed and
it is interesting to note how the phenomena known as "shortage"
awakens scientific interest and catalyses research work.

In particular, the concept of heat gives rise to a fair amount
of confusion. People talk abundantly of deterioration of energy,
of energy losses, wastage and efficiency, but the present estimation methods do not match such concepts. For example,
people often add up without any differentiation X calories supplied
at 20°C and Y calories supplied at 500°C, whereas thermodynamics showed a long time ago that their energetic values are quite
different. Statistics are even published in which people merrily
add up "heat calories" and "mechanical or electrical calories".
In short, it is easy to demonstrate that present energy accounting
methods are not satisfactory.

I am going to try to outline a new tool known under different names and in particular as <u>exergy</u>. I will confine myself to summarising a few leading ideas, so as to deal as soon as possible with the practical consequences, discarding intentionally the scientific demonstrations. I will therefore handle one aspect only of the very vast problem of energy economics.

If he wants to, the reader will find the necessary theoretical bases in the survey I have published, entitled "Fonctions d'etat, bilans de travail, pertes et rendement thermodynamiques"[9].

I should like to express my thanks here to Mr. Georges Yanni, Engineer, Lausanne Federal Polytechnic, for his invaluable assistance.

2.   EXERGY AND EXERGETIC EFFICIENCY

If we apply the first and second principles of thermodynamics to any system, we obtain the exergetic balance in power according to:

$$\dot{E}^- + \dot{E}_q^- + \dot{E}_w^- = \dot{E}^+ + \dot{E}_q^+ + \dot{E}_w^+ - \dot{L} \qquad (1)$$

where:

$\dot{E}$ is mechanical or electric power, (mechanical or electric exergy is the same as mechanical or electric energy) in W

$\dot{E}_q$ is heat-exergy (or heat-copower)   in W

$\dot{E}_w$ is transformation-exergy (or transformation copower)   in W

$\dot{L}$ is exergy losses (or thermodynamic losses)   in W

The exponents + or - indicate that exergy is received by the system from outside, respectively given by the system to the outside.

Heat-exergy $\dot{E}_q$ is defined by the relation:

$$\dot{E}_q = \Theta \cdot \dot{Q} \qquad (2)$$

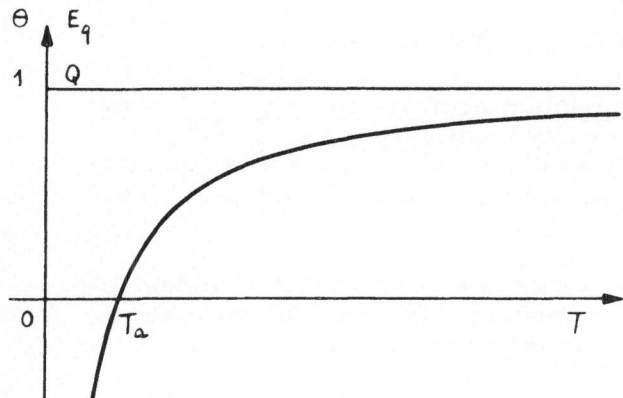

Fig. 1.   Carnot coefficient Θ and heat-exergy $\dot{E}_q$ as a
function of the transfer temperature T

where:

$\dot{Q}$ is heat-power received or given by the system to the
outside

Θ is the CARNOT coefficient defined by:

$$\Theta = 1 - \frac{T_a}{T} \qquad (3)$$

In this definition:

T_a is the temperature of the atmosphere

T   is the temperature at which heat-power $\dot{Q}$ is transferred.

Fig. 1 shows that:

-    Where $T = \infty$, the Carnot coefficient is equivalent to 1
and heat-exergy is equivalent to heat power $\dot{Q}$

-    Where $T_a < T < \infty$, these values diminish as T diminishes

-    Where $T = T_a$, these values are zero

- Where $0 < T < T_a$, they change sign and increase as T diminishes

- Where $T = 0$, they are equivalent to $-\infty$.

Heat-exergy $\dot{E}_q$ is the maximum equivalent mechanical power that can be obtained reversibly from heat power $\dot{Q}$. This heat-exergy can therefore only be obtained by means of a perfectly reversible driving cycle operating between the temperatures T and T_a

The concept of heat-exergy is a good quantitative demonstration of the observations made qualitatively by specialists, according to which a heat-power received:

- is all the more valuable because it is transferred at a higher temperature (thermo-electric power stations);

- does not present any direct interest when it is transferred at atmospheric temperature (cooling of the water of a lake);

- necessitates an expense of energy when it is transferred at a temperature lower than atmospheric temperature (refrigeration facilities);

- necessitates an even greater expense of energy if it is transferred at a lower temperature (gas liquefaction).

Transformation-exergy $\dot{E}_w$ is the maximum equivalent mechanical power that can be obtained reversibly from the thermodynamic transformation of a material. This concept gives rise to more complex considerations than heat-exergy. In view of the scope of this paper, I do not intend to expand it and will thus formulate the two following assumptions:

- energy transformations are effected by means of closed systems;

- the system operates on a steady-state (or cycle) basis.

Exergy losses $\dot{L}$ are exergy reductions due to the imperfections of the system. In the energy sense, any imperfection gives rise to a thermodynamic irreversibility and any irreversibility results in a deterioration of energy. This sequence of events can be expressed quantitatively as a reduction of exergy. The main causes of irreversibility, and therefore of exergy losses, are as follows:

- dissipation (friction);

- heat-transfer under temperature reduction;

- chemical reactions (combustion);

- physical mixtures.

Considering the exergy balance expressed by relation (1), we can define the exergetic efficiency of any system by the relation:

$$\eta = \frac{\dot{E}^- + \dot{E}_q^- + \dot{E}_w^-}{\dot{E}^+ + \dot{E}_q^+ + \dot{E}_w^+} \qquad (4)$$

which can be expressed as follows:

$$\eta = \frac{N}{D} = \frac{N}{N + \dot{L}} = \frac{D - \dot{L}}{D} = 1 - \frac{\dot{L}}{D} \qquad (5)$$

when N and D are the numerator and the demoninator of relation (4).

By virtue of the second principle of thermodynamics, we have:

$$\dot{L} \geqslant 0$$

$$\eta \leqslant 1$$

Thus, the exergetic efficiency of any system is always 100% or less. It would only be equivalent to 100% in the limit of a perfectly reversible system.

In the following paragraphs, I will deal with the problem of the combined supply of electrical energy and heat-energy for space heating. I will first examine the cases where these two forms of energy are supplied independently, and also the performance characteristics of the different parts of the system.

To give an example, I will assume that:

- the premises have to be kept at an average temperature of $T_1 = 20°C$;

- the atmosphere is at the average temperature of $T_a = 0°C$.

56

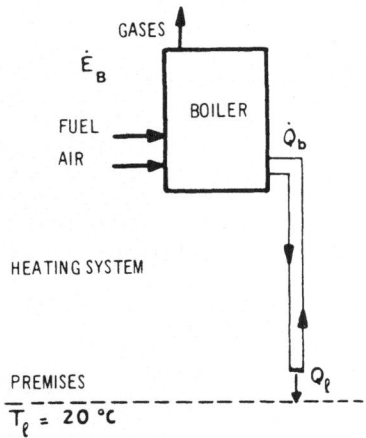

Fig. 2a.   Diagram of an individual heating system.

3.   INDIVIDUAL HEATING (Fig. 2a)

The heat power $\dot{Q}_1$ supplied to the premises is given by the energy balance:

$$\dot{Q}_1 = \dot{Q}_b = \mathcal{E}_b \; \dot{E}_B = 0.75 \; \dot{E}_B \qquad (6)$$

where:

$\dot{Q}_b$ is heat power supplied by the boiler;

$\mathcal{E}_B$ is the efficiency of the boiler's heating which is assumed here to be 75%;

$\dot{E}_B$ is the calorific power received by the boiler, i. e. the chemical power linked to the combustion phenomenon, given by the relation:

$$\dot{E}_B = \dot{M}_B \; q_{pi} \qquad (7)$$

where:

$\dot{M}_B$ is the weight capacity of fuel;

q_{pi} is the lower isobar calorific power of the fuel.

The Carnot coefficient relative to temperatures T_1 and T_a is:

$$\theta_1 = 1 - \frac{T_a}{T_1} = 0.07 \qquad (8)$$

By virtue of definition (4), and assuming that the transformation exergy received $\dot{E}_w$ is equivalent to the calorific power $\dot{E}_B$, the <u>exergetic efficiency</u> of the system is:

$$\eta_b = \frac{\dot{E}_{ql}}{\dot{E}_w} = \frac{\theta_1 \, \mathcal{E}_b \, \dot{E}_B}{\dot{E}_B} = \theta_1 \, \mathcal{E}_b \qquad (9)$$

$$\boxed{\eta_b = 5\%}$$

This extremely low value may seem, at first sight, quite surprising to the uninitiated reader. Having deliberately decided to regrain from scientific speculation, I have tried to make it plausible by means of the cartoon in fig. 2b.

I apologise for the puerile nature this measure might have in the eyes of those readers already conversant with the concepts of energy. Its only claim is to illustrate the sequence "thermodynamic irreversibility $\longrightarrow$ deterioration of energy $\longrightarrow$ exergetic losses" by depicting exergetic reduction.

The cartoon in fig. 2b must be interpreted on the basis of the following code.

- energy = fuel (fuel-oil) or little men;

- temperature level = altitude;

- exergy = combination of the above elements.

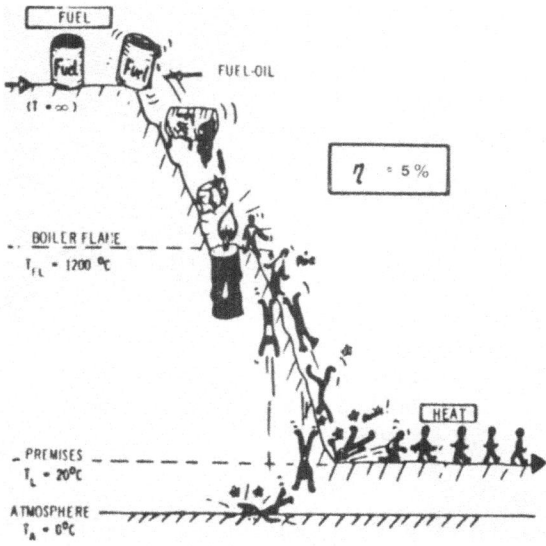

Fig. 2b.   Exergetic interpretation of the individual heating
           shown in fig. 2a.

It will easily be seen that the major exergetic losses are
caused by the three following phenomena:

-   combustion;

-   heat transfer from 1,200°C to 20°C;

-   energy losses (chimney, radiation, unburnt matter).

4.   COLLECTIVE BOILER

The schematic diagram of a collective boiler and its
exergetic interpretation are the same as for an individual boiler.

Since a large collective boiler has better performance
characteristics than a small individual boiler, we have assumed
the heating efficiency here to be 90%.

In order to be able to use the heat supplied in a district
heating system, I have assumed the average outlet temperature of
the boiler to be $T_h = 70°C$, which gives the Carnot coefficient:

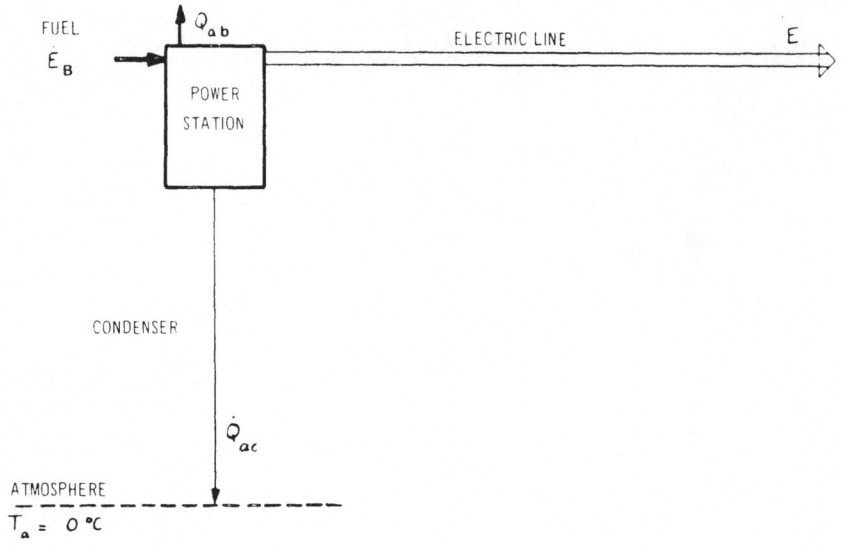

Fig. 3a. Succinct diagram of a thermo-electric power station.

$$\theta_h = 1 - \frac{T_a}{T_h} = 0.20$$

The boiler's exergetic efficiency is then:

$$\eta_h = \frac{\dot{E}_{qh}}{\dot{E}_w} = \frac{\theta_h \, \mathcal{E}_b \, \dot{E}_B}{\dot{E}_B} = \theta_h \, \mathcal{E}_b$$

$$\boxed{\eta_h = 18\%}$$

5.   THERMO-ELECTRIC POWER STATION (Fig. 3a)

The electric power E furnished by the power station is given by the energy balance:

$$\dot{E} = \dot{E}_B - \dot{Q}_{ab} - \dot{Q}_{ac} = \mathcal{E}_c \, \dot{E}_B = 0.40 \, \dot{E}_B \qquad (12)$$

60

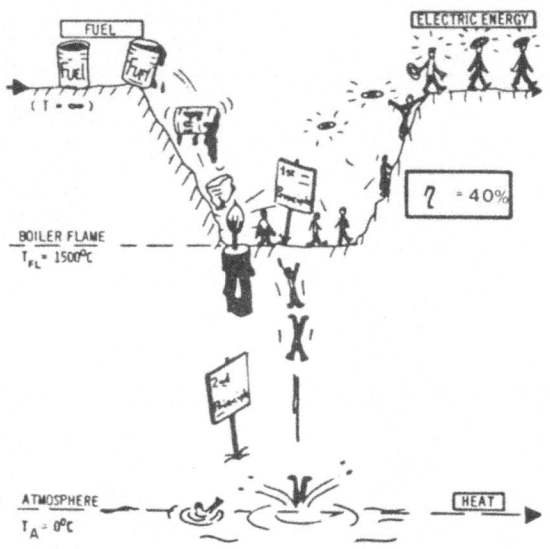

Fig. 3b.   Exergetic interpretation of the thermo-electric
power station in fig. 3a.

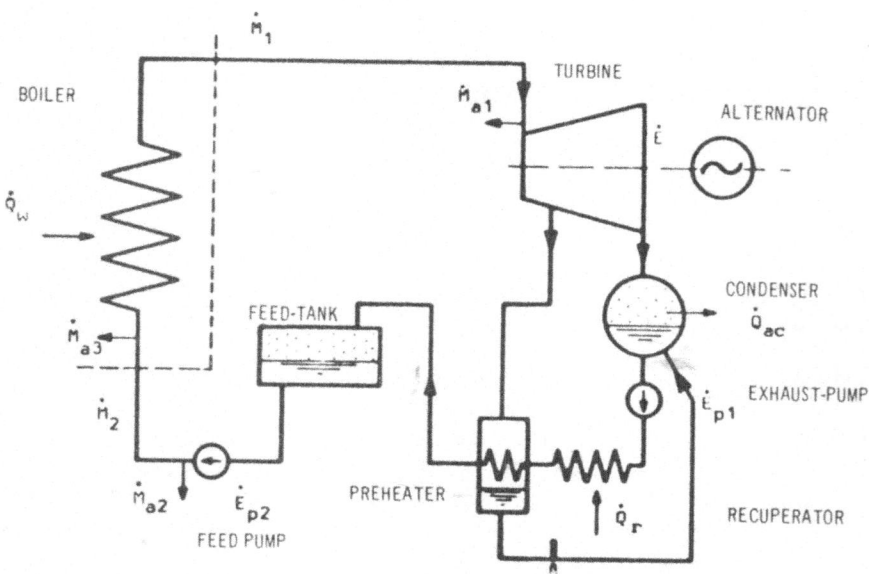

Fig. 4.   Schematic diagram of a thermo-electric power
station producing electric energy.

where:

$\dot{E}$ is the electric power supplied by the power station;

$\dot{Q}_{ab}$ is the power lost in the boiler;

$\dot{Q}_{ac}$ is the heat power supplied to the cold source (atmosphere);

$\mathcal{E}_c$ is the driving efficiency of the power station, assumed here to be 40%.

The power station's <u>exergetic efficiency</u> is:

$$\eta_c = \frac{\dot{E}}{\dot{E}_B} = \mathcal{E}_c$$

$$\boxed{\eta_c = 40\%}$$

The cartoon in fig. 3b depicts the balance in accordance with the first principle, the switching imposed by the second principle, the exergy losses as above and the exergetic equivalent of electric energy, the "noble" nature of the latter being interpreted by haloes!

Fig. 4 shows the schematic diagram of a thermo-electric power station, while fig. 5 presents three photographs of the Vouvry thermo-electric power station.

6.   ELECTRIC RADIATOR (Fig. 6a)

The heat power Q_{lr} supplied to the premises is given by the exergy balance:

$$\dot{Q}_{lr} = \mathcal{E}_r \ \dot{E}_r = 1 \ \dot{E}_r \qquad (14)$$

where:

$\mathcal{E}_r$ is the heating efficiency of the radiator, which is equivalent to 100%;

$\dot{E}_r$ is the electric power consumed.

(a)

(c)

(b)

Fig. 5.   Vouvry thermo-electric power station (Valais - Switzerland)
   (a)   General view
   (b)   Engine Room (300 MW)
   (c)   Turbo-unit open (150 MW)

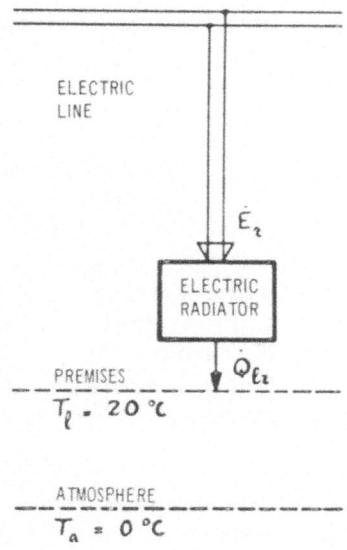

Fig. 6a.   Diagram of an electric radiator.

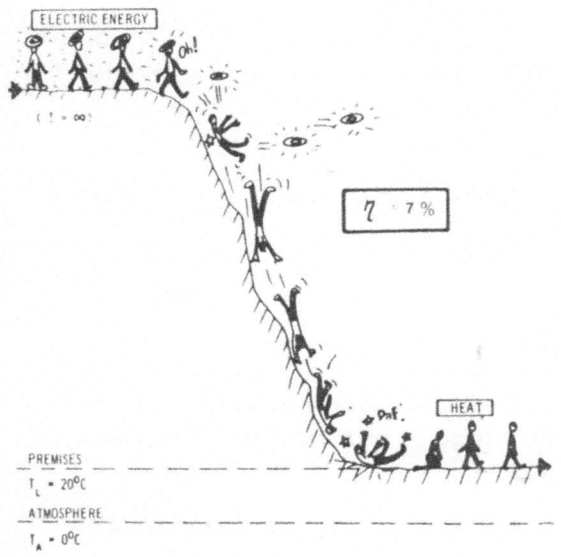

Fig. 6b.   Exergetic interpretation of the electric radiator
           in fig. 6a.

The radiator's <u>exergetic efficiency</u> is:

$$\eta_r = \frac{\dot{E}_{ql}}{\dot{E}_r} = \frac{\theta_1 \; \mathcal{E}_r \; \dot{E}_r}{\dot{E}_r} = \theta_1 \; \mathcal{E}_r$$

$$\boxed{\eta_r = 7\%} \tag{15}$$

This value is extremely low. At the risk of traumatising the people who - with extensive advertising back-up material - recommend electric heating, the cartoon in fig. 6b depicts the very irreversible nature of the transformation "electric energy ———→ heat", the enormous energy deterioration which it causes and the resulting high exergetic losses.

It is important to point out that when specialists support electric heating, saying in particular that the transformation efficiency is 100%, they exploit the fact that the efficiency $\mathcal{E}_r$ = 100%, but they do not know or pretend they do not know that this figure is deceptive, since it is not at all representative in terms of energy economy in its overall aspect.

7. HEAT PUMP (Fig. 7a)

The heat power $\dot{Q}_p$ furnished by the heat pump is given by the energy balance:

$$\dot{Q}_p = \dot{E}_p + \dot{Q}_{ap} = \mathcal{E}_p \; \dot{E}_p = 4 \; \dot{E}_p \tag{16}$$

where:

$\dot{E}_p$ is the electric power consumed;

$\dot{Q}_{ap}$ is the heat power received from the environment, for example from the water of a lake or a river, or even from air;

$\mathcal{E}_p$ is the heating efficiency of the heat pump, assumed here to be 400%.

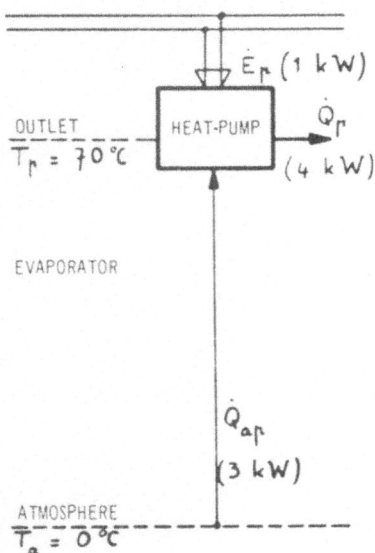

Fig. 7a.   Succinct diagram of a heat pump.

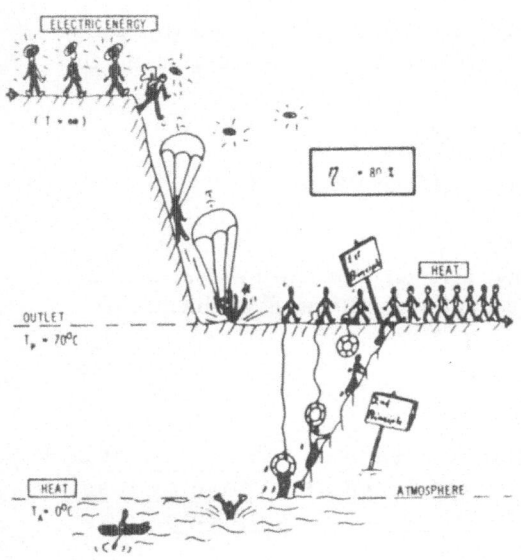

Fig. 7b.   Exergetic interpretation of the heat pump in fig. 7a.

In order to use the heat supplied in a district heating system, I have assumed that the average temperature at the outlet of the heat pump was $T_p = 70^\circ C$ (as for the collective boiler), which gives the Carnot coefficient again of:

$$\theta_p = 1 - \frac{T_a}{T_p} = 0.20$$

The exergetic efficiency of the heat pump is then:

$$\eta_p = \frac{\dot{E}_{qp}}{\dot{E}_p} = \frac{\theta_p \, \mathcal{E}_p \, \dot{E}_p}{\dot{E}_p} = \theta_p \, \mathcal{E}_p$$

$$\boxed{\eta_p = 80\%} \tag{17}$$

This value already appears to be very interesting.

Fig. 7b depicts the fact that everything is very satisfactory from the thermodynamics point of view.

Fig. 8 shows the schematic diagram of a heat pump, while fig. 9 is a photograph of an installation operating according to the principle of the heat pump.

8. COMBINED THERMO-ELECTRIC POWER STATION (Fig. 10a)

The sum of the electric power $\dot{E}$ and of the heat power $\dot{Q}_c$ supplied by the power station is given by the energy balance:

$$\dot{E} + \dot{Q}_c = \dot{E}_B - \dot{Q}_{ab} - \dot{Q}_{ac} \tag{18}$$

The supply of heat, operated by means of a forced draft and heat exchanger system, may be considered as a more or less complete recovery of the power station's heat rejection.

In order to be able to use the heat supplied in a district heating system, I have assumed the average temperature at the power station outlet to be $T_c = 70^\circ C$ (as for the collective boiler

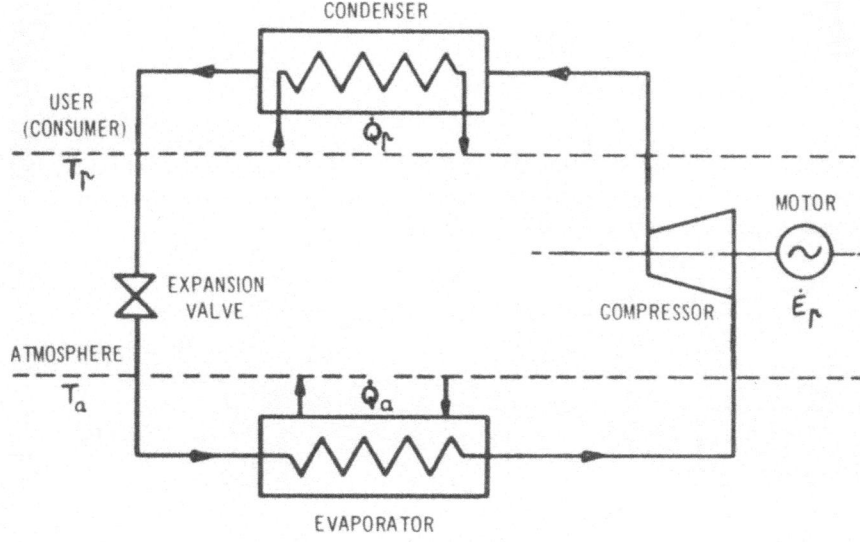

Fig. 8.   Schematic diagram of a heat pump.

Fig. 9.   View of the engine room of a heat pump installation.

and the heat pump), which gives the Carnot coefficient:

$$\theta_c = 1 - \frac{T_a}{T_c} = 0.20$$

If we assume that the quality of the combined power station is the same as that of the single power station as seen in section 4, its exergetic efficiency is also the same, that is:

$$\eta_c = \frac{\dot{E} + \dot{E}_{qc}}{\dot{E}_B} = \frac{\dot{E} + \theta_c \, \dot{Q}_c}{\dot{E}_B}$$

$$\boxed{\eta_c = 40\%} \tag{19}$$

The cartoon in fig. 10b is the same as in fig. 3b, but also depicts the heat exergy supplied at the temperature $T_c = 70^{\circ}C$.

Fig. 11 shows the schematic diagram of a combined thermo-electric power station.

9.   DISTRICT HEATING LINE (Fig. 12a)

The heat power $\dot{Q}_d$ supplied to a building by means of a heat exchanger is given by the energy balance:

$$\dot{Q}_d = \dot{Q}_c - \dot{Q}_{ad} = \mathcal{E}_d \, \dot{Q}_c = 0.85 \, \dot{Q}_c \tag{20}$$

where:

$\dot{Q}_{ad}$ is the heat power lost due to lack of insulation;

$\mathcal{E}_d$ is the transmission efficiency of the line, assumed here to be 85% (value assumed to be intentionally pessimistic).

I have assumed that the heat power $\dot{Q}_d$ is supplied at the average temperature $T_d = 65^{\circ}C$, which gives the Carnot co-efficient:

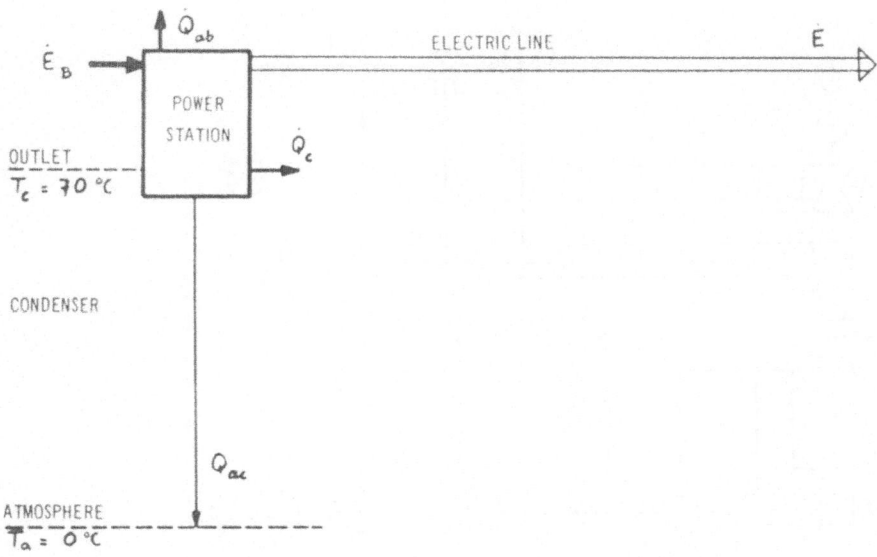

Fig. 10a. Succinct diagram of a combined thermo-electric
power station.

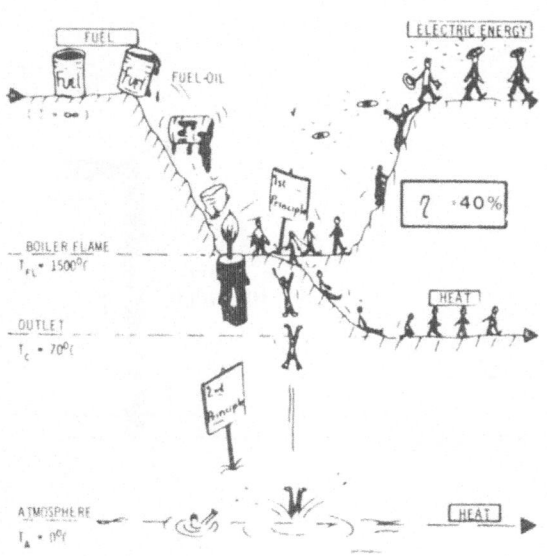

Fig. 10b. Exergetic interpretation of the combined thermo-
electric power station of fig. 10a.

70

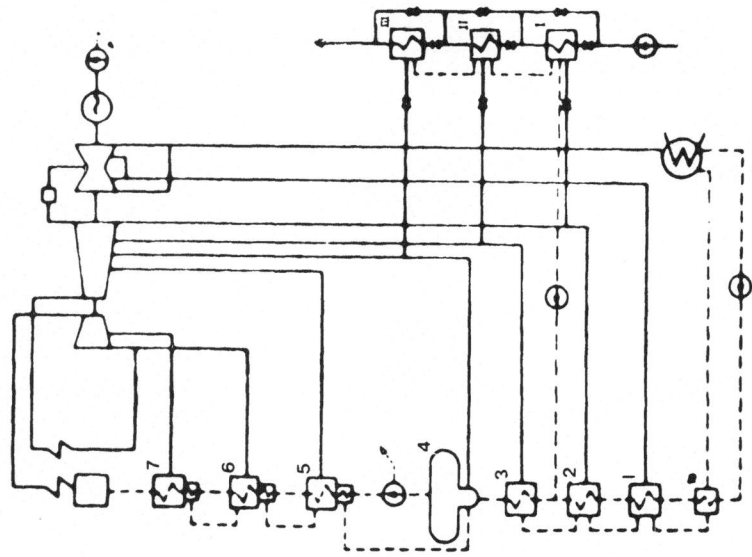

Fig. 11.   Schematic diagram of a combined thermo-electric
power station.

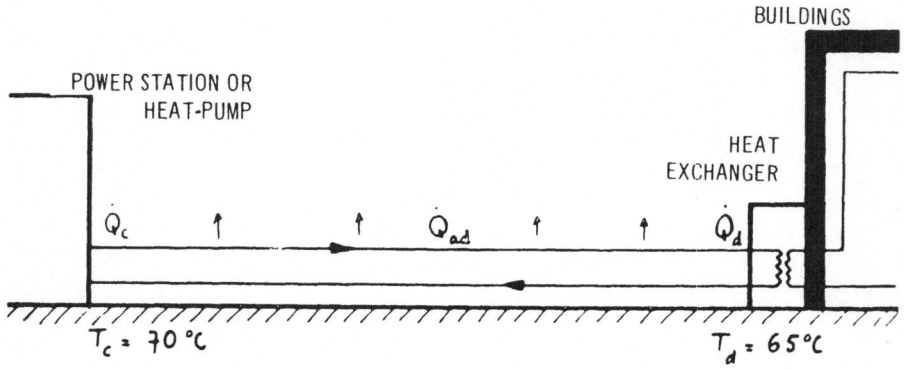

Fig. 12a.   Diagram of the line designed for the district heating
of a building.

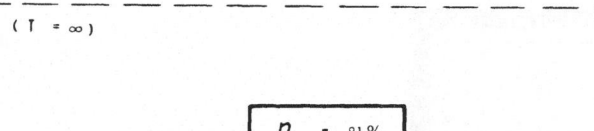

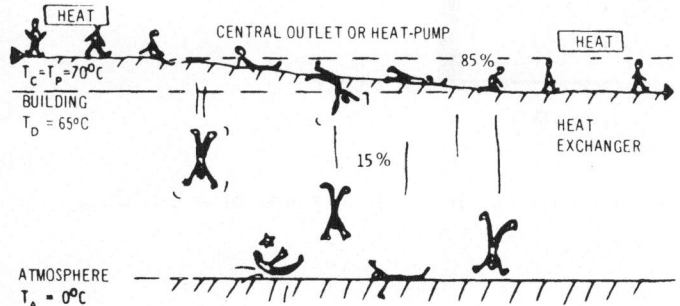

Fig. 12b. Exergetic interpretation of the district heating line in fig. 12a.

$$\Theta_d = 1 - \frac{T_a}{T_d} = 0.19$$

The line's exergetic efficiency is therefore:

$$\eta_d = \frac{\dot{E}_{qd}}{\dot{E}_{qc}} = \frac{\Theta_d \, \mathcal{E}_d \, \dot{Q}_c}{\Theta_c \, \dot{Q}_c} = \frac{\Theta_d}{\Theta_c} \, \mathcal{E}_d \qquad (21)$$

$$\boxed{\eta_d = 81\%}$$

Fig. 12b depicts the exergetic losses due to the lack of insulation and to the reduced temperature level of the water in the pipe.

72

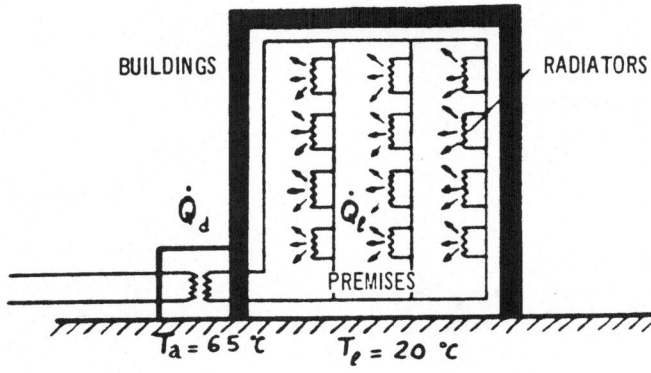

Fig. 13a. Diagram of the internal heating system of a building.

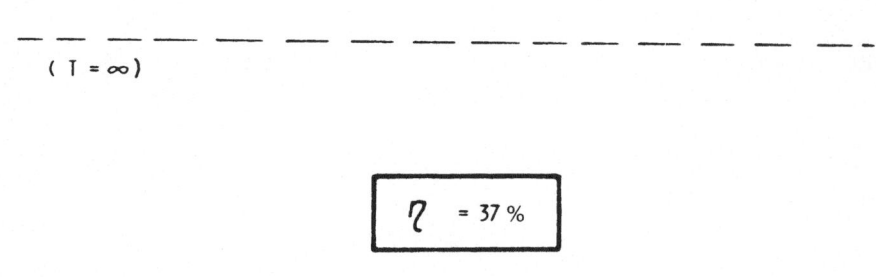

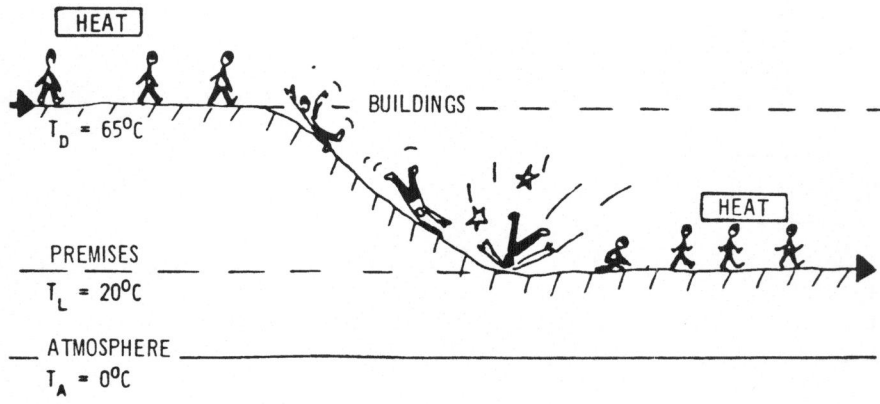

Fig. 13b. Exergetic interpretation of the internal heating system of fig. 13a.

10. INTERNAL HEATING SYSTEM (Fig. 13a)

The heat power $\dot{Q}_1$ supplied to the premises is:

$$\dot{Q}_1 = \dot{Q}_d \tag{22}$$

The system's exergetic efficiency is:

$$\eta_1 = \frac{\dot{E}_{ql}}{\dot{E}_{qd}} = \frac{\Theta_1 \dot{Q}_1}{\Theta_d \dot{Q}_d} = \frac{\Theta_1}{\Theta_d}$$

$$\boxed{\eta_1 = 37\%} \tag{23}$$

Fig. 13b depicts the exergetic losses due to the reduction of the temperature level in the heat exchanger, in the pipes and in the radiators.

11. HEAT ENERGY CONSUMPTION (Fig. 15a)

The heat power $\dot{Q}_{al}$ which is finally transferred to the atmosphere through the walls of the building is:

$$\dot{Q}_{al} = \dot{Q}_1 \tag{24}$$

The heat exergy corresponding to $\dot{Q}_1$ is:

$$\dot{E}_{ql} = \Theta_1 \dot{Q}_1 = 0.07 \dot{Q}_1$$

The Carnot coefficient relative to the temperature T_a of the atmosphere is obviously:

$$\Theta_a = 1 - \frac{T_a}{T_a} = 0$$

74

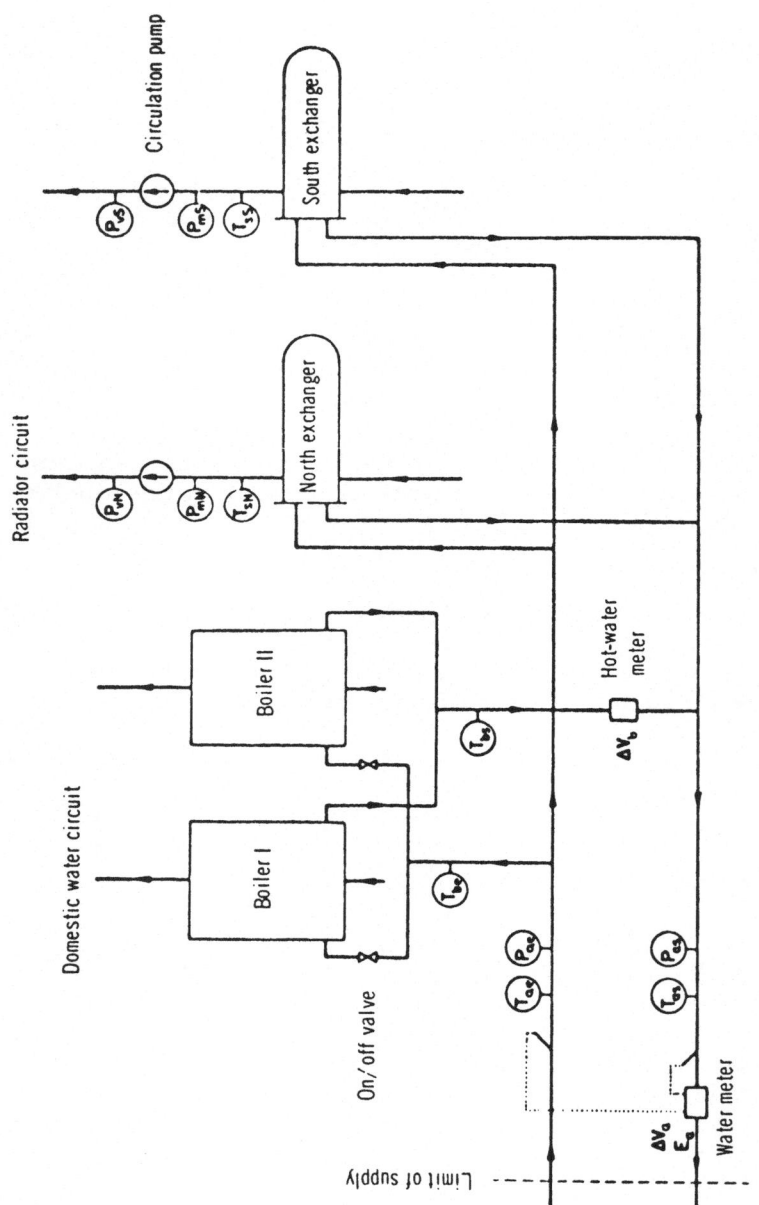

Fig. 14.   Diagram of the transformation station of the district heating of a building (Cité du Lignon, Geneva)

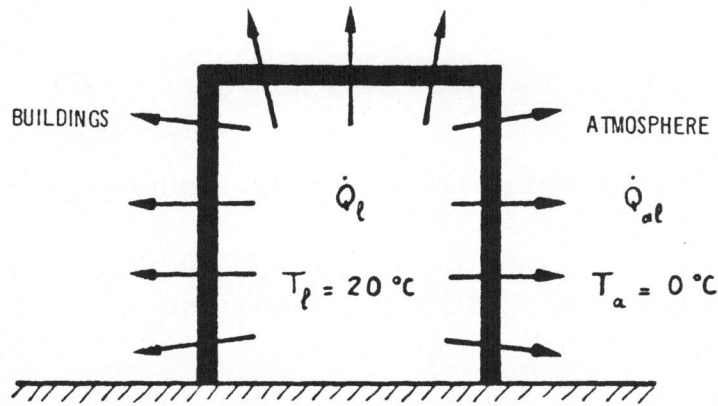

Fig. 15a. Heat energy consumption of a building.

Fig. 15b. Exergetic interpretation of the consumption of
heat energy made by the building in fig. 15a.

so that the heat exergy corresponding to $\dot{Q}_{al}$ is finally:

$$\dot{E}_{qal} = \theta_a \, \dot{Q}_{al} = 0 \tag{25}$$

Fig. 15b truly depicts the fact that the heat exergy has become zero.

12. COMBINED SUPPLY OF ELECTRIC ENERGY AND HEAT ENERGY

Fig. 16 represents very diagrammatically the combination of the different possible methods of supplying electric energy and heat energy from a combined thermo-electric power station.

To characterise the relative importance of the heating energy as compared with the total energy consumed, I am introducing the heating rate u_q defined by the relation:

$$u_q = \frac{\dot{Q}_1}{\dot{E}_m + \dot{Q}_1} \tag{26}$$

where $\dot{E}_m$ is the electric power supplied for purposes other than heating.

Heat energy can be supplied by one of the five following methods:

- individual heating;

- collective heating;

- heating by a heat pump;

- heating by forced draft (recovery).

The selection of one method of distribution must be based in particular on the three following criteria:

(a) exergetic performance characteristics;

(b) harmful effects;

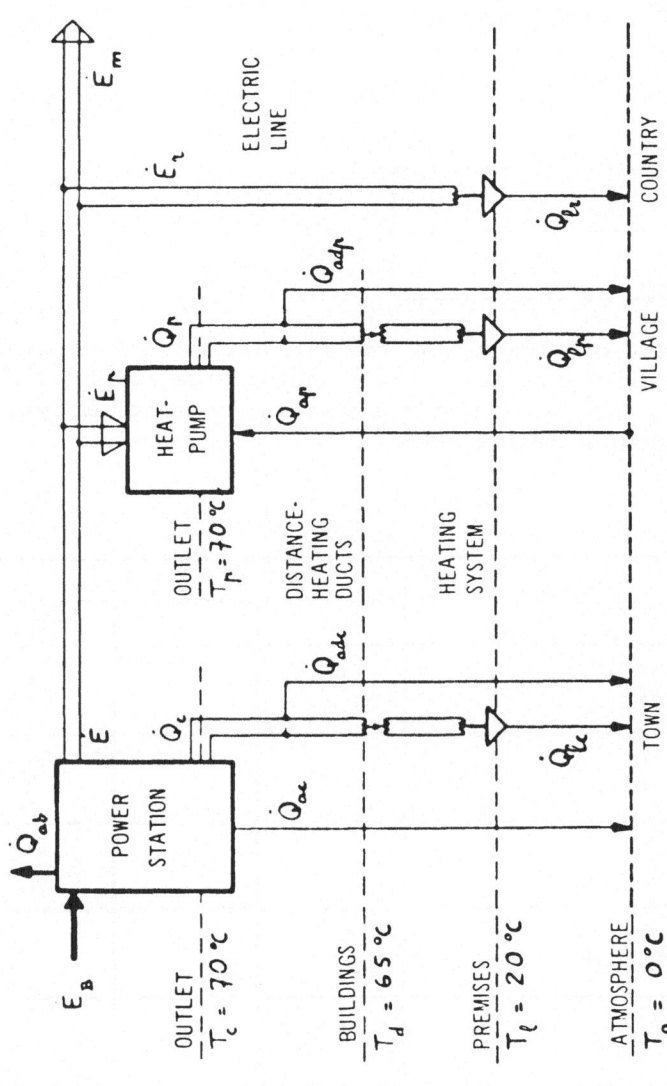

Fig. 16. Succinct model of the distribution of electric energy and heat energy from a combined thermo-electric power station.

(c) technical and economic aspects.

(a) Exergetic Performance Characteristics

By virtue of general definition (4), the system's overall exergetic performance is characterised by the <u>overall exergetic efficiency</u>:

$$\eta = \frac{\dot{E}_m + \dot{E}_{q1}}{\dot{E}_w}$$

$$\eta = \frac{\dot{E}_m + \theta_1 \cdot \dot{Q}_1}{\dot{E}_B} \qquad (27)$$

Table I shows the results of the calculation of the overall exergetic efficiency η , carried out for the supply of electric energy and the five methods of heating examined above, for the three following heating rates:

METHOD OF DISTRIBUTION OF ENERGY	HEATING RATE u_q		
	0 %	65 %	100 %
ELECTRIC ENERGY INDIVIDUAL HEATING	40 %	23 %	5 %
ELECTRIC ENERGY COLLECTIVE HEATING	40 %	23 %	5 %
ELECTRIC ENERGY ELECTRIC HEATING	40 %	16 %	3 %
ELECTRIC ENERGY HEATING BY HEAT-PUMP	40 %	30 %	10 %
ELECTRIC ENERGY FORCED-DRAFT HEATING	40 %	32 %	--

Table I   Overall exergetic efficiency η of the different methods of energy distribution.

u_q = 0% Electric energy only

u_q = 65% Electric energy + heating

u_q = 100% Heating only

In the most realistic case of the 65% heating rate, we observe that:

- forced-draft heating leads to the overall exergetic efficiency η of 32%, which is the best;

- heat-pump heating has an efficiency of 30%, which is also very good;

- individual and collective heating, both with an efficiency of 23%, are less favourable;

- electric heating with an efficiency of 16% is the worst.

This means that, under the conditions envisaged, "all electric" heating requires about twice as high a fuel consumption as heat-pump or forced-draft heating.

The cases examined above are extreme cases in which all the heating would be provided by one heating method. Of course, in practice, the heating of an area or of a country will be provided by a suitable combination of the different heating methods.

This combination is particularly essential if one has a condensation and forced-draft type thermo-electric power station, where the heating rate is high. This is because when the production of the forced-draft heat increases, the heat rejection linked to condensation drops, and when the latter becomes zero, a limit is reached where the condensation-type power station becomes a back-pressure station. If the heating rate continues to increase, this must be supplemented by another heating method, for example by a direct collective heating system. It can be assumed that the latter would be provided either by the boiler of the back-pressure power station, or by an auxiliary boiler.

As an example, I have shown in figs. 17a and 17b the influence, the energy and exergy balances of a combined thermo-electric power station, of the heating rate u_{qc} defined by the relation:

$$u_{qc} = \frac{\dot{Q}}{\dot{E} + \dot{Q}}$$

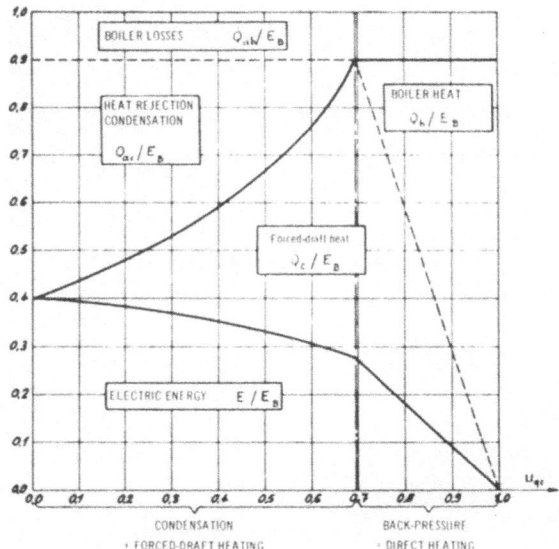

Fig. 17a. Influence of the heating rate u_{qc} on the energy balance of a combined (forced-draft and direct heating) power station.

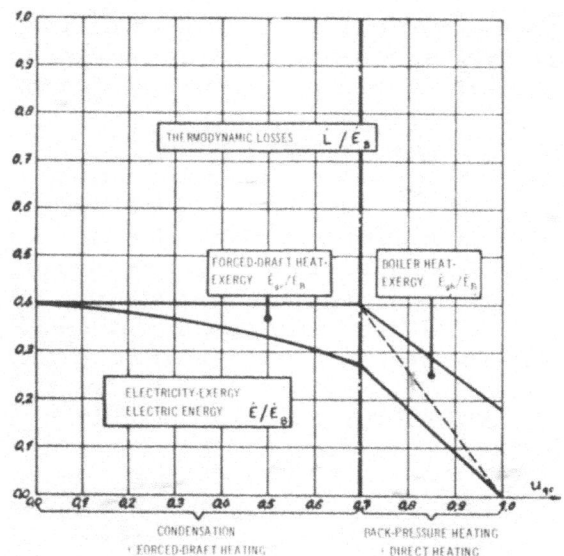

Fig. 17b. Influence of the heating rate u_{qc} on the exergetic balance of a combined (forced-draft and direct heating) power station.

in which:

$$\dot{Q} = \dot{Q}_c + \dot{Q}_h$$

$\dot{Q}$ is the total heating heat power

$\dot{Q}_c$ is the forced-draft heating power

$\dot{Q}_h$ is the direct heating heat power

Figure 17a shows that the energy balance is not representative of the system's energy economics. On the other hand figure 17b shows that the exergy balance is a determining factor, because:

-   the exergetic efficiency of the condensation and forced-draft power station is still good, since it remains at its first approximation constant and equivalent to 40%, for heating rates ranging from 0 to 69%;

-   the system's exergetic efficiency drops from 40% to 18% when the heating rate rises from 69% to 100%.

(b)   Harmful Effects

In view of the scope of this paper, I will confine myself to the pollution due to <u>heat rejection</u>, that is, unused heat energy forced out to the atmosphere (water or air) because of the requirements related to the operation of the boilers or the design of the heat cycles.

To characterise the relative importance of the pollution as compared with the total energy consumed, I am introducing the <u>coefficient of thermal pollution</u> Ψ defined by the relation:

$$\Psi = \frac{\dot{Q}_a}{\dot{E}_m + \dot{Q}_1} \tag{28}$$

in which $\dot{Q}_a$ is the unused heat power transferred to the atmosphere.

METHOD OF DISTRIBUTION OF ENERGY	HEATING RATE u_q		
	0 %	65 %	100 %
ELECTRIC ENERGY INDIVIDUAL HEATING	150 %	75 %	33 %
ELECTRIC ENERGY COLLECTIVE HEATING	150 %	73 %	31 %
ELECTRIC ENERGY ELECTRIC HEATING	150 %	150 %	150 %
ELECTRIC ENERGY HEAT-PUMP HEATING	150 %	36 %	−26 %
ELECTRIC ENERGY FORCED-DRAFT HEATING	150 %	27 %	--

Table II   Coefficient of pollution Ψ of the different methods
of energy distribution.

Table II shows the results of the calculation of the pollution
coefficient Ψ , carried out for the supply of electric energy and
the five methods of heating examined above, for the three heating
rates u_q = 0%, 65% and 100%.

In the particularly interesting case of the 65% heating rate,
we observe that:

- forced-draft heating is accompanied by a thermal pollution
  coefficient Ψ of 27%, which is the lowest;

- heat-pump heating, by a coefficient of 36%, also fairly low;

- collective heating, by a coefficient of 73%, average;

- individual heating, by a coefficient of 75% which is also
  average;

- electric heating, by a coefficient of 150%, which is by far
  the highest.

This means that, <u>when we consider, as we should, the whole</u> <u>system,</u> "all-electric" heating is accompanied by a thermal pollution which is by far the highest.

(c)   Technical and Economic Aspects

First, there is - around a thermo-electric power station or a heat pump - a circle with a limit radius beyond which the principle of distance heating is no longer interesting, either because the line energy losses become too high, or because the cost of the pipes becomes prohibitive.   In this respect, thermo-electric power stations and heat pumps <u>should therefore be placed</u> <u>as close as possible to the centre of urban developments.</u>

Second, without going into detail, it is reasonable to assume <u>that the investments likely to be envisaged for heating purposes</u> <u>may be even higher if the urban centre considered is large.</u>

13.   MODEL OF ENERGY DISTRIBUTION

On the basis of the criteria examined above, we can imagine a <u>model of energy distribution,</u> taking into account the size of the urban centre to be heated (see fig. 16).

<u>Case of the Large Urban Centres (Conurbations)</u> - In any case, the most interesting heating method for the larger urban centres is <u>district heating by the forced-draft system,</u> the characteristics of which are:

-    best exergetic efficiency;

-    lowest thermal pollution coefficient;

-    high investment and field of use limited to a given radius around the power station.

Fig. 19 depicts the whole process from the exergy point of view and shows the rational nature of the supply of heat.

<u>Case of the Medium-sized Urban Centres (Localities)</u> - Because of criterion (c), it is not reasonable to use forced-draft distance heating, unless there is a thermo-electric power station nearby for other reasons.

Generally, the method of heating most suitable for medium-

84

sized urban centres is heat pump distance heating, the features of which are:

- very good exergetic efficiency;

- fairly low thermal pollution coefficient;

- medium-sized investment and field of use limited to a given radius around the heat pump.

Fig. 21 depicts the whole process from the exergetic point of view and shows the satisfactory nature of the supply of heat.

Case of the Small Urban Centres (Countryside) - Still because of criterion (c), it is not reasonable to use distance heating. Apart from other heating methods which this study has not dealt with (solar energy, individual heat pump, etc.), the methods left for the heating of small urban centres are individual heating and electric heating, the features of which are:

- average and bad exergetic efficiency;

- average and high pollution coefficient;

- low investment and field of use not limited.

Fig. 18.   View of a large conurbation.

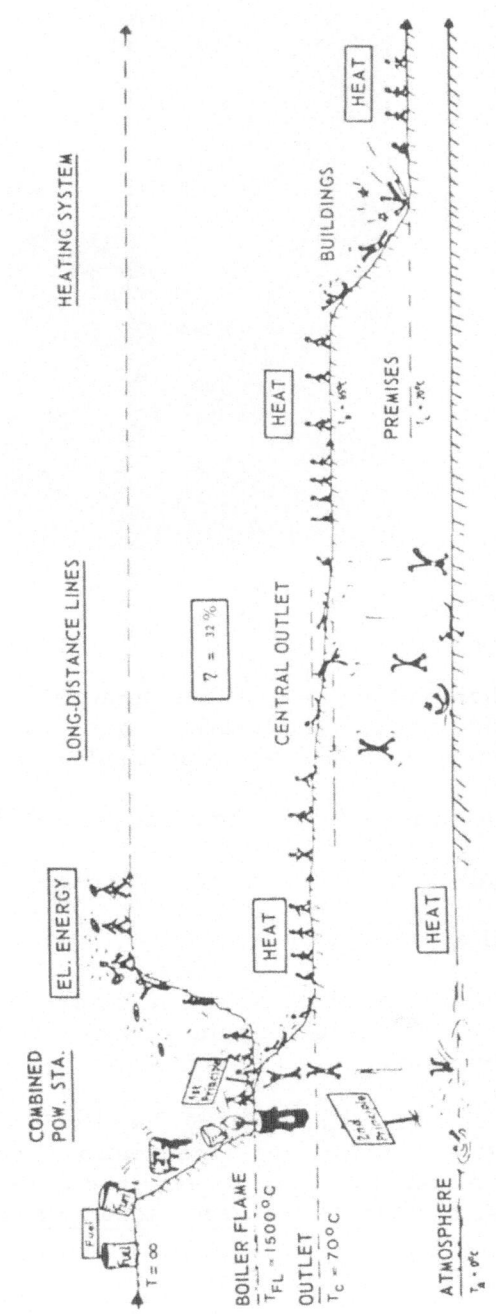

Fig. 19. Exergetic interpretation of the method of distribution "electric energy + forced-draft heating".

Fig. 20.   View of a locality.

The choice of any one of these methods of heating must be guided by additional considerations which are out of the scope of this paper and which are, in particular, as follows:

- other harmful effects (chemical pollution);

- adjustment of the system (combustion);

- flexibility.

Fig. 23 depicts, in the case of electric heating, the whole process from the exergetic point of view and shows its not very satisfactory nature.

These considerations justify the efforts being made at present with a view to using unconventional sources of energy, such as, for example, solar energy and to developing individual heat pumps, together with energy accumulation systems.

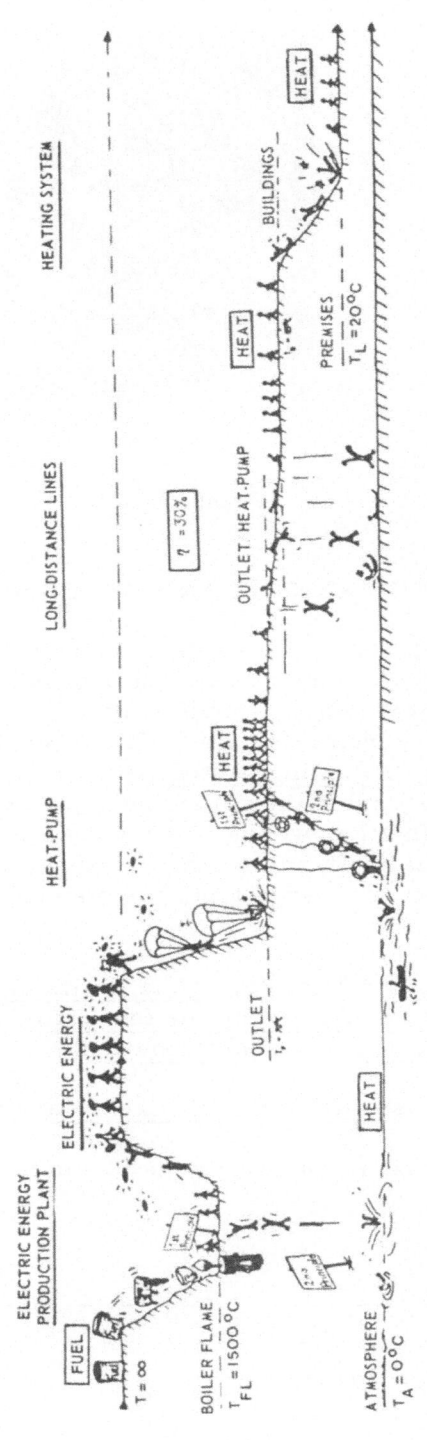

Fig. 21. Exergetic interpretation of the distribution method "electric energy + heating by heat pump".

Fig. 22.   View of the countryside.

14.   COST OF EXERGY AND OF ENERGY

In contrast to the studies conducted at present, in which heat energy is considered as a byproduct of electric energy, I feel that heat energy should play a full part in energy optimisations and economic studies.   It is therefore necessary to adopt an <u>exergy accounting system,</u> that is to say an energy accounting system observing the equivalences expressed in the theory of exergy.   In other words, we must assume that, at the limits of supply, heat or transformation exergy is equivalent to mechanical or electric energy from the standpoint of the energy service performed.

Thus, in principle, <u>at the limits of supply, the cost of exergy should not be affected by the form it is available in</u> (mechanical, electrical, heat or transformation).

<u>We must therefore sell exergy and not energy.</u>

The cost P_e of exergy may be determined using the relation:

$$P_e = \frac{D_{an}}{\displaystyle\int_{1\,year} (\dot{E}^- + \dot{E}_q^- + \dot{E}_w^-)\,dt} \tag{29}$$

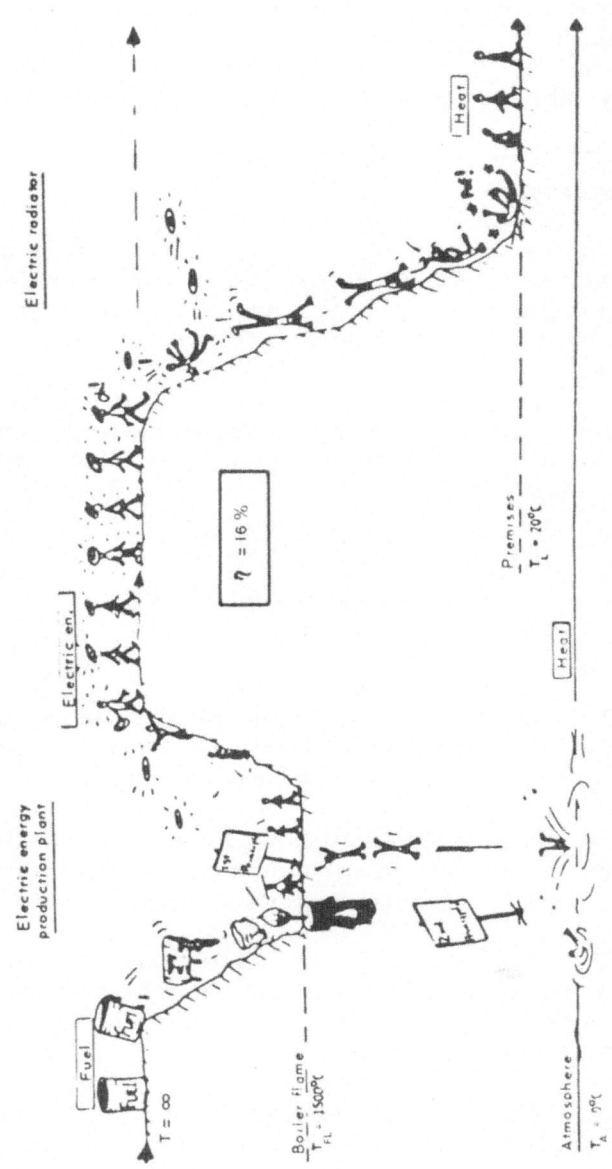

Fig. 23. Exergetic interpretation of the method of distribution "electric energy + electric heating".

90

in which:

D_{an} are the total annual expenses:

- fixed expenses (amortisation, maintenance, personnel, etc.);

- variable expenses (fuel, etc.).

This price P_e may, for example, be expressed in Francs per kW-h of exergy.

It would always be possible, of course, to take into account certain constraints of a technical, economic or business-cycle order, by placing in front of $\dot{E}_q^-$ and $\dot{E}_w^-$ the coefficients k_q and k_w which would soften the rigour of the principle of exergy equivalence.

With regard to heat exergy, it is interesting to switch from the unified cost of exergy to the price P_q of heat energy, following the relation:

$$P_q = \frac{\dot{E}_q^- \cdot P_e}{\dot{Q}^-} = \frac{\theta \cdot \dot{Q}^- \cdot P_e}{\dot{Q}^-} = \theta \, P_e \qquad (30)$$

in which θ is the Carnot coefficient relative to the temperature T of supply of the heat energy. Price P_q may, for example, be expressed in Francs per kW-h of heat energy.

Fig. 24, showing the evolution of price P_q as a function of temperature T, with the temperature T_a of the atmosphere as a parameter, gives rise to the following remarks:

- price P_q changes exactly like coefficient θ , since price P_e is a fixed price. It well reflects the fact that the energetic value of heat depends mainly on the temperature level T at which it is supplied.

- the sign of P_q has been changed for T included within 0 and T_a, in other words in the case of refrigeration. This reflects the fact that the supplier actually performs an exergetic service, in spite of the fact that he receives heat energy from the consumer.

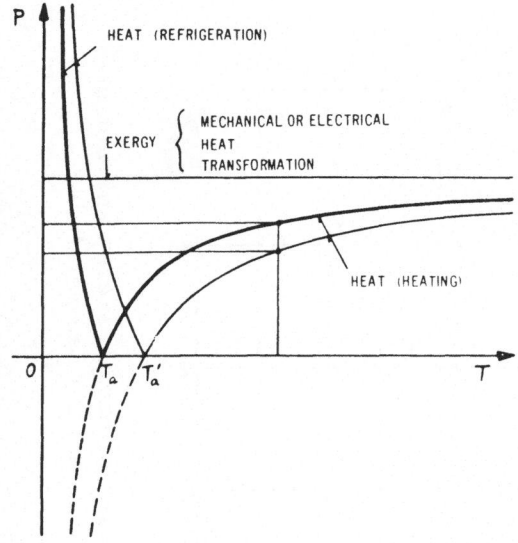

Fig. 24.   Cost of exergy and of heat energy.

-    price P_q tends in principle to move towards ∞ for T
     tending to 0.   This corresponds to the fact that it is very
     difficult to get to absolute zero.

-    in the case of heating, price P_q is, for a given value of T,
     as high as the temperature of the atmosphere T_a is low
     $(T_a < T_a')$.   This well expresses the fact that heat is more
     valuable in winter than in summer.

-    in the case of refrigeration, the opposite is true.   This
     truly expressed the fact that refrigeration is more difficult
     in summer than in winter.

Figures 25a and 25b show practical results derived from an
exergetic survey conducted by the EPF-Lausanne Thermodynamics
Institute.   This survey, which concerns the Electrical Energy
heat Power Station of South-West Lausanne, showed how valuable
exergetic considerations are when it comes to assessing a development in terms of energy economics.

92

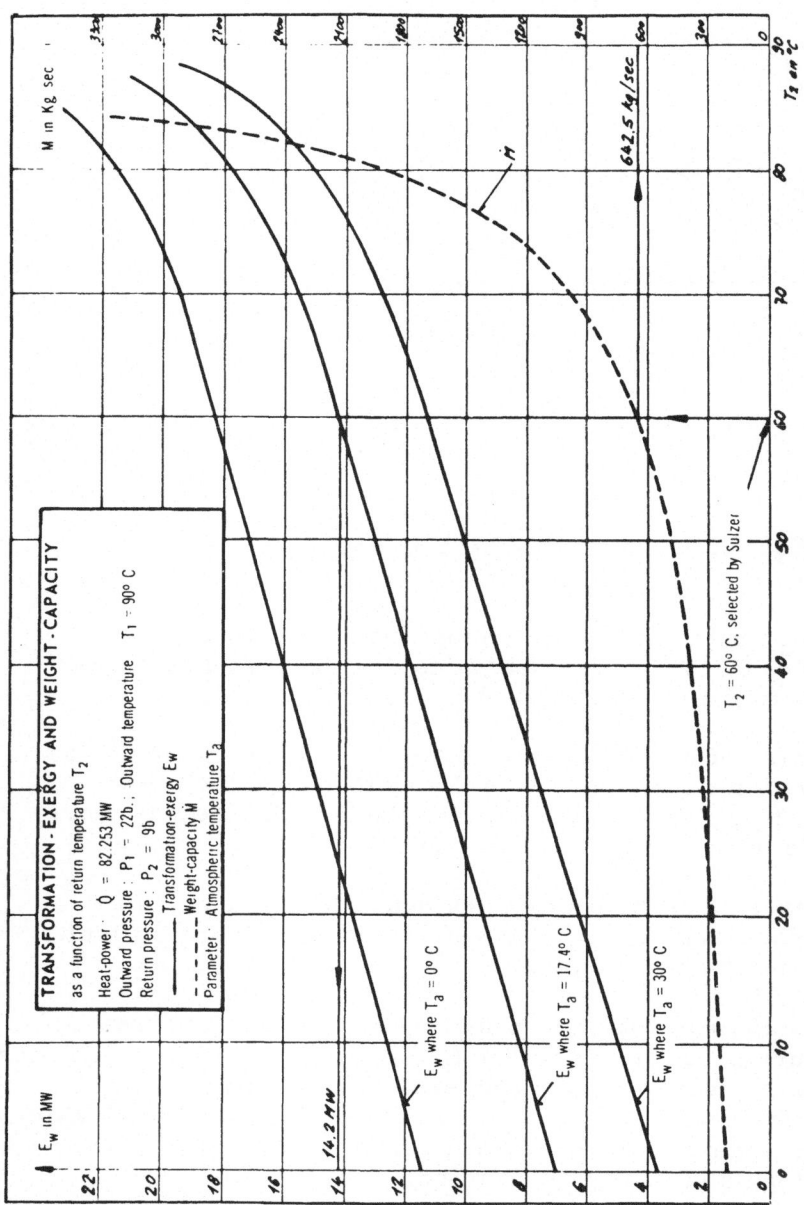

Fig. 25a. South-West Lausanne Power Station. Transformation-exergy and weight-capacity as a function of return temperature and air temperature.

Outward temperature: $T_l = 90°C$.

93

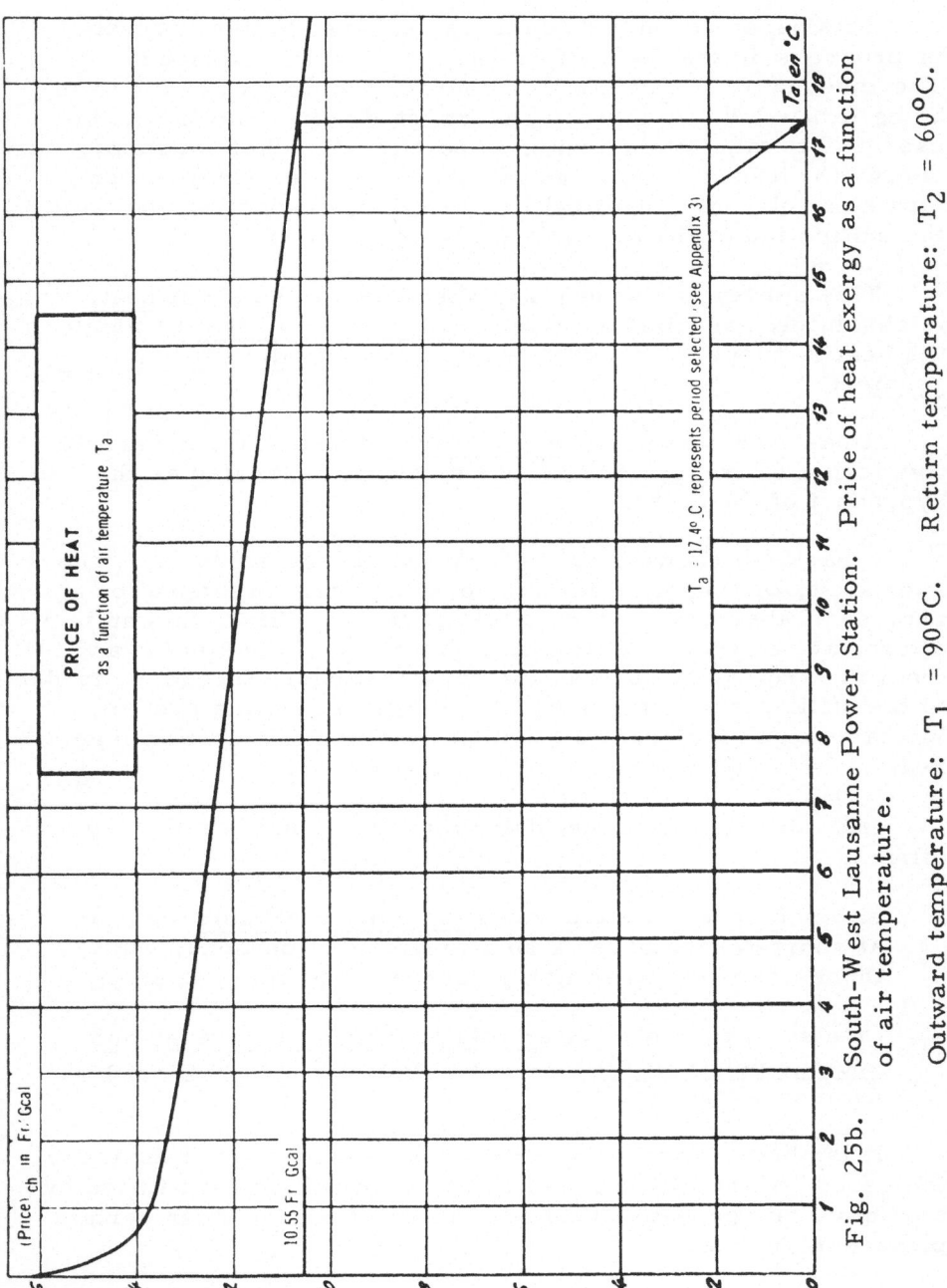

Fig. 25b.   South-West Lausanne Power Station.   Price of heat exergy as a function of air temperature.

Outward temperature: $T_1 = 90°C$.   Return temperature: $T_2 = 60°C$.

15. CONCLUSION

This paper is only a small part of the studies currently in progress in the field of energy at the EPF-Lausanne Thermodynamics Institute. Of course, the subject is far from being exhausted. In a more general study, it would in fact be essential to take into account, in particular, other forms of energy (hydraulic, solar, aeolian, geothermic, etc.), other forms of pollution (chemical, radioactive, accoustic, etc.) and the exhaustion of the primary energy resources.

This survey has shown that when one speaks about heat, it is absolutely essential to state the temperature level at which the heat is supplied, together with the temperature of the environment.

It has also shown that electricity and heat are two correlative forms of energy and that one must not be treated as the byproduct of the other.

It has also shown that, in terms of energy economics, the specialist, the supplier and the consumer are threatened by a number of aberrant prospects, if not traps. First, this state of affairs arises from the fragmentary nature of the studies and from the sectorial aspect of the optimisations. Second, it is due to the utilization of an inadequate energy accounting system. We have seen, for example, how the method of a certain type of heating is artificial and illusory in given cases.

Our society should now determine itself on the following dilema:

- either to continue to develop the most profitable solutions at short-term and for a small number of persons, while inconsiderately wasting the sources of primary energy;

- or else to promote better solutions from the energy and ecology points of view, at long term and for the greater number of persons.

It is obvious that, if we are far-sighted, we must envisage the second of the two above choices, even if it is conditioned by a painful, and perhaps cruel awareness of the facts for certain private interests.

This choice automatically leads to the need to use a correct energy accounting system, within the scope of an overall

conception of energy. In this connection, the exergy accounting
system should be effective. It should facilitate the achievement
of the following objectives:

- better optimisation and more rational exploitation of the
  facilities

- fight against waste, by urging manufacturers and operators
  to carry out operations that are as reversible as possible,
  and to start by eliminating the most obvious irreversibilities;

- preparation of a very general model for the distribution of
  energy in the different forms.

It is clear that any measure involving exergetic estimation
must be accompanied by concomitant measures involving economic
estimation and estimation of impacts on the environment.

The concept of exergy must, lastly, help to clarify the situation and to assume diagnoses from the standpoint of energetic
economics. It is a precious tool for the establishment of an
overall conception of energy, with a view to planning it. The
exergetic accounting system is likely to facilitate the choice of
the energy options and to guide certain decisions on the economics
and political level.

The task is considerable taking into account the choices
already made.

REFERENCES

1. Seippel, C. and Bereuter, R. , Contribution à la technique
   de l'emploi combiné des turbines à vapeur et à gaz. Revue
   BBC tome 47, décembre 1960.
2. Dzung, L.S. , Fundamental Concepts of Thermodynamics in
   Classical Treatment. Cours donné au MIT (Cambridge),
   décembre 1960.
3. Roegener, H. , Anwendung des Energiebegriffs auf den
   Dampfkraftprozess. Elektrizitätswirtschaft 60, 1961.
4. Grassmann, P. , Energie und Exergie. BWK 13, 1961.
5. Tratscher, W. , Exergetische Beurteilung technischer
   Verbrennungsreaktionen. Energietechnik 12, 1962.
6. Van Lier, J.J.C. , Thermodynamische processen in de
   centrale en mogelijkheden tot het verbeteren van deze
   processen. Argus, Amsterdam, 1963.

96

7. Baehr, H. D. and Schmidt, E. F., Definition und Berechnung von Brennstoffexergien. BWK 15, 1963.

8. Baehr, H. D., Bergmann, E., Bošnjaković, F., Grassmann, P., Van Lier, J. J. C., Rant, Z., Roegener, H., Schmidt, K. R., Energie und Exergie. Die Anwendung des Exergiebegriffs in der Energietechnik. VDI-Verlag, Düsseldorf, 1965.

9. Borel, L., Fonctions d'état, bilans de travail, pertes et rendement thermodynamiques. Bulletin technique de la Suisse romande. No. 19 et 20 des 18 septembre et 2 octobre 1965.

10. Berchtold, M., Zur Bewertung thermodynamischer Prozesse. Schweizerische Bauzeitung, No. 12, mars 1970.

11. Rant, Z. and Gaspersic, B., Ein allgemeines Temperatur-Enthalpie - Exergie - Diagramm für Verbrennungsgase, VDI-Verlag, Düsseldorf, 1972.

12. Hohl, R., Suisse chauffée à distance? Publication BBC, juin, 1973

13. Sulzer, Geb, Städtefernheizung. Verlag EDMZ, Bern, Januar 1974.

14. Dzung, L. S., Raumheizung, Exergie und Licht. Publication interne BBC, Februar 1974.

15. Berg, C. A., A technical basis for energy conservation. Mechanical Engineering, No. 30, May 1974.

16. Bidard, R., Exergie, rendements de cycles, rendements de machines. Revue générale de thermique. No. 150-151, juin-juillet 1974.

BASIC PRINCIPLES OF THERMODYNAMICS AS APPLIED
TO HEAT PUMPS:

THERMODYNAMIC CYCLES IN HEAT PUMPS

M. Duminil,

Institut Français du Froid Industriel,
(Conservatoire National des Arts et Métiers, Paris),
Ecole Centrale des Arts et Manufactures, and
Ecole Nationale du Génie Rural, des Eaux et
des Forêts.

1. GENERAL REMARKS AND DEFINITIONS

1.1 A heat pump is a thermodynamic system which enables heat
to be transferred from one medium to another medium which is
at a higher temperature. The second law of thermodynamics
makes it clear that a system which performs such an "unnatural"
transfer must necessarily be supplied with energy.

There is no difference between the principle of a cooling
machine and that of a heat pump. The only difference between
them is their purpose. A cooling machine is essentially designed
to maintain a given medium (a cold chamber, a liquid which is
to be cooled or frozen etc.) at a temperature lower than the
ambient temperature by continuously absorbing heat from this
medium. The heat thus extracted is transferred to a warmer
medium (atmospheric air or water), which evacuates it from the
system. By virtue of the first law of thermodynamics, the energy
supplied to operate the machine is added to the transferred heat
and it is this total which is given up to the external medium
(fig. 1(a)).

A heat pump, on the contrary, is designed to give up heat to
a medium (e.g. air or water) which needs to be heated in order
to maintain it at a temperature higher than the ambient temperature. The greater part of this heat is derived from a medium at
a lower temperature:

98

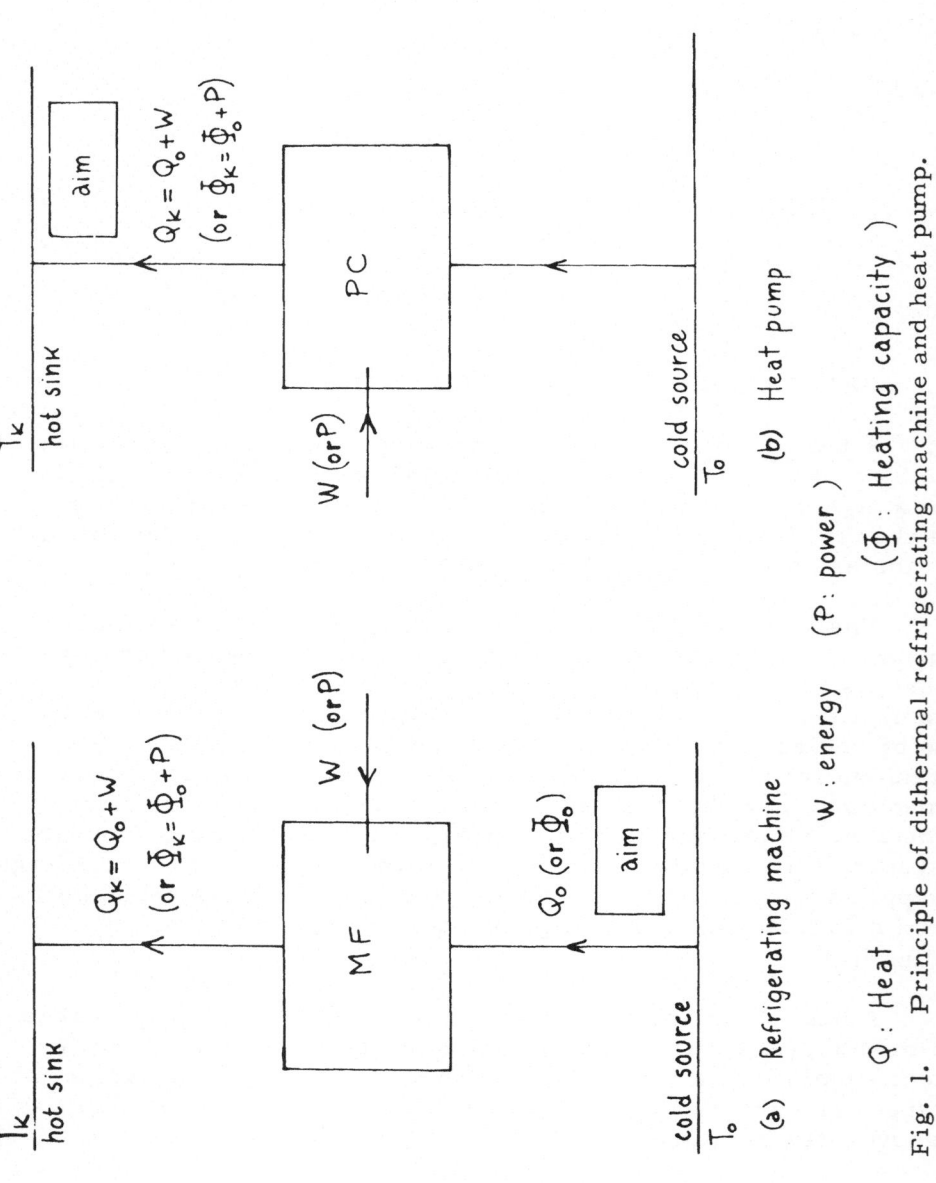

(a) Refrigerating machine   W : energy   (P : power)   (b) Heat pump

Q : Heat   (Φ : Heating capacity)

Fig. 1.   Principle of dithermal refrigerating machine and heat pump.

- air extracted from a building;
- river or geothermic water
- a chemical reactor
- the condenser of a rectification column, etc.
- a fluid heated by solar energy,
  etc.

Certain systems can be used both for refrigeration and for heating.

Like all thermal engines, the heat pump involves at least two heat sources (two-temperature heat pump):

- a <u>cold source (extraction)</u>, at temperature θ_o (or at absolute temperature T_o Kelvin);
- a <u>hot sink</u> (<u>supply</u>), at temperature θ_k (or $T_k°K$) (fig. 1(b)).

It should be noted that a distinction must be made between the temperature (which, incidentally, is generally variable) of the medium which is to be cooled and which is <u>outside the thermodynamic circuit of the pump</u> and the temperature of the cooling medium, the <u>working fluid in the pump</u> (generally known as the <u>refrigerant</u>), which is obviously lower than that of the medium which is to be cooled.

What we shall henceforward call the <u>cold source at T_o</u> is the temperature of the refrigerant at the lower thermal level of the system.

Similarly a distinction must be made between the temperature of the external medium which is to be heated and that of the working fluid which does the heating. The refrigerant temperature at the higher thermal level of the system will henceforward be known as the <u>hot sink at T_k</u>.

1.2 Heating capacity ϕ_k

This represents the amount of heat given up, per unit of time, to the hot sink T_k of the heat pump.

1.3 Coefficient of performance (COP) ε_c

This is defined as the ratio of the heat Q_k (or the heating capacity ϕ_k) supplied by the heat pump at the heat sink to the energy W (or the power P) absorbed by the pump in order to perform this operation.

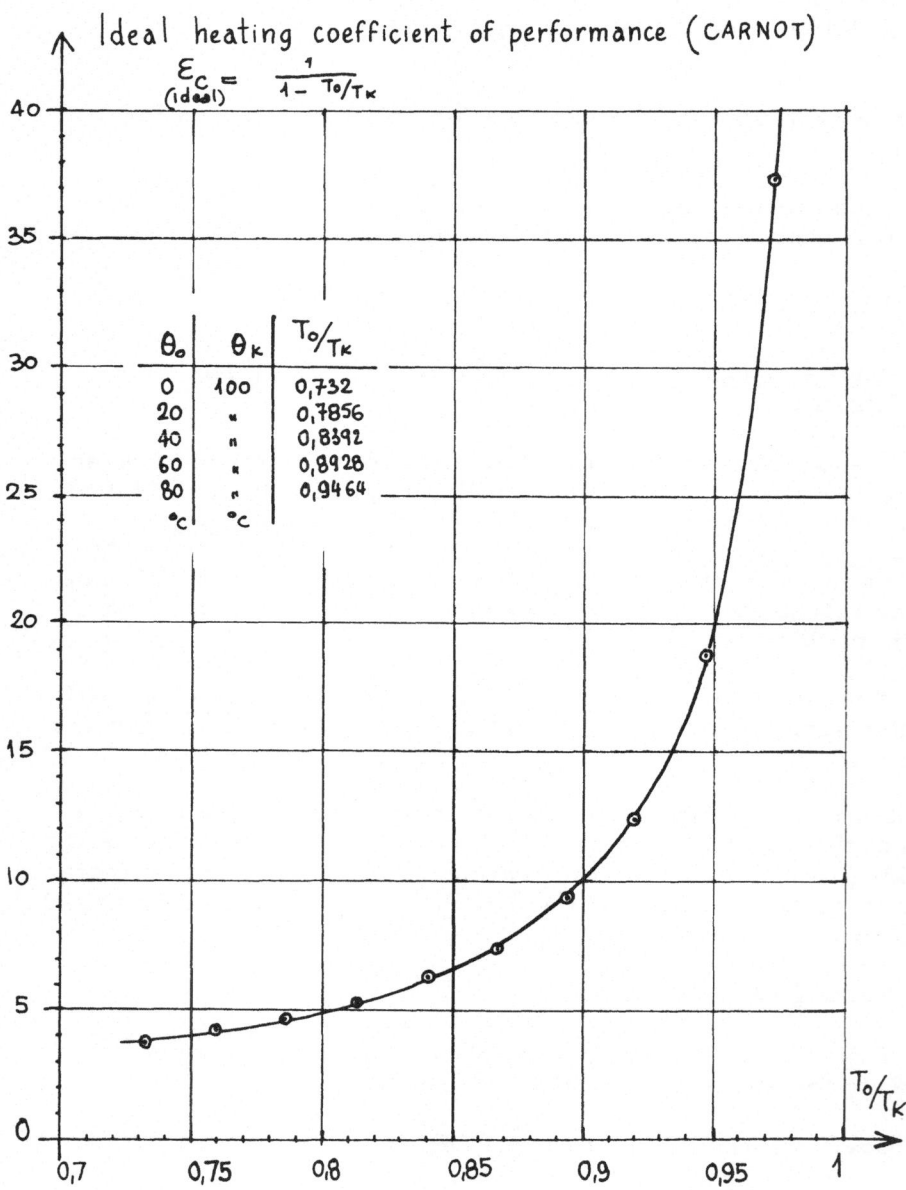

The table within the figure:

θ_o	θ_k	T_o/T_k
0	100	0,732
20	"	0,7856
40	"	0,8392
60	"	0,8928
80	"	0,9464
°C	°C	

Ideal heating coefficient of performance (CARNOT)

$$\varepsilon_{C \atop (ideal)} = \frac{1}{1 - T_o/T_k}$$

Fig. 2.   Ideal heating coefficient of performance of dithermal heat pump.

In the general case of a two-temperature heat pump:

$$\varepsilon_C = \frac{|Q_K|}{|W|} = \frac{|Q_K|}{|Q_K - Q_O|}$$

$$= \frac{|\Phi_K|}{|P|} = \frac{|\Phi_K|}{|\Phi_K - \Phi_O|} \qquad (1)$$

In the case of a reversible two-temperature heat pump (Carnot cycle):

$$\varepsilon_{C\,(ideal)} = \frac{T_K}{T_K - T_O} = \frac{1}{1 - T_O/T_K} \qquad (2)$$

This latter ratio tends to infinity when T_O tends towards T_K. Fig. 2 represents the changes in the ideal heating coefficient of performance $\varepsilon_{C\,(ideal)}$ depending on the ratio T_O/T_k.

1.4 Efficiency compared with the ideal cycle η_c

This is defined as the ratio of the real COP ε_C to the ideal value $\varepsilon_{C\,(ideal)}$ for a reversible heat pump operating within the same temperatures:

$$\eta_C = \frac{\varepsilon_C}{\varepsilon_{C\,(id)}} \qquad (3)$$

This efficiency, which will obviously always be less than unity, largely depends, as we shall see, on the operating conditions of the real heat pump considered.

1.5 Various types of heat pump

As we know, there is a wide variety of refrigeration systems:

-   refrigeration systems using the cold produced by vaporizing a liquid refrigerant:

    compression cooling machines, in which the refrigerant is moved by means of a compressor

    absorption cooling machines, in which the refrigerant is moved by the circulation of a suitable absorbent solution

    vapour-jet cooling machines, in which the refrigerant is moved by an ejector

    (these last two types operate almost exclusively on heat);

-   refrigeration systems using the cold produced by the expansion of compressed gas;

-   thermo-electric refrigeration systems, etc.

A fairly wide variety of heat pumps can therefore be used. The simplest and most efficient however are compression heat pumps. We shall now examine these various types of heat pump.

2. SINGLE-STAGE LIQUEFIABLE VAPOUR COMPRESSION HEAT PUMPS

2.1 Diagram and description

Fig. 3(a) is a diagrammatic representation of the circuit of such a pump, which is used to transfer heat from a circuit containing water to be cooled, e.g. geothermic water, to a circuit containing water which is to be heated.

It comprises:

(1) a water cooler-evaporator E, in which the refrigerant circulates from 5 to 1. This vaporizes at the lower temperature θ_o and at the vapour pressure p_o and then superheats at the same pressure from θ_o to θ_1. This vaporization and superheating use the heat extracted from the water circuit E_g, those temperature falls from θ_{ge} to θ_{gs}.

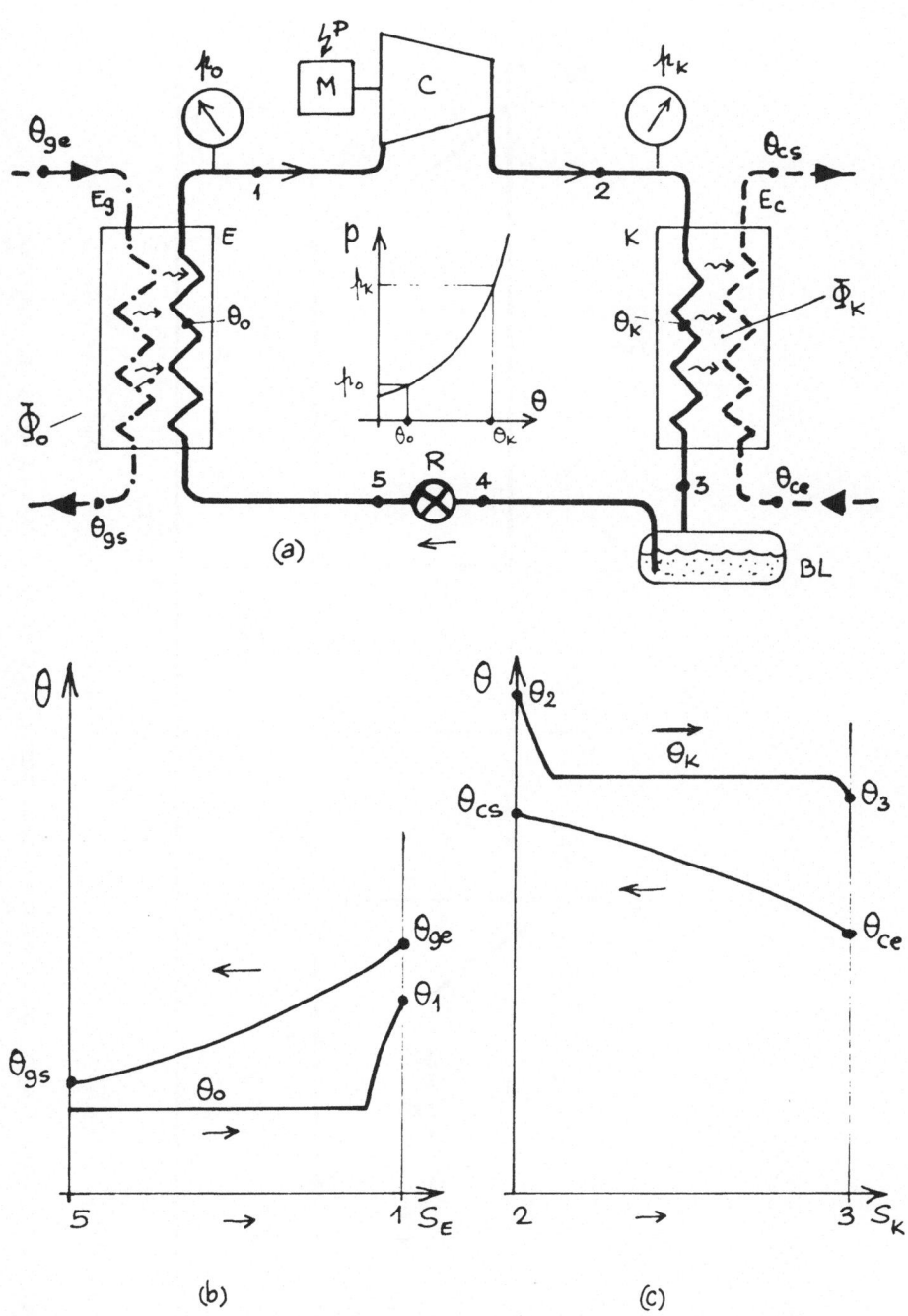

Fig. 3. One stage compression mechanical heat pump.

104

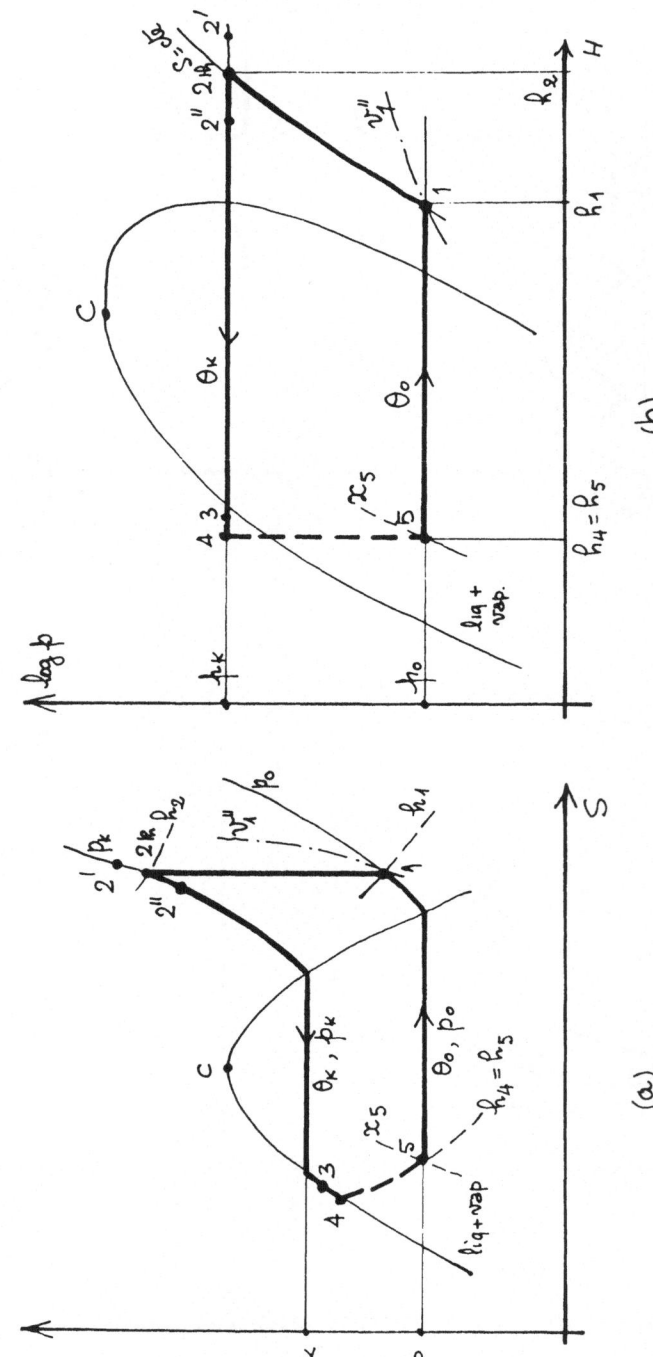

Fig. 4.   One stage compression heat pump cycles in T-S and log p-H diagram.

The graph in fig. 3(b) represents the changes in water and refrigerant temperatures along the heat exchange surface S of the evaporator;

a mechanical compressor C driven by a motor M which sucks in refrigerant vapour from the evaporator at 1 at the evaporation pressure p_o and compresses it as far as 2 at the condensation pressure p_K corresponding to the condensation temperature;

(3) a water heater-condenser K in which the refrigerant is de-superheated, liquefied and sub-cooled from 2 to 3 at the constant condensation pressure p_K. Condensation occurs at the temperature θ_K. Cooling and liquefaction use the heat transferred to the working water circuit E_c, whose temperature rises from θ_{ce} to θ_{cs}. The graph in fig. 3(c) shows the changes in the temperatures of the working water and refrigerant along the heat exchange surface S of the condenser.

(4) a liquid receiver BL which constitutes a reservoir enabling the system to function smoothly.

(5) a regulator or expansion valve R, which is an automatic valve whose purpose is to supply the evaporator E with refrigerant as necessary. The fluid expands from p_k to p_o with no addition or subtraction of external energy and hence at constant enthalpy.

2.2 Presentation of the operating cycle of a single-stage compression heat pump in the entropic graph (T-S) and the enthalpic graph (H-log p)

Fig. 4(a) represents this cycle plotted in the entropic diagram (T-S) which is generally better known.

Fig. 4(b) represents this cycle plotted in the enthalpic diagram (H-log p) which is more widely used.

From 1 to 2: Compression. If this were adiabatic and reversible and hence isentropic, the transformation 1→ 2 would follow the isentrope S=cte(constant) (vertical in the entropy graph). The extremity 2=th (theoretical) of such an ideal transformation lies at the intersection of the isentrope passing through 1 and the isobar p_k.

With a real compressor, point 2:

- is at 2' if the compressor gives up less heat to the external medium than would appear to be the case as a result of the degradation of part of the mechanical energy supplied;

- is at 2" if the compressor gives up more heat than would appear to be the case as a result of internal thermodynamic losses;

- coincides with the theoretical point 2 if the heat losses are exactly equivalent to the mechanical energy degraded during compression.

From 2 (or 2' or 2 ") to 3: de-superheating of the vapour, condensation and sub-cooling at the constant pressure p_K (if the pressure losses in the high-pressure circuit are ignored). The representative point evolves according to the isobar p_k which, in the entropy graph, presents a horizontal segment in the liquid-vapour area.

From 3 to 4: the sub-cooling is completed in the receiver and in the liquid line.

From 4 to 5: isenthalpic expansion. The representative point follows the isenthalp H=cte (vertical in the enthalpy graph). It will be noted that at 5 part of the refrigerant has already been vaporized, making it possible to cool the fraction which remains liquid from θ_4 to θ_0.

From 5 to 1: vaporization and then superheating of the vapour formed at the constant pressure p_0. The representative point varies according to the corresponding isobar.

2.3 Equations for the single-stage compression heat pump

(1) Quantity of heat given up at the hot sink per unit of mass of refrigerant in circulation

$$q_C = h_2 - h_3 \qquad (\frac{j}{K_g}) \qquad (4)$$

where h represents the specific enthalpy of the refrigerant at the points indicated by the subscript.

(2) Mass flow rate of refrigerant to be moved to ensure the production of Φ_k thermal watts at the hot sink

$$m = \frac{\Phi_K}{h_2 - h_3} \qquad (\frac{Kg}{s}) \qquad (5)$$

(3) Volume flow rate to be sucked in by the compressor to produce $\Phi_k w$

$$q_v = m \, v''_1 \qquad (\frac{m^3}{s}) \qquad (6)$$

where v_1 is the specific volume of fluid sucked in.

(4) Effective power (at the shaft) absorbed by the compressor to produce $\Phi_k w$

$$P = \frac{m \cdot (h_{2th} - h_1)}{\eta_i \cdot \eta_m} \qquad (w) \qquad (7)$$

where η_i is the indicated efficiency of the compressor compared with the isentropic compression and η_m the mechanical efficiency of the compressor.

(5) Coefficient of performance ε_c

$$\varepsilon_C = \frac{|\Phi_K|}{|P|} = \frac{(h_2 - h_3)}{(h_{2th} - h_1)} \, \eta_i \cdot \eta_m \qquad (8)$$

(6) Cycle efficiency compared with the ideal cycle η_c

$$\eta_C = \frac{(h_2 - h_3)}{(h_{2th} - h_1)} \cdot \eta_i \cdot \eta_m / \varepsilon_{C(ideal)} \tag{9}$$

$$= \frac{(h_2 - h_3)}{(h_{2th} - h_1)} \cdot \eta_i \cdot \eta_m \left(1 - \frac{T_0}{T_K}\right) \tag{10}$$

This latter value depends on the operating conditions of the heat pump, particularly the ratio T_0/T_k.

Figs. 5, 6 and 7 represent, for three halogenated refrigerants

R 11: $(C\ Cl_3\ F)$

R 12: $(C\ Cl_2\ F_2)$

R 22: $(C\ Cl\ H\ F_2)$

and various condensation temperatures θ_k , the graphs for the changes in the efficiency η_C of the single-stage compression heat pump compared with the ideal cycle, depending on the ratio T_0/T_k.

To calculate η_C from the equation (10) the following assumptions are made:

- $h_2 = h_{2th}$

- mechanical efficiency of the compressor $\eta_m = 0.9$

- indicated efficiency of the compressor η_i variable with the compression ratio $\tau = p_k/p_0$ according to the equation:

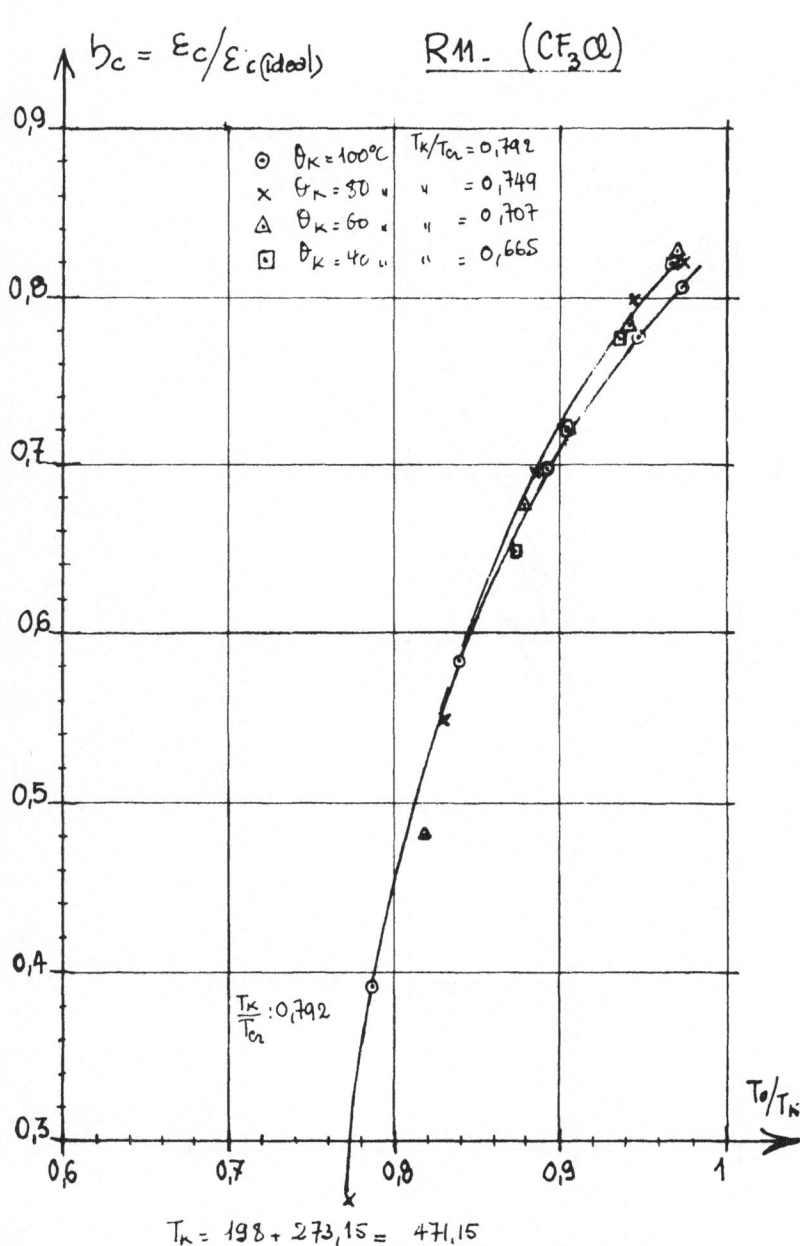

$b_c = \varepsilon_c / \varepsilon_{c(ideal)}$ $R11 - (CF_3Cl)$

$\frac{T_k}{T_{cr}} : 0{,}792$

$T_k = 198 + 273{,}15 = 471{,}15$

Fig. 5. $n_c = f\,(T_O\,/\,T_k)$ for R 11 (CF_3Cl)

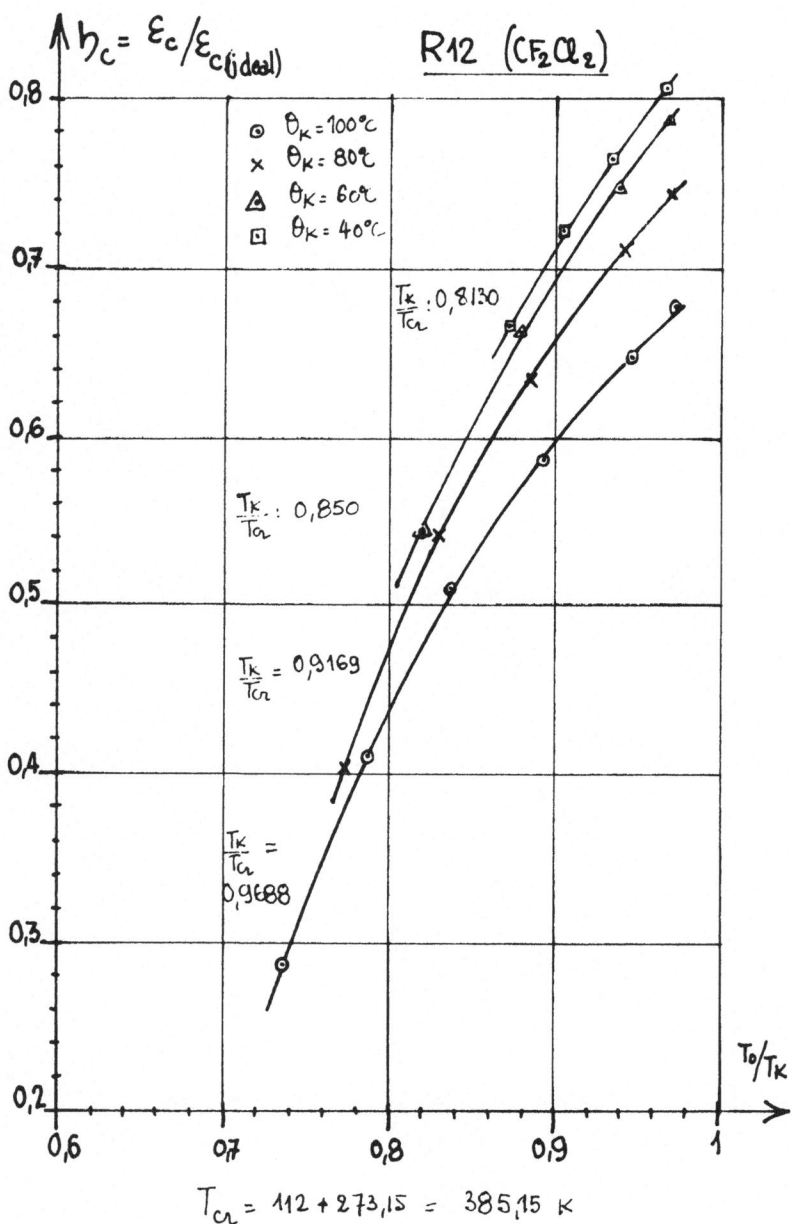

Fig. 6. $\eta_c = f(T_0/T_k)$ for R 12 (CF_2Cl_2)

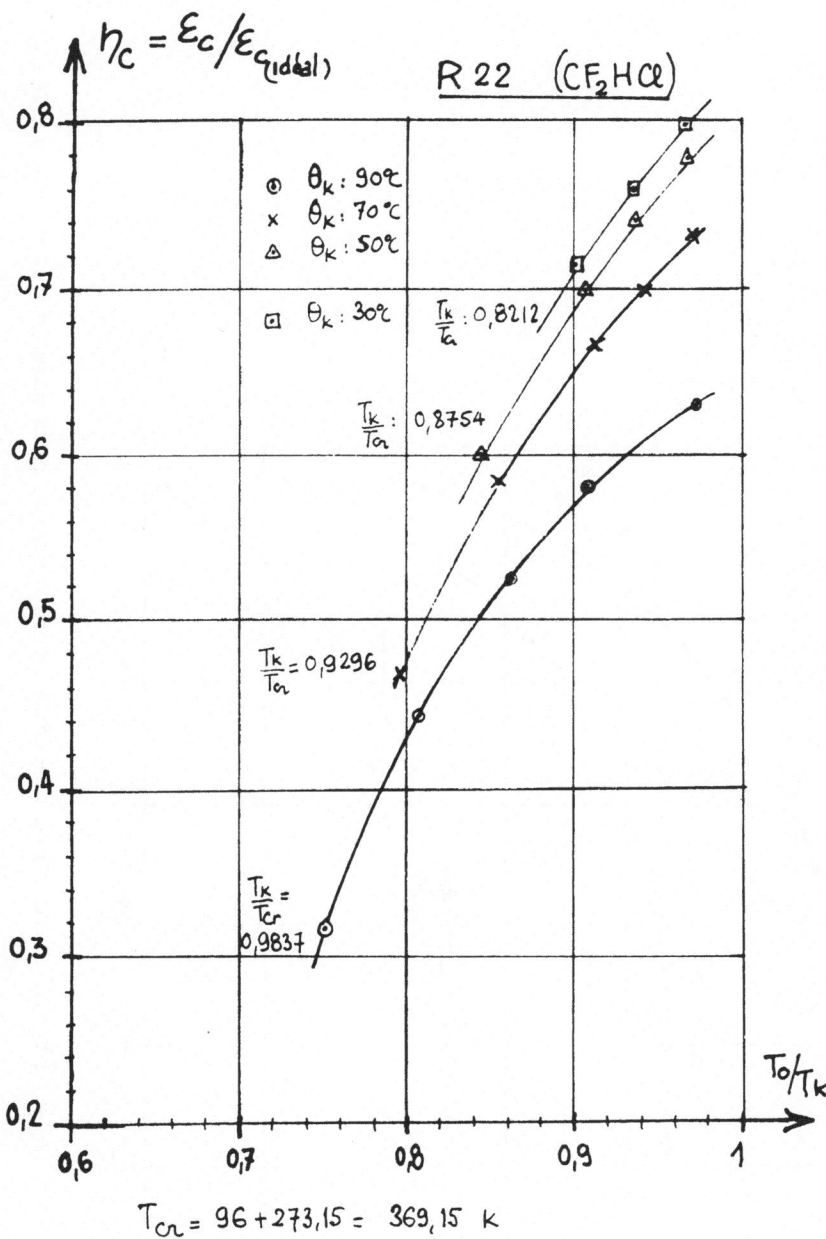

Fig. 7. $n_c = f(T_o/T_k)$ for R 22 (CF$_2$HCl)

112

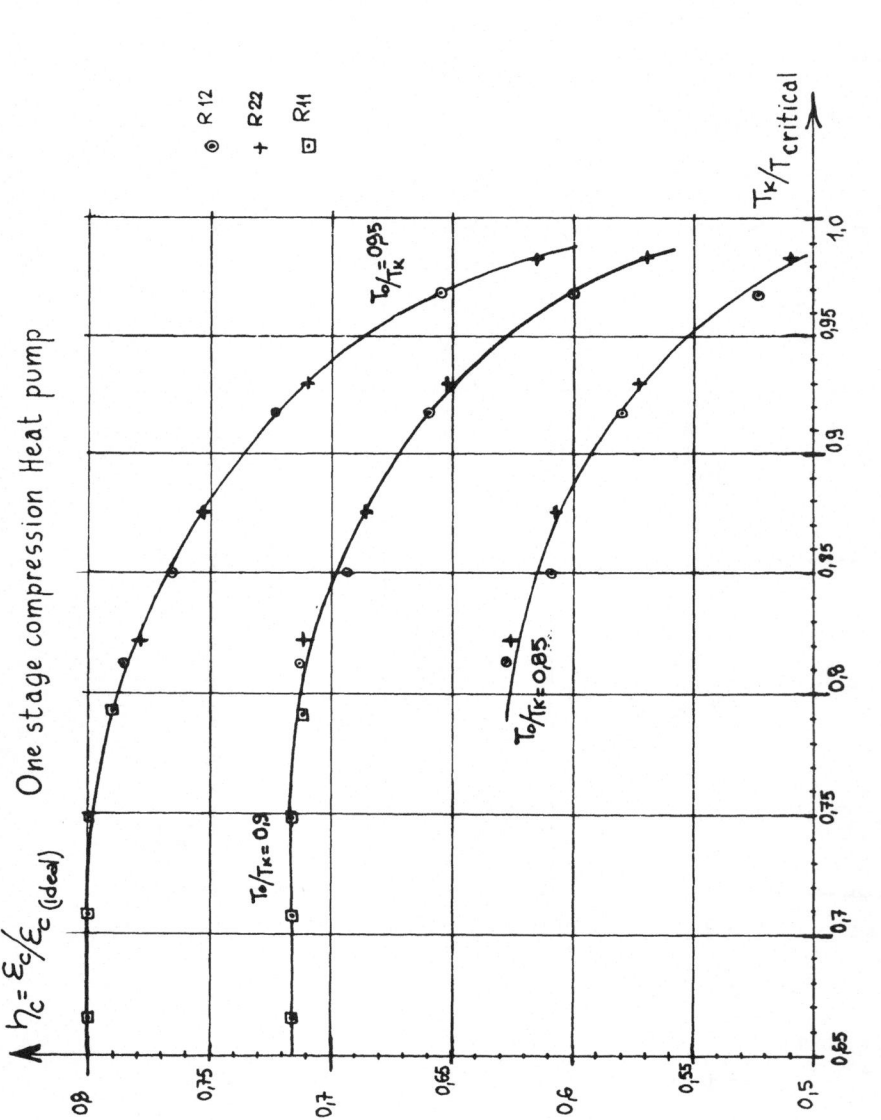

Fig. 8. $\eta_c = f(T_K/T \text{ critical})$ of one stage compression heat pump for various values of $\dfrac{T_0}{T_k}$

$$n_i = 1 - 0,05\ \tau \tag{11}$$

which represents fairly accurately the evolution of this value, particularly for reciprocating compressors.

- superheating of the vapour sucked in at 1 by the compressor

$$\theta_1 - \theta_0 = 10^\circ C$$

sub-cooling of the liquid leaving the condenser:

$$\theta_K - \theta_3 = 5^\circ C$$

In figs. 5, 6 and 7, study of the curves shows that:

the efficiency n_c diminishes rapidly when the relation T_0/T_k decreases, i.e. when the temperatures of the cold source and hot sink of the machine diverge from one another. This is due to the considerable decrease in the indicated efficiency n_i of the compressor when the compression ratio $\tau = p_k/p_0$ rises;

for the same value of T_0/T_k the efficiency n_c decreases as the condensation temperature approaches the critical temperature of the refrigerant, in other words as the ratio $T_k/T_{critical}$ approaches 1.

The graphs in fig. 8 show, for different values of the ratio T_0/T_k and for the three refrigerants envisaged above, how n_c decreases as T_k/T_{crit} increases.

Whatever refrigerant is used, the efficiency of the heat pump cycle diminishes very rapidly when the temperature of the hot sink approaches the critical temperature.

It is therefore clear that to obtain, with such a heat pump, a high enough efficiency n_c and hence acceptable COPs, we must:

use a cold source whose temperature θ_0 is as high as possible;

avoid approaching the critical point of the cycle fluid too closely. The values 0.8 or 0.85 for T_k/T_{crit} appear to be suitable upper limits, which as far as possible should not be exceeded.

(7) Heating effect per unit of swept volume q_{cv}

The volume flow rate q_v which must be sucked in by the the compressor to enable the heat pump to produce a heating capacity of Φ_k at its hot sink will decrease as the ratio

$$q_{cv} = \frac{\Phi_K}{q_v} = \frac{h_2 - h_3}{v''_1} \qquad (\frac{j}{m^3}) \qquad (12)$$

increases. This value, known as the heating effect per unit of swept volume depends on:

the nature of the fluid: it will increase as the latent heat of liquefaction at θ_k rises and as the specific volume of vapour sucked in decreases. Other things being equal, the higher the vapour pressure of the fluid and therefore the lower its normal boiling point, the smaller will be v''_1. The drawback with such fluids is that, although they offer the advantage of requiring smaller compressors, their critical temperatures are lower and higher condensation pressures are required;

the operating conditions of the heat pump:
- condensation temperature
- evaporation temperature
- superheating of vapour sucked in, etc.

For the purposes of comparison, the following values of q_{cv} are found for the three refrigerants envisaged above (R 11, R 12 and R 22) and for the superheating and the supercooling mentioned, a condensation temperature of $\theta_k = 80°C$ and a vaporization temperature $\theta_o = 20°C$:

R 11:   859.2 Kj/ m^3 (or 205.2 Kcal/ m^3)

R 12:   3,581.8 Kj/ m^3 (or 855.6 Kcal/ m^3)

R 22:   5,633.4 Kj/ m^3 (or 1,545.8 Kcal/ m^3)

115

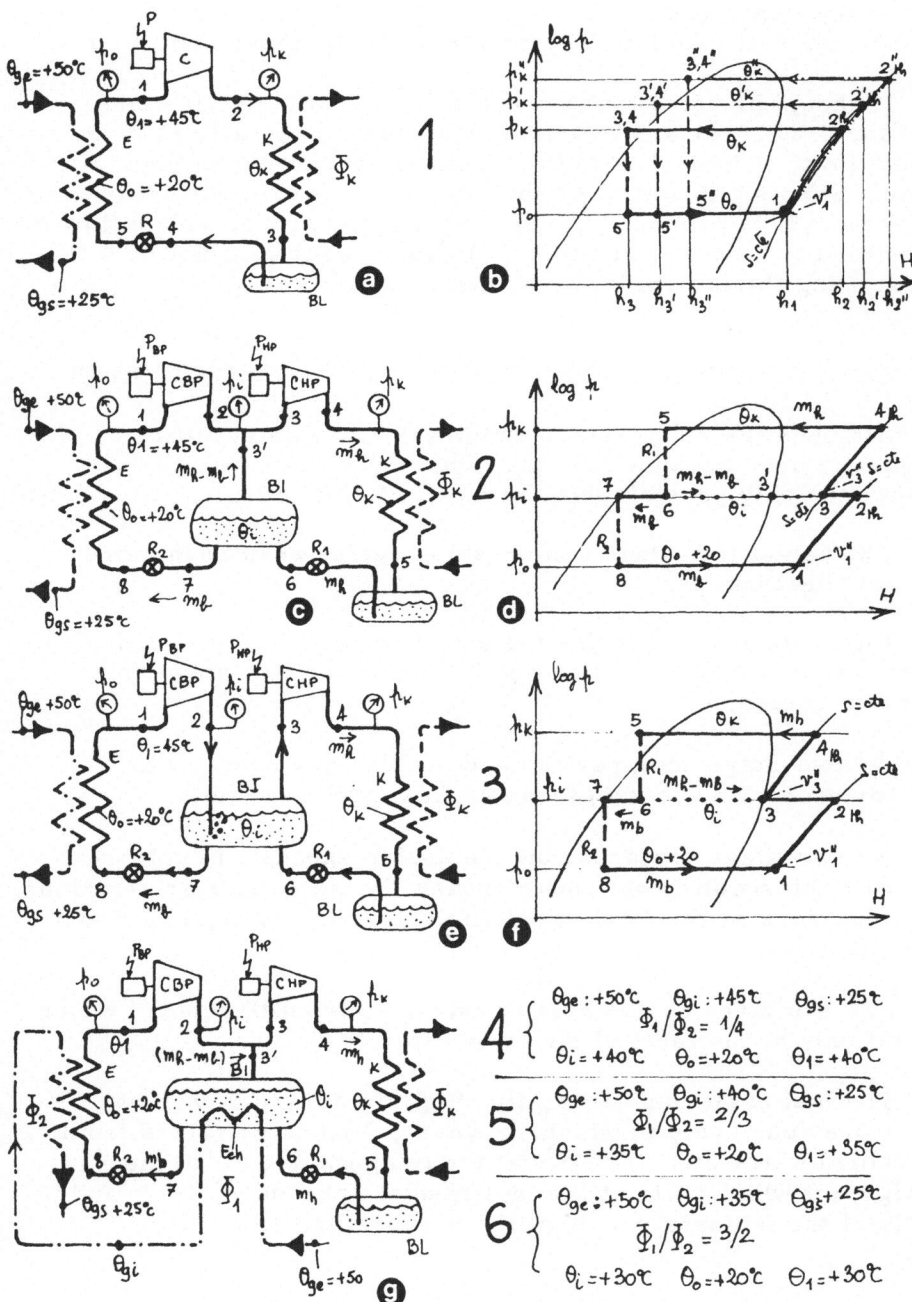

Fig. 9.   Various cycles of compression heat pump with one
or two stage.

To produce the same heating capacity it will therefore be necessary to move a vapour volume flow-rate 6. 55 times greater for R 11 than for R 22 or 4.17 times greater than for R 12.

It should be noted that if we wish to use volumetric compressors, whose possible volume flow rates generally have an upper limit, it is desirable to select refrigerants with high heating effects per unit of swept volume. The problem is different if it is proposed to use turbocompressors, whose flow rates tend to have a lower limit. In such cases it might be worth using fluids with lower values of q_{cv}.

3. COMPARISON BETWEEN OTHER COMPRESSION CYCLES

Although the single-stage compression cycle is most frequently used because it is the simplest to operate, it is neither the only possible nor the most efficient one.

The use of two-stage compression cycles results in more efficient heat pumps.

Fig. 9 shows, alongside the single-stage cycle just considered (case 1, figs. 9(a) and 9(b)), two other two-stage compression cycles:

- the two-stage compression and double expansion cycle (or economizer cycle) (case 2, figs. 9(c) and 9(d));

- the two-stage compression and double expansion cycle with intercooling of vapour, generally known to refrigeration engineers as the total injection cycle (case 3, figs. 9(e) and 9(f)).

In these three cycles heat is extracted from the cooled water circuit only at the level of the evaporator E.

The last cycle envisaged, fig. 9(g), is simply a variant of the previous cycle in which, however, heat is extracted from the cooled water circuit partly at the temperature θ_i of the refrigerant which boils at the intermediate pressure, Φ_1 , and partly at the temperature θ_o of the evaporator, Φ_2 .

Cycle 2: two-stage compression and double expansion (fig. 9(c))

The vapour leaving the evaporator E is compressed from the evaporation pressure p_o to the intermediate pressure p_i (1→ 2) in the low-pressure compressor (CBP). At 3 the high-pressure compressor (CHP) sucks in this vapour, to which is added the vapour taken from the interstage drum (BI) which results from the first expansion in the valve (R1). It compresses these vapours from p_i to p_k, the condensation pressure (3→ 4). As before, from 4 to 5 the refrigerant is de-superheated, liquefied and sub-cooled, giving up heat to the working circuit. The liquid formed expands from p_k to p_i in the first expansion valve (R1) (5→ 6). The refrigerant liquid cools from θ_5 to θ_i, the saturation temperature at the intermediate pressure p_i. The The second expansion valve expands this saturating liquid from p_i to p_o (7 →8). The liquid-vapour mixture at 8 enters the evaporator.

The compression ratios being lower, the indicated efficiency is higher. On the other hand, the vapour produced by the first cooling of the liquid from θ_5 to θ_i is compressed only by the high-pressure compressor. For these reasons the efficiency of this cycle is higher than that of a single stage cycle.

The heating capacity Φ_k is:

$$\Phi_K = m_h \cdot (h_4 - h_5) \qquad (w) \tag{13}$$

where m_h is the high-pressure mass flow rate.

The actual power absorbed by the two stages is as follows:

low-pressure $\quad P_b = \dfrac{m_b \cdot (h_{2th} - h_1)}{(\eta_i \eta_m)_{BP}} \qquad (w) \tag{14}$

high-pressure $\quad P_h = \dfrac{m_h \cdot (h_{4th} - h_3)}{(\eta_i \eta_m)_{HP}} \qquad (w) \tag{15}$

The COP ε_C then becomes

$$\varepsilon_C = \frac{\Phi_K}{P_b + P_h} = \frac{(h_4 - h_5)}{\dfrac{(h_{2th} - h_1) \cdot m_b}{(\eta_i \cdot \eta_m)_{BP} \cdot m_h} + \dfrac{(h_{4th} - h_3)}{(\eta_i \cdot \eta_m)_{HP}}} \tag{16}$$

The ratio m_b/m_h is deduced from the energy balance of the interstage drum BI ($h_6 = h5$: isenthalpic expansion):

$$+ \, m_h \, h_5 - m_b \, h_7 - (m_h - m_b) \, h_{3'} = 0 \tag{17}$$

$$\frac{m_b}{m_h} = \frac{(h_{3'} - h_5)}{(h_{3'} - h_7)} \tag{18}$$

The specific enthalpy at point 3 can be obtained by calculating the energy balance for the node comprised by 2. 3' and 3:

$$+ \, m_b \, H_2 + (m_h - m_b) \, h_{3'} = m_h \, h_3 \tag{19}$$

$$h_3 = h_{3'} + \frac{m_b}{m_h} \, (h_2 - h_{3'}) \tag{20}$$

The curve for this cycle in the enthalpy graph is shown in fig. 9(d).

Cycle 3: two-stage compression with double expansion and intercooling of vapour (fig. 9(e)).

This differs from the previous cycle only by the more marked cooling of the vapour compressed by the low-pressure stage. The vapour sucked in at 3 by the high-pressure compressor is virtually saturated at the intermediate pressure p_i.

The equations (13), (14), (15) and (16) above are equally applicable in this case. The ratio m_b/m_h is however different in form. The balance for the interstage drum gives us:

$$+ \ m_b \ h_2 - m_h \ h_3 + m_h \ h_5 - m_b \ h_7 = 0 \tag{21}$$

$$\frac{m_b}{m_h} = \frac{(h_3 - h_5)}{(h_2 - h_7)} \tag{22}$$

The curve for this cycle in the H-log p graph is presented in fig. 9(f).

Cycles 4, 5 and 6

These are similar to cycle 2 but with an additional heat load ϕ_1 at the intermediate temperature. The heat cycles 4, 5 and 6 differ from one another by the ratio ϕ_1/ϕ_2 between the heat loads given up at the heat pump to the cold sources at the temperatures θ_1 and θ_0. Since ϕ_1 varies from one cycle to another, the intermediate temperature θ_{gi} of the cooled water takes different values. This leads to a corresponding change in:

-   the intermediate temperature θ_i of the refrigerant so that this is always less than 5°C (in the example chosen) at θ_{gi};

-   the suction temperature θ_1 which is also assumed to be less than 5°C at θ_{gi}.

Here again equations (13), (14), (15) and (16) apply. The relation m_b/m_h takes yet another form

$$\Phi_2 = m_b \cdot (h_1 - h_7) \qquad (w) \qquad (23)$$

and the balance for the interstage cooler gives us:

$$+ \ m_h \cdot h_5 + \ \Phi_1 - m_b \cdot h_7 - (m_h - m_b) \cdot h_{3'} = 0 \qquad (24)$$

From (23) and (24) we derive:

$$+ \ m_h \cdot h_5 + \frac{\Phi_1}{\Phi_2} \cdot m_b \cdot (h_1 - h_7) - m_b \cdot h_7 - (m_h - m_b) h_{3'} = 0 \ (25)$$

giving

$$\frac{m_b}{m_h} = \frac{(h_{3'} - h_5)}{(h_{3'} - h_7) + \dfrac{\Phi_1}{\Phi_2} \cdot (h_1 - h_7)} \qquad (26)$$

Comparison between the various cycles

To make this comparison, we have calculated, for four condensation temperatures θ_k= 60, 70, 80 and 90°C, the mechanical power required by heat pumps using the above-mentioned cycles to produce 1 thermal kw at the hot sink, from the heat taken from a water circuit which is cooled from 50 to 25°C.

For cycles 1, 2 and 3, the evaporation temperature assumed is θ_o = 20°C and the temperature of the vapour sucked in at 1 by the compressor θ_1 = 45°C. These values are compatible with the temperatures of the water circuit to be cooled.

For cycles 4, 5, and 6, the same vaporization temperature θ_o= 20°C has been retained and, as stated above, values of θ_1 compatible with the evaporator water input temperatures θ_{gi} have been adopted.

For cycles 2 and 3, we have adopted an intermediate pressure p_i which is the geometric mean between the extreme values:

$$p_i = \sqrt{p_O \cdot p_K} \qquad (27)$$

For cycles 4, 5 and 6, on the other hand, the intermediate temperatures at the selected value of θ_{gi} have been adopted. For all these cycles it has been assumed that the state of the refrigerant fluid when compressed by the various compressors corresponds to that obtained after isentropic compression: $4 = 4_{th}$, $2 = 2_{th}$. For the indicated output, the same variation law $\eta_i = f(\tau)$ as before has been adopted.

All these calculations were made for R 12 (CCl_2F_2)

The graphs in fig. 10 show the results of these calculations.

It will be noted that:

(1) the power required by the single-stage compression heat pump (1) is considerably higher than that absorbed by the other machines;

(2) the efficiency of cycles 2 and 3 is approximately the same. Considerable energy is saved by using these two-stage cycles, especially at high condensation temperatures ($\approx$ 11% at 60°C, $\approx$ 25% at 90°C);

(3) the usefulness of the various cycles, 4, 5 and 6 largely depends on the hot sink temperature required. They can be recommended for the lowest θ_k temperatures but are of no value for the highest temperatures. This is due to the fact that θ_i is limited by the temperature of the external water to be cooled and by the fact that, for high condensation temperatures, the compression ratios of the high-pressure stage increase, causing a corresponding decrease in the indicated efficiency. This is particularly true for cycle 6.

To conclude:

- the two-stage cycles are more efficient;

- the optimization of each cycle constitutes a special case, since the main factor when choosing the cycle to be adopted is the hot sink temperature θ_k.

Note: The use of a two-stage compression cycle does not necessarily entail the use of two separate compressors. Some compressors are designed with two stages (compound reciprocating compressors, turbo-compressor with two or more impellers, etc.).

122

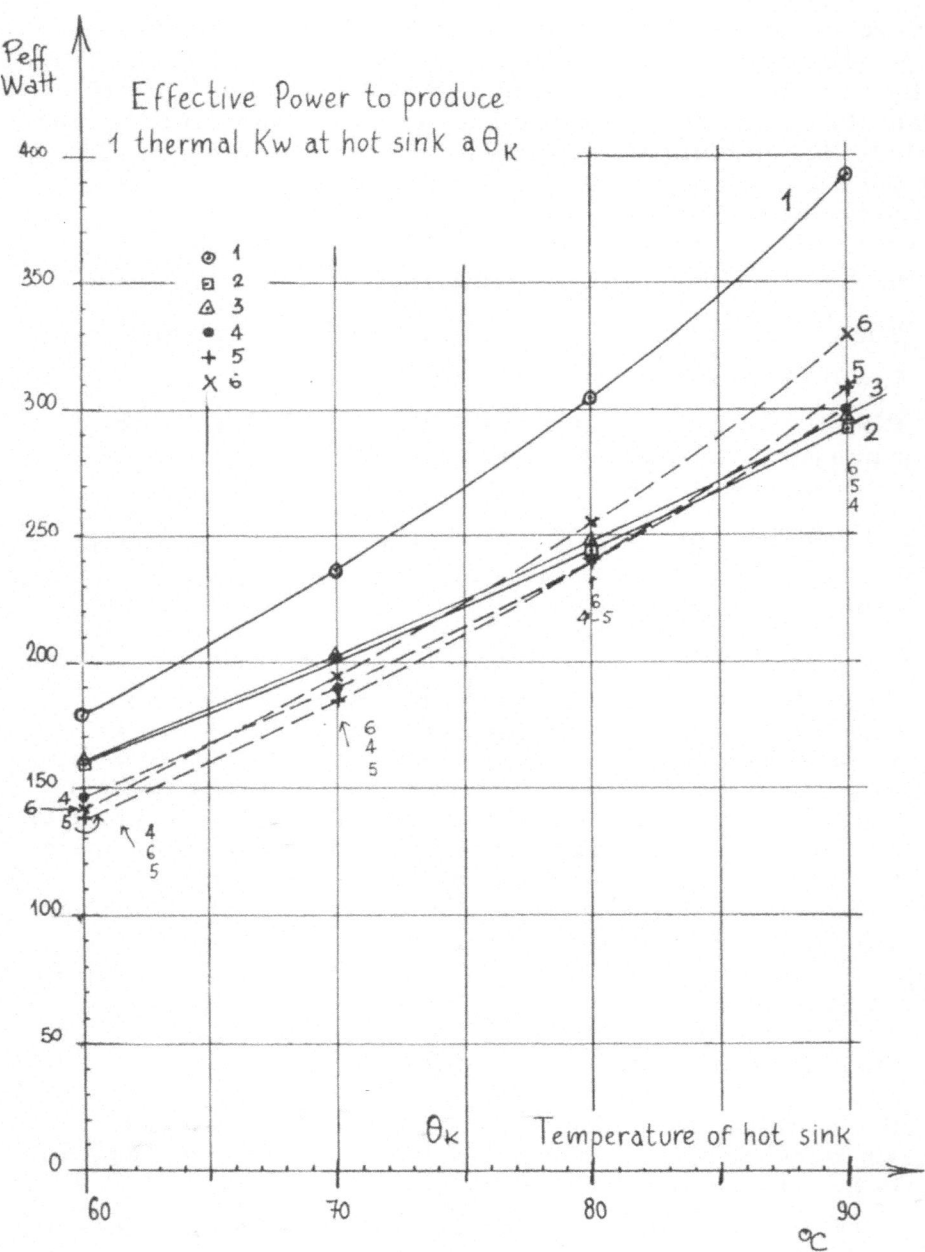

Fig. 10   Energetic consumption of the various heat pumps
of fig. 9.

123

4. WORKING FLUIDS (REFRIGERANTS)

4.1 Thermodynamic criteria

Critical temperature θ_{cr}. We have already noted the importance of this factor. For compounds of a given chemical family, this temperature increases with the molar mass.

Normal boiling point θ_{eb}. If the evaporation temperature θ_0 is less than θ_{eb}, the low-pressure part of the heat pump is under a pressure less than that of the atmosphere, and this may facilitate the penetration of non-condensable gas into the cooling circuit.

Latent heat of condensation. This conditions the mass flow rate of refrigerant to be moved.

Heating effect per unit of swept volume q_{cv}. The volume flow rate of the compressor depends directly on this. Finally, we shall examine the importance of the condensation pressure at the desired hot sink temperature.

4.2 Safety criteria

Inflammability (or otherwise)

Toxicity (or otherwise)

Chemical stability of the fluid. This depends largely on the temperature and varies considerably according to the nature of the materials with which the refrigerant is in contact.

4.3 Technical criteria

Action on metals

Action on material used for joints, plastics and elastomers.

Action on lubricating oils
   physical: mixing properties
   chemical

Heat exchange efficiency, especially as regards boiling and condensation.

Liability to leak

TABLE 1

Refrigerant	Chemical formula	Molecular Weight	Boiling point θ_{eb} °C	Critical point θ_{cr} °C	Critical pressure h_{cr} bar	Vapories pressure at $\theta_K=90°c$ $h_{K_{90}}$ bar	Latent heat of vaporiz. at θ_{eb} Kj/Kg	
R-113	$C_2Cl_3F_3$	187,39	47,57	214,1	34,36	3,44	146,8	*
R-114 B2	$CB_2F_2-CB_2F_2$ $C_2F_4Br_2$	259,85	47,26	214,5	34,67	3,5	104,6 (est.)	
R-216	$C_3Cl_2F_6$	221	35	180	27,54	4,84	117,63	
R-11	CCl_3F	137,38	23,77	198	44,09	6,66	180,24	**
R-21	$CHCl_2F$	102,93	8,92	178,5	52	10,56	242,20	
R-114	$CClF_2-CClF_2$ $C_2Cl_2F_4$	170,94	3,77	145,7	32,8	11,477	137	*
R-12 B1	$CClF_2Br$	165,4	-4	153,8	42,42	11,26	135,12	
R.C318 e	C_4F_8	200,04	-6,07	115,34	28,12	17,12	116,95	
R-12	CCl_2F_2	120,9	-29,8	112	41,15	27,88	165,34	**
R-22	$CHClF_2$	86,5	-40,8	96	49,8	44,42	233,6	**

*   often used
* * widely used

Halocarbon refrigerants for heat pumps.

4.4 Economic criteria

Price

Availability

The refrigerants used in cooling techniques are as follows:

- ammonia
- hydrocarbons (for petrochemistry)
- halogenated (fluorinated, chlorinated, brominated) hydrocarbon derivatives

The major drawbacks of ammonia, which is an excellent, well-known and widely-used refrigerant, are its toxicity and the difficulty of employing high pressures. The hydrocarbons propane, butane, etc. can be used but are dangerous owing to their inflammability. These refrigerants are cheap.

Table 1 shows some of the thermodynamic properties of the halogenated hydrocarbon derivatives that can be used in heat pumps.

These refrigerants are non-inflammable and practically non-toxic: R 11, R 22, R 114 B2 have about the same toxicity as CO_2; R 12, R 114 and R C 318 are still less toxic.

Their chemical stability is good. Tests made at fairly high temperatures (315-537°C in the presence of steel){5} have shown the particularly high stability of R 114 and R 216.

They are compatible with all current metals. Zinc, magnesium and its alloys should, however, be avoided. Certain metals exert a catalytic action which favours the thermal decomposition of the refrigerant.

As regards the elastomers, most of the halogenated fluids can be used with Buna-N (butadiene acrylonitrile), except R 21 and R 22 (for the latter, neoprene is preferable). They are either fully miscible with oil (R 12) or partly miscible, depending on the temperature, the nature of the oil and the concentration of the mixture (R 22). Oils chemically compatible with these refrigerants can very easily be found.

The convective heat transfer with these fluids is generally poor. Their low surface tension makes them very liable to leak from the circuits in which they are contained. Special care must be taken to make the circuits leakproof.

In 1975 R 12 cost F. fr. 5. 5 per kg (excluding taxes) in France. The prices of the various halogenated refrigerants, compared with that of R 12, are shown in Table 2 below:

TABLE 2

Refrigerant	Price of refrigerant considered / Price of R 12
R 113	1. 26
R 11	0. 57
R 21	6. 13
R 114	1. 62
R 12 B1	3. 54
R 12	1
R 22	1. 80

5.   OTHER COMPRESSION CYCLES

5.1 Three or more stage compression cycles

These are an extrapolation of the two-stage cycles. Their efficiency is still greater, but so is their complexity. They are seldom used except with multiple impeller turbo-compressors.

5.2 Cascade cycles

Cascade heat pumps consist of several single or multi-stage machines arranged in series (fig. 11(a)). The evaporator of the "low temperature" machine, which uses a refrigerant F_1 , absorbs heat from the medium to be cooled in the evaporator E and gives it up to the second machine, which follows it in increasing order of temperature, via the evaporator-condenser KE '. This second pump, using a refrigerant F'_1 which is different from the first one, finally yields heat to the circuit to be heated via its condenser K '.

The characteristics of the fluids F_1 and F'_1 are suitable for the temperature range covered by this machine. In particular fluid F'_1 can be at a low vapour pressure and a high critical temperature.

127

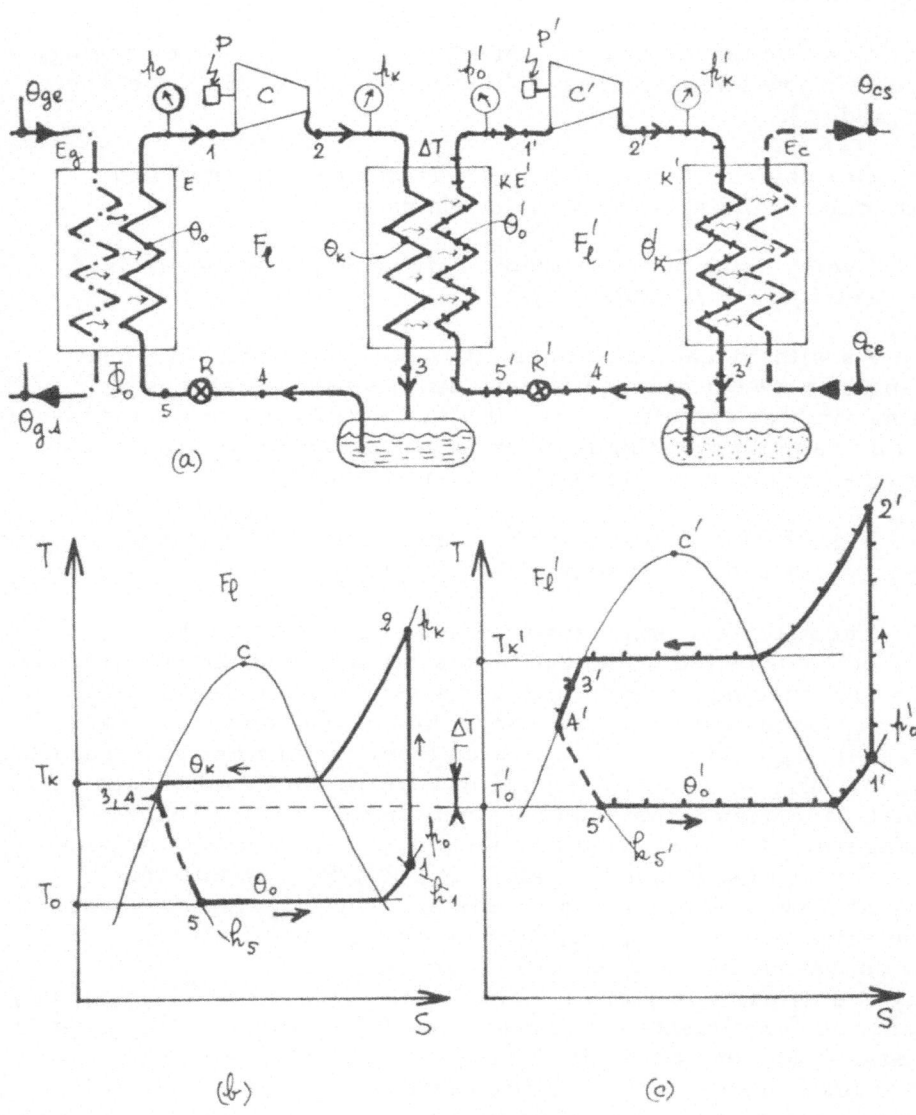

Fig. 11 - Compression heat pump in cascade.

To produce high temperatures it is quite possible to use water, for instance, as the F'_1 fluid.

Fig. 11(b) and (c) shows the cycles for the two fluids F_1 and F_1 in the entropy graph T-S. It will be noted that the condensation temperature θ_k of the fluid F_1 at KE' must be ΔT(from 5 to 10°C in general) higher than the vaporization temperature θ'_o of the fluid F'_1 in this heat exchanger.

Cascade machines are currently in refrigeration technology to obtain low temperatures (differences of 120 to 150°C between θ_o and θ_k).

One obvious drawback is their relative complexity and particularly their use of several compressors.

5.3 Cycles using non-azeotropic refrigerant mixtures (mixed refrigerant cascade cycle)

As with cascade machines, it is possible to obtain a heat transfer between sources at very different temperatures by using simpler circuits, especially those with only one compressor. In this case the working medium is a non-azeotropic mixture of refrigerants (binary, ternary, etc. mixtures).

Fig. 12(a) represents a heat pump based on the mixed refrigerant cascade system.

The machine comprises a series of evaporators E_1, E_2, E_3, etc. connected to the common suction line of the compressor at a pressure p_o. It also includes a series of condensers K_1, K_2, K_3 connected to the discharge side of the compressor at a pressure p_k. These are separated from one another by separators B_1, B_2, B_3, which are designed to receive the liquid phases which have been formed successively in the various heat exchangers. It is these liquids which supply the various evaporators through the expansion valves R_1, R_2, R_3. Within this circuit flows a non-azeotropic mixture of refrigerants, whose concentrations m_1, m_2, m_3, change throughout the circuit. In the condenser K1, between temperatures θ_{k1} and θ'_{k1}, the mixture m_1 partly liquefies. The liquid phase formed is obviously richer in constituents with lower vapour pressures. The gaseous mixture m_2, enriched with lighter constituents, partially liquefies from θ_{k2} to θ'_{k2} in the second condenser, where the temperatures are lower. The mixture m_3, which is still richer in light constituents, liquefies between θ_{k3} and θ'_{k3} at still lower temperatures. Similarly, these various expanded mixtures

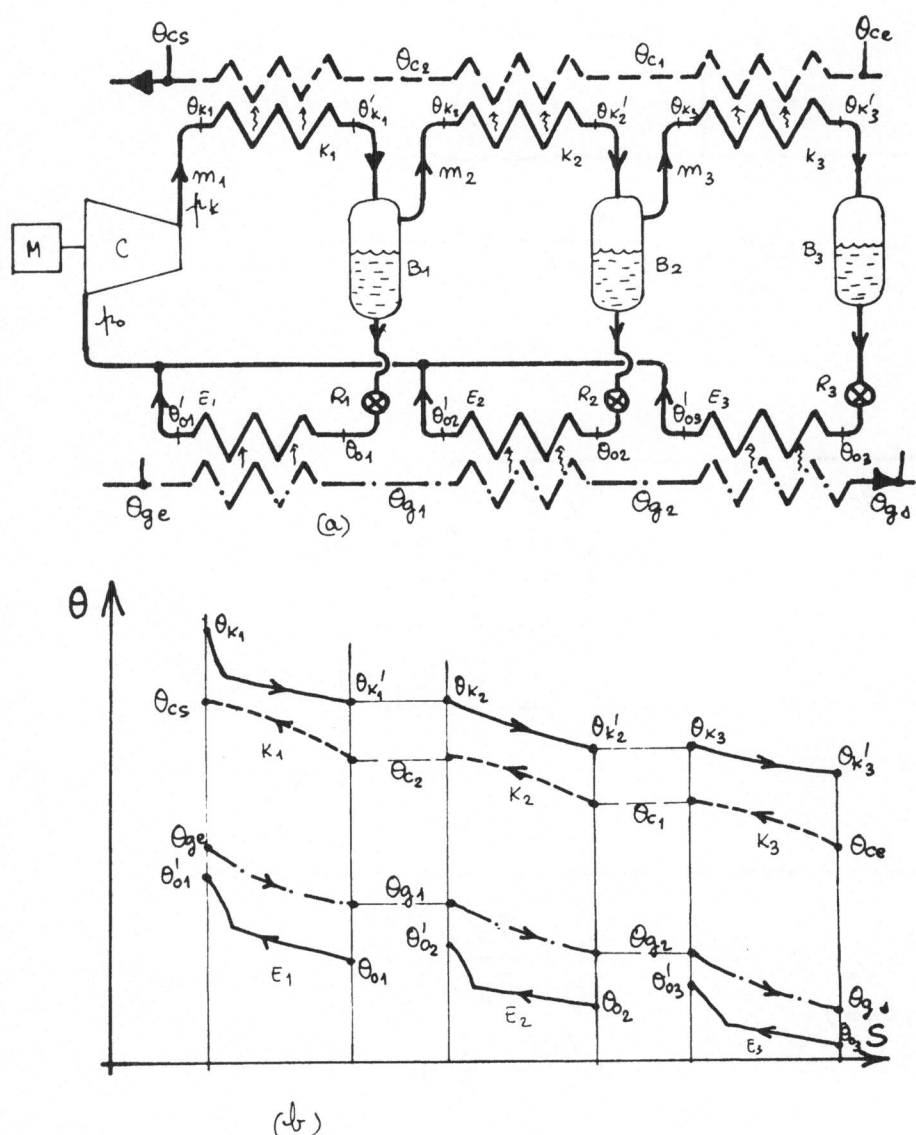

Fig. 12.   Mixing fluids heat pump.

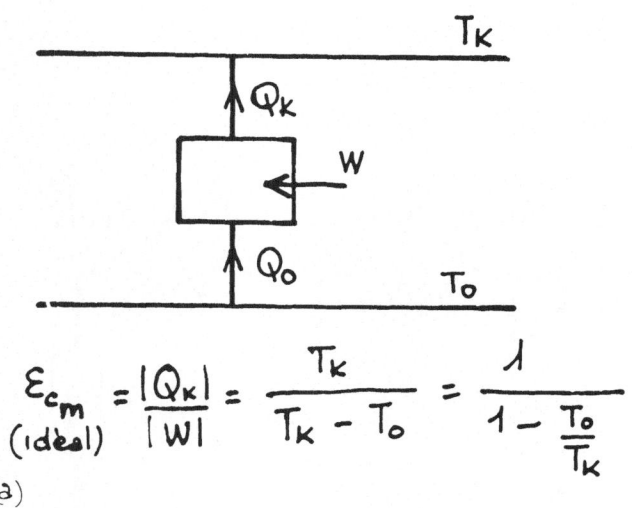

$$\mathcal{E}_{c_m} = \frac{|Q_k|}{|W|} = \frac{T_k}{T_k - T_0} = \frac{1}{1 - \frac{T_0}{T_k}}$$
(ideal)

(a)

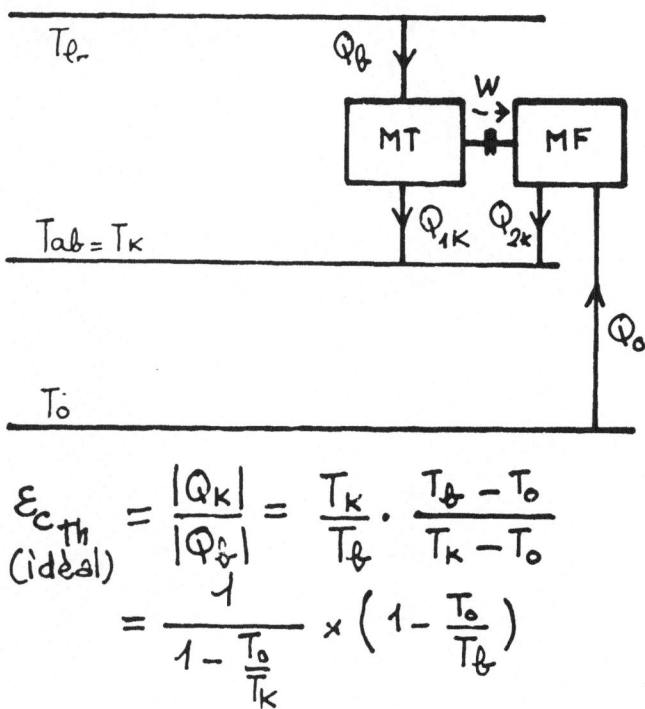

$$\mathcal{E}_{c_{th}} = \frac{|Q_k|}{|Q_\theta|} = \frac{T_k}{T_\theta} \cdot \frac{T_\theta - T_0}{T_k - T_0}$$
(ideal)

$$= \frac{1}{1 - \frac{T_0}{T_k}} \times \left(1 - \frac{T_0}{T_\theta} \right)$$

(b)

Fig. 13.   Dithermal (a) and trithermal (b) heat pump.

will vaporize and superheat at varying temperatures, $\theta_{03} \to \theta'_{03}$, $\theta_{02} \to \theta'_{02}$, $\theta_{01} \to \theta'_{01}$, in the different evaporators, progressively taking heat from the heat-giving fluid, which thus cools from θ_{ge} to θ_{gs}. It is thus possible to cause the temperatures of the fluids, which are to be cooled from θ_{ge} to θ_{gs} and heated from θ_{ce} to θ_{cs}, to vary considerably, while maintaining reasonable differences of temperature between these fluids and the refrigerant (fig. 12(b)).

Mixed refrigerant cascade systems are currently used for liquefying natural gas (differences of approx. 200°C between the lowest and the highest temperatures). They do not appear to have been used to produce fairly high temperatures in heat pumps.

6. HEAT PUMPS USING HEAT FOR THEIR OPERATION

These are:

- compression heat pumps driven by thermal engines;
- absorption heat pumps;
- ejection heat pumps.

6.1 Heating coefficient of performance of such an ideal heat pump

As compared with the single-stage mechanical compression pump (see fig. 13(a) for the ideal machine), these systems need at least three heat sources in order to function. The tri-thermal cycle is shown on figure 13(b):

- the cold source at T_o, where heat is taken from the fluid to be cooled;

- the hot sink at T_k, where heat is given up to the fluid to he heated;

- the high temperature source T_b, from which the machine receives the thermal energy necessary to transfer heat from T_o to T_k.

An ideal heat pump assembly consists of a two-temperature, reversible thermal engine, MT, functioning between the sources T_b and T_k and driving a reversible mechanical heat pump which functions between the sources T_o and T_k.

We must first note that the circuit to be heated receives, at T_k, the following heat:

$$Q_K = Q_{1K} + Q_{2K} = Q_0 + Q_b \tag{27}$$

(first law of thermodynamics)

Q_k is higher than Q_b but at a lower temperature.

To determine the ideal heating coefficient of performance ε_{cth} (ideal) let us apply the second law of thermodynamics (Clausius):

$$+ \frac{|Q_b|}{T_b} - \frac{|Q_K|}{T_K} + \frac{|Q_0|}{T_0} = 0 \tag{28}$$

From (27) we obtain:

$$+ \frac{|Q_b|}{T_0} - \frac{|Q_K|}{T_0} + \frac{|Q_0|}{T_0} = \tag{29}$$

From (28) and (29) the ideal coefficient of performance can easily be deduced by eliminating $\frac{|Q_0|}{T_0}$

$$\varepsilon_{C_{th(ideal)}} = \frac{|Q_K|}{|Q_b|} = \frac{T_K}{T_K - T_0} \cdot \frac{T_b - T_0}{T_b} \tag{30}$$

It will be noted that this is equal to the coefficient of performance of a heat pump absorbing mechanical energy multiplied by the reduction coefficient:

$$\left(1 - \frac{T_0}{T_b}\right)$$

$$\varepsilon_{C_{th(ideal)}} = \varepsilon_{C_{m(ideal)}} \cdot \left(1 - \frac{T_0}{T_b}\right) \tag{31}$$

This reduction becomes more marked as T_o and T_b are closer. To clarify the subject, let us calculate this coefficient of performance where:

$$\theta_O = + 20^\circ C \qquad\qquad T_O = 293,15 \text{ K}$$

$$\theta_K = + 50^\circ C \qquad\qquad T_K = 323,15 \text{ K}$$

$$\theta_b = + 100^\circ C \qquad\qquad T_b = 373,15 \text{ K}$$

$$\varepsilon_{C_{m}(ideal)} = \frac{323,15}{323,15 - 293,15} = 10,77$$

$$\varepsilon_{C_{th}(ideal)} = 10,77 \times (1 - \frac{293,15}{373,15}) = 10,77 \times 0,214 = 2,31$$

In this case the coefficient of performance is 4 or 5 times lower. As the system receives unco-ordinated thermal energy it consumes more.

6.2 Compression machine driven by a thermal engine

- Reciprocating compressor → driven by diesel or gas engine.
- Centrifugal turbocompressor driven by steam turbine.
- Centrifugal turbocompressor driven by gas turbine.

These different versions have been used in total energy plants.

6.3 Absorption heat pump

This system uses a couple consisting of two substances:

- a refrigerant as the working medium of the heat pump;
- a solvent or absorbent.

The refrigerant is transported by the circulation of the solutions from the low-pressure section to the high-pressure section of the machine.

Fig. 14(a) represents diagrammatically the circuit of an absorption heat pump. As with compression heat pumps, we find, in order, between points 2 and 1:

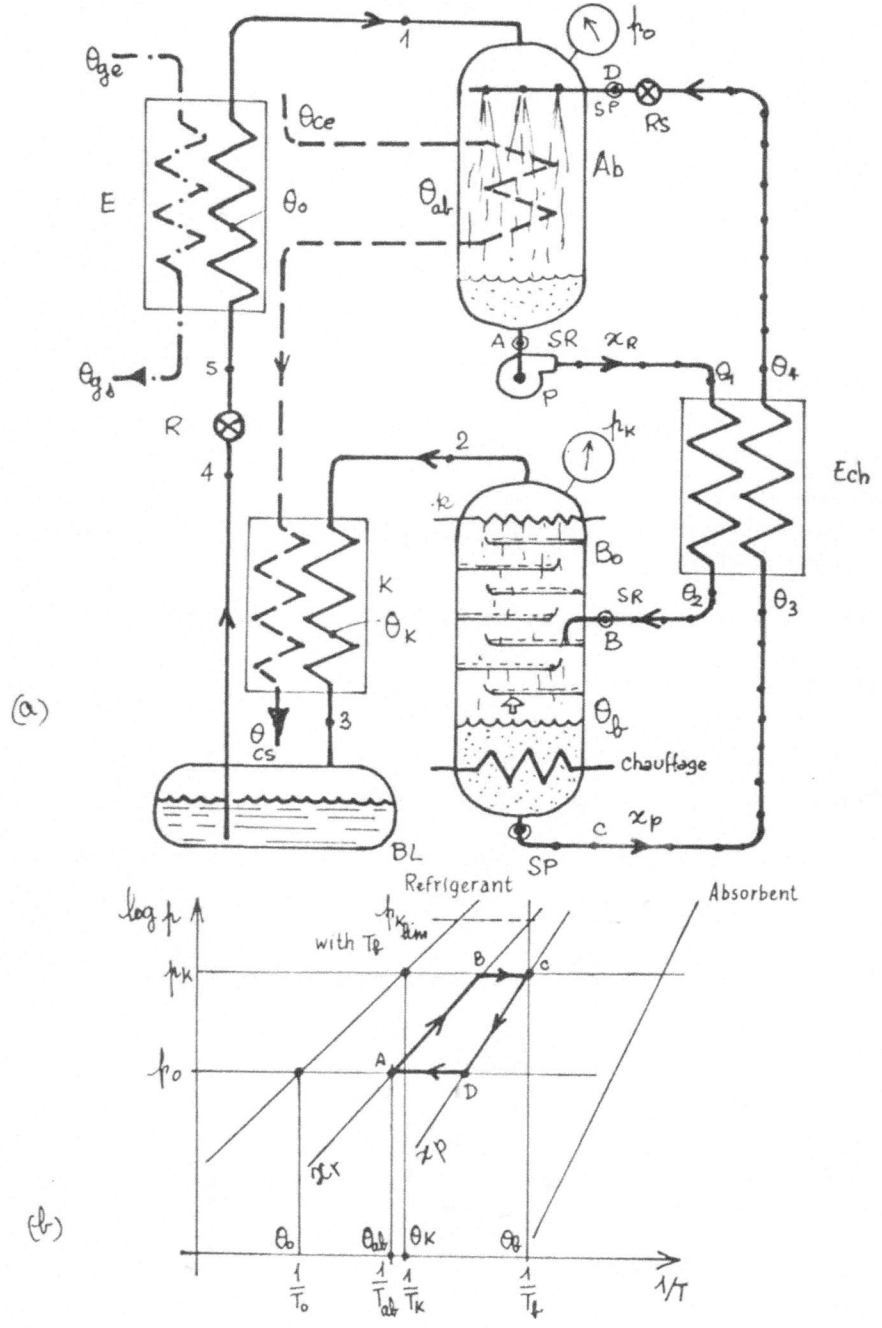

Fig. 14.   Absorption heat pump.

- the <u>condenser K</u>, in which the refrigerant condenses at supplying heat to the circuit containing fluid to be heated which emerges at θ_{cs};

- the <u>liquid receiver BL</u>;

- the <u>expansion valve R</u>;

- the <u>evaporator E.</u>

Here, the compressor is replaced by the following assembly:

- the <u>absorber</u> Ab;

- the <u>pump</u> P;

- the <u>heat exchanger</u> Ech;

- the <u>generator</u> Bo;

within which the solutions carrying the refrigerant circulate permanently.

In the absorber Ab the refrigerant vapour from the evaporator and a weak solution of refrigerant SP, in the form of a fine spray, are brought together. This solution dissolves and absorbs these vapours and becomes stronger. The latent heat of dissolution is extracted from this solution by means of a heat exchanger through which the fluid to be heated flows and which enters at θ_{ce}. The absorber is at the vaporization pressure p_0. The strong solution of refrigerant, SR, is received by the pump P, which pumps it, via the heat exchanger Ech, where it is heated from θ_1 to θ_2, to the <u>generator</u> B_0, in which the condensation pressure prevails.

This frequently takes the form of a rectification column equipped with baffles or adequate packing. The base of the boiler is heated (to θ_b) and the machine receives the thermal energy it needs in order to function. The upper part comprises a small condenser k to provide the return flow in the column. Below the baffle which receives, at B, the strong solution, the liquid is enriched with solvent and the solution becomes weaker in refrigerant. Above, the vapour is enriched with refrigerant. The weak, hot solution at C returns to the absorber and flows through the heat exchanger Ech, where it cools from θ_3 to θ_4 and heats the strong solution. An expansion valve RS regulates the arrival of the weak solution in the absorber.

Fig. 14(b) represents, in the thermodynamic graph $\log p - \dfrac{1}{T}$ of the absorbent-refrigerant couple considered, the theoretical development cycle for the solutions. "Refrigerant" represents the vapour pressure curve for the pure refrigerant; "absorbent" those of the pure solvent, whose pressures are much lower. If the evaporation temperature θ_0 (or T_0) and condensation temperature θ_k (or T_k) are known, it is possible to plot the isobars p_0 and p_k (in relation to the refrigerant vapour pressure curve). The concentration of the strong solution is obtained from the isotitric curve x_R(1) passing through the point of intersection A of the isotherm θ_{ab} (or T_{ab}) of the absorber and the isobar p_0.

Similarly the concentration of the weak solution x_p is obtained from the isotitric curve passing through the point of intersection C of the isotherm θ_b (or T_b) of the generator and the isobar p_k.

In order that the machine may function it is obviously necessary that $x_p < x_R$. It will be noted that, for a given absorber temperature θ_{ab} and a cold source temperature θ_0, the temperature of the generator θ_b does not permit the heat pump to give up heat at a hot sink temperature equal to or greater than $\theta_{k(lim)}$ corresponding to $p_{k(lim)}$; for this pressure we should find $x_R = x_p$. To produce heat at θ_k(lim) or a higher temperature we must use a generator temperature θ_b which is itself higher, or an absorber temperature θ_{ab} which is lower. Few absorbent-refrigerant couples can be used at the present time. Halogenated refrigerants, which are chemically almost inert, are scarcely compatible with current absorbents. The only couples actually used in absorption cooling machines are: Ammonia (refrigerant) - water (absorbent); water (refrigerant) - lithium bromide solution (absorbent).

The absorption machine, which is frequently used for air-conditioning, appears to be very seldom employed at the present time as a pure heat pump.

It should be noted that, in total energy plants, assemblies have been used consisting of centrifugal turbocompressors driven by back-pressure steam and absorption machines heated by exhaust steam.

Note: The mechanical energy absorbed by the pump P is generally very small compared with the thermal energy used in the various exchangers.

---

(1) x represents the strength of the solution in refrigerant;

$$x = \frac{m \text{ refrigerant}}{m \text{ refrigerant} + m \text{ solvent}}$$

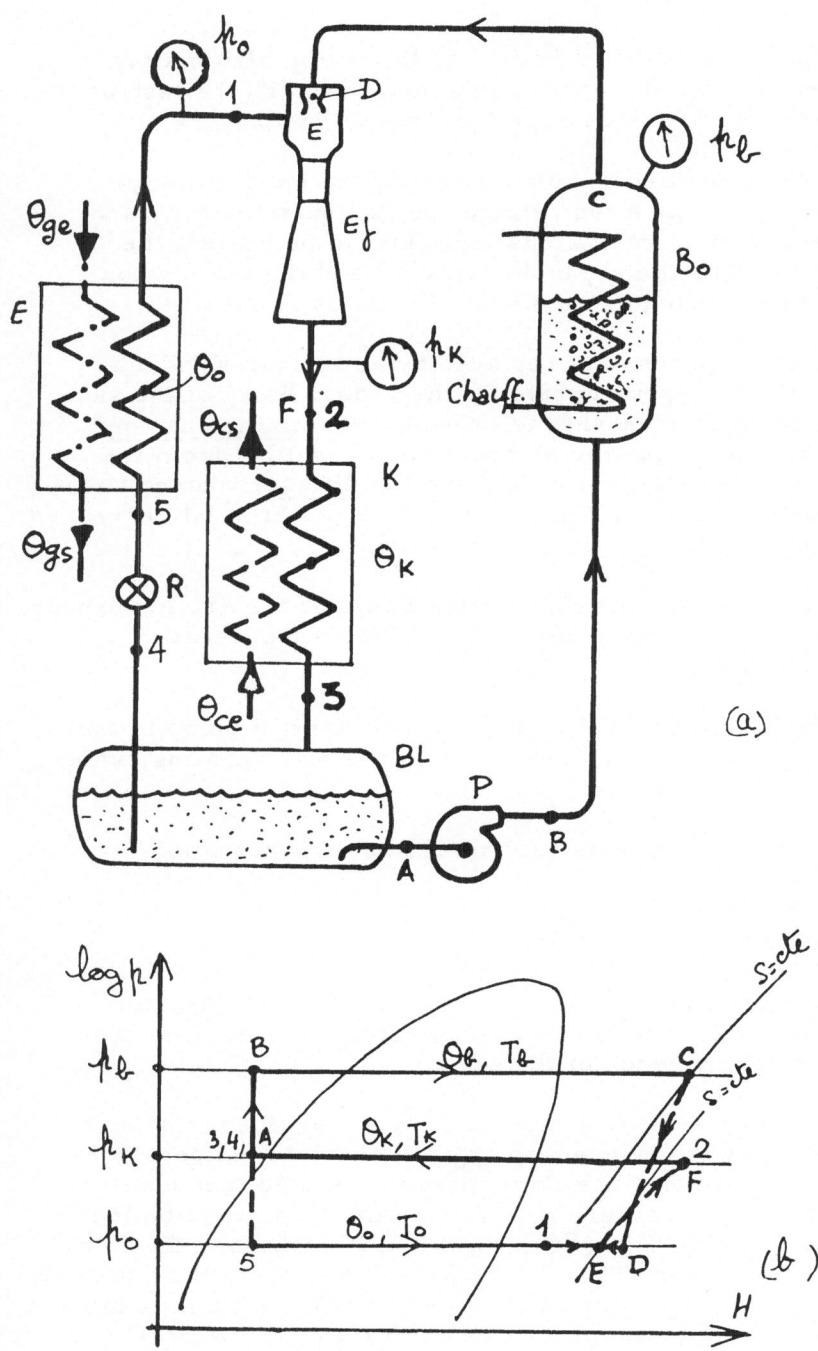

Fig. 15.   Ejection heat pump.

6.4 Ejection heat pump

Like those previously mentioned, this pump also mainly receives heat energy in order to function. Fig. 15(a) illustrates diagrammatically the circuit for this type of pump.

Here again, we find the classic refrigeration assembly: a condenser K, in which the refrigerant is liquefied at θ_k and gives up heat to the circuit containing fluid to be heated, the liquid receiver BL, the expansion valve R and the evaporator E in which the fluid providing the heat circulates.

Here the compressor is replaced by an ejector Ej which includes a driving vapour nozzle D, the vapour being of the same type as the cycle refrigerant, followed by the mixing chamber, in which the driving vapour meets the vapour resulting from the evaporation of the refrigerant in E and the diffuser whose cross-section widens and in which the kinetic energy of the jet decreases so that the pressure may rise.

The condenser liquifies the refrigerant and the driving vapour. The pump P supplies the generator B_O (boiler-superheater assembly) with liquid.

Fig. 15(b) represents the cycle for this type of pump in the graph H-log p. 1, 2, 3, 4, 5 represents the refrigerating cycle; ABCDEFA the motor cycle.

The ejection machine is seldom used as a heat pump.

7. GAS CYCLE HEAT PUMPS

In these systems, the state of the working fluid does not change. The refrigerating and heating effects result from the expansion or compression of the gas.

Fig. 16 represents diagrammatically a machine using such a cycle. The gas at low pressure p_o and absolute temperature T_1 °K (or θ_1°C) enters the compressor C, where it is compressed to the high pressure p_k. During this compression it is heated from T_1 to T_2 °K (θ_2°C). In the heat exchanger E_c, the working gas cools from T_2 to T_3 °K (θ_3°C) and heats the external fluid from θ_{ce} to θ_{cs}. During this operation the fluid maintains its high pressure p_k, except for pressure losses, in a real machine. It is then admitted into the expansion engine T, e.g. a turbine. The pressure falls from p_k to p_o and the temperature

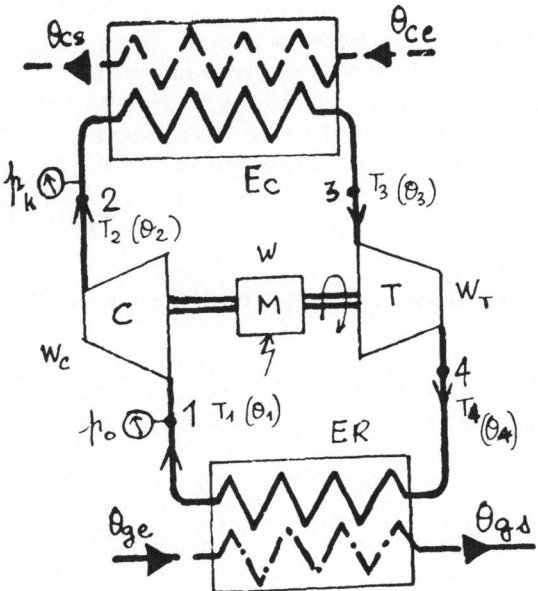

Fig. 16

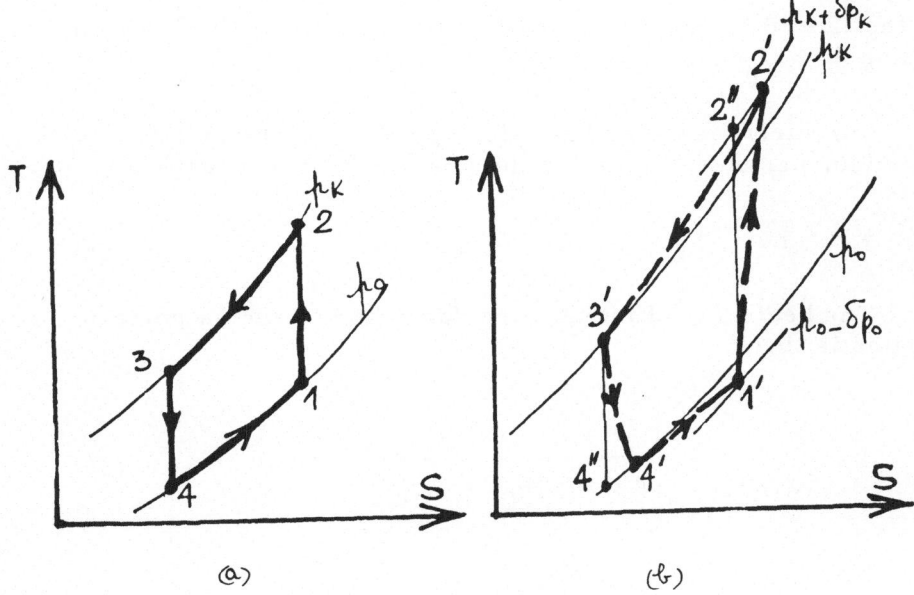

(a)

(b)

Fig. 17 (a)

Fig. 17(b)

Gas cycle heat pump.

from T_3 to T_4 °K (θ_4 °C). The cold gas is then introduced into the second heat exchanger ER, where it is heated from T_4 to T_1 and cools the external fluid, from which heat is taken, from θ_{ge} to θ_{gs}. Since the compressor consumes more mechanical energy than is provided by the turbine, an auxiliary motor M is obviously necessary. Per unit of mass of fluid in circulation, the amount of heat given up at the hot sink of the pump is:

$$Q_K = C_p \cdot (T_2 - T_3) \tag{32}$$

where C_p is the specific heat at constant pressure of the working gas.

The compressor C, operating isentropically, requires the following mechanical energy per unit of mass of fluid:

$$W_C = h_2 - h_1 \tag{33}$$

$$= C_p \cdot (T_2 - T_1) \tag{34}$$

if the fluid is ideal (in accordance with Joule's second law, in which $dh = C_p dT$).

Similarly, if the turbine also operates isentropically it will provide, per unit of mass of fluid, a mechanical energy of:

$$W_T = h_3 - h_4 = C_p \cdot (T_3 - T_4) \tag{35}$$

The heating coefficient of performance of such a perfect engine is then:

$$\varepsilon_{C(ideal)} = \frac{|Q_K|}{|W|} = \frac{|Q_K|}{|W_C - W_T|} \tag{36}$$

$$= \frac{C_p \cdot (T_2 - T_3)}{C_p(T_2 - T_1) - C_p(T_3 - T_4)}$$

This of course gives:

$$\frac{T_1}{T_2} = \frac{T_4}{T_3}$$

since, in both cases, these ratios are equivalent to $(\dfrac{p_o}{p_k})^{\frac{\gamma-1}{\gamma}}$

where $\gamma = \dfrac{C_p}{C_v}$ is the ratio between specific heat at constant

pressure and specific heat at constant volume. Hence, according to (36):

$$\varepsilon_{C\,(ideal)} = \frac{1}{1 - \dfrac{(T_1 - T_4)}{(T_2 - T_3)}} \tag{37}$$

or again

$$\varepsilon_{C\,(ideal)} = \frac{T_2}{T_2 - T_1} = \frac{T_3}{T_3 - T_4} \tag{38}$$

From this aspect, the coefficient of performance is high and such an engine appears to be fully competitive as compared with vapour heat pump systems. Fig. 17(a) represents the cycle for such an ideal heat pump plotted in the entropy graph T - S.

Unfortunately the real machines have certain drawbacks;

- pressure losses in the high-pressure (δp_k) and low-pressure (δp_o) circuits of the machine (fig. 17(b));

- thermodynamic and mechanical losses in the compressor, resulting in an indicated efficiency η_c defined, of course, in relation to the isentropic compression and a mechanical efficiency η_{mc} for the compressor;

thermodynamic and mechanical losses in the turbine, resulting in an indicated efficiency n_t and a mechanical efficiency n_{mt}.

Fig. 17(b) represents the cycle for a real machine.

Hence, the real heating coefficient of performance of such a heat pump, the gas still being considered as perfect and the cooling operation $2' \rightarrow 3'$ as approximately isobaric, is as follows:

$$\varepsilon_{C_{(reel)}} = \frac{(T_{2'} - T_{3'})}{\dfrac{(T_{2''} - T_{1'})}{n_C \cdot n_{mC}} - (T_{3'} - T_{4''})n_T \cdot n_{mT}} \tag{39}$$

where $\varepsilon_{C_{(reel)}} << \varepsilon_{C_{(ideal)}}$

The coefficients of performance of such real heat pumps are considerably lower than those of vapour heat pumps, which explains why they are used only in specific cases where they offer decisive advantages.

When the temperature differences between the hot exchanger EC and the cold exchanger ER increase, it becomes advantageous to insert between them an intermediate recovery heat exchanger EI, fig. 18. The gas at high pressure, which has cooled in the working circuit EC from θ_2 to θ_3, completes its cooling in the exchanger EI from θ_3 to θ_4. The heat thus given off is transferred to the low-pressure fluid, which is heated from θ_6 to θ_1 before entering the compressor.

8.   THE RANQUE-EFFECT HEAT PUMP

The Ranque or vortex tube is a thermal separator which operates as follows.

A flow of gas m_t at a pressure p_k and temperature θ_2 is directed tangentially onto the inner surface of a tube T (fig. 19).

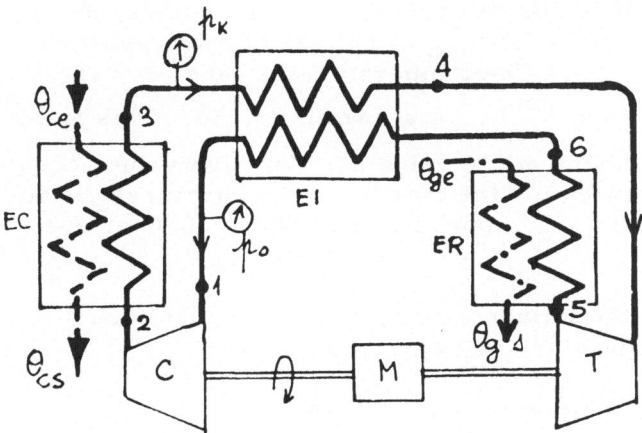

Fig. 18.   Gas cycle heat pump with regenerator

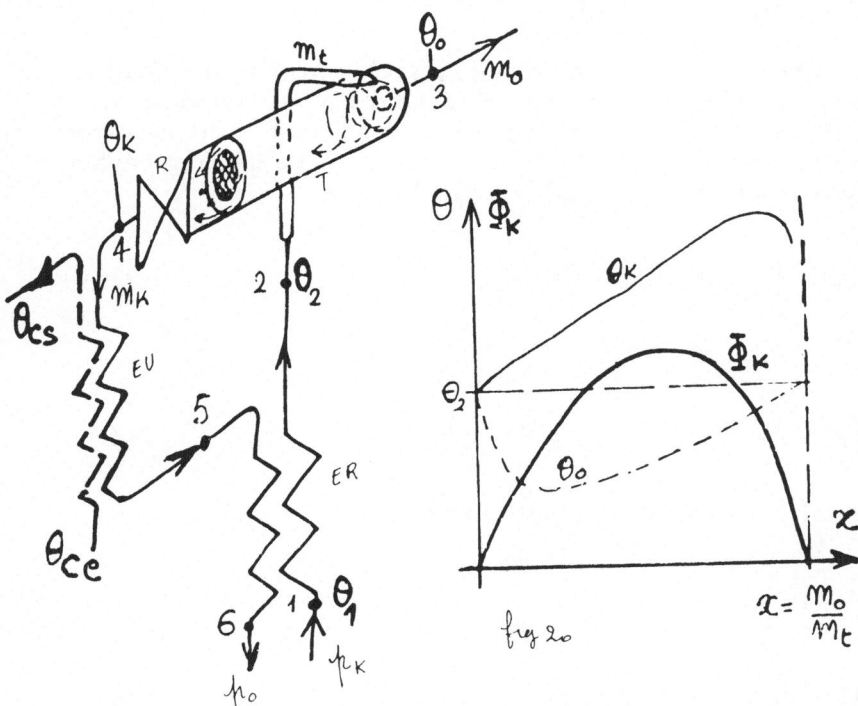

Fig. 19.   Fig. 20.

Ranque tube as heat pump.

At the centre of the vortex thus formed, a flow of gas m_o, cooled to a temperature $\theta_o < \theta_2$, is led off at 3 through a suitable central orifice. At the periphery of the vortex, downstream from a valve R, a flow m_k of hot gas at a temperature $\theta_k > \theta_2$ is extracted at 4. The temperatures θ_o and θ_k vary according to the ratio $x = \dfrac{m_o}{m_t}$, as shown in fig. 20. This ratio can be adjusted by means of the valve R. For high values of x, i. e. for a low flow rate m_k, the maximum temperature θ_k attained can be relatively high (e. g. approximately 50 to 100°C or even more).

In addition to the influence of x, which has just been mentioned, the temperature θ_k depends on:

-       the injection temperature θ_2 of the gas;

-       the pressure p_k
        (θ_k increases when p_k and θ_2 increase);

-       the nature of the gas.

The hot gas, at a pressure p_o, yields heat to the fluid to be heated in the heat exchanger EU, where it is cooled from θ_k to θ_5. A recovery heat exchanger ER may enable the gas from the compressor to be heated from θ_1 to θ_2. The heating capacity produced, in the case of fig. 19, will then be:

$$\Phi_K = m_K \cdot C_p \ (\theta_K - \theta_5) \tag{40}$$

$$\Phi_K = m_t \cdot (1-x) C_p \cdot (\theta_K - \theta_5) \tag{41}$$

If the gas is cooled to the input temperature θ_2

$$\Phi_K = m_t (1-x) C_p \cdot (\theta_K - \theta_2) \tag{42}$$

The heating capacity is nullified when $x = 1$ and $x = o$, for in that case the entire flow m_t passes through EU and if the gas is ideal $\theta_K = \theta_2$. The heating capacity attains a maximum, which unfortunately is not identical with the maximum heating of the gas (fig. 20).

145

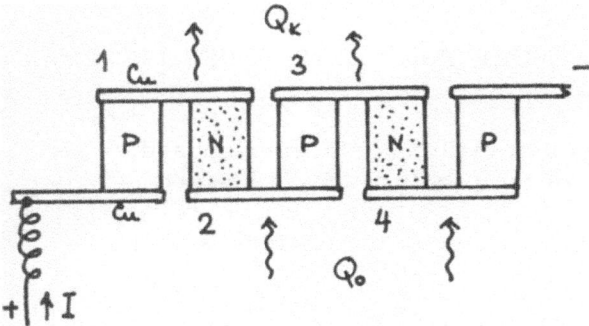

Fig. 21.   Principle of thermo-electric heat pump.

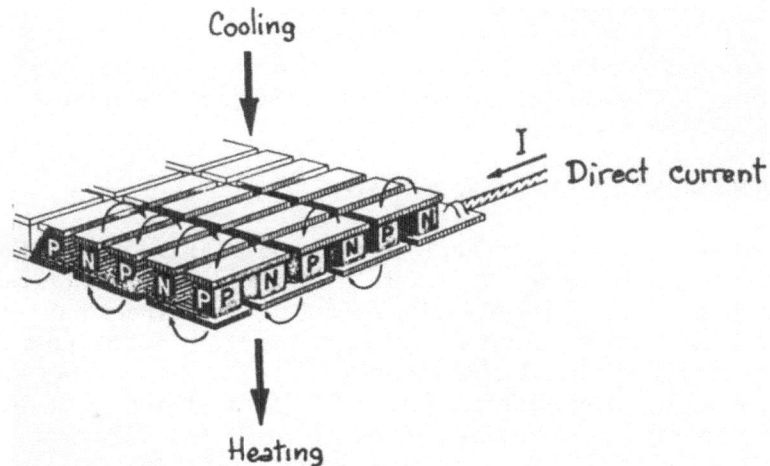

Fig. 22.   Thermo-electric module.

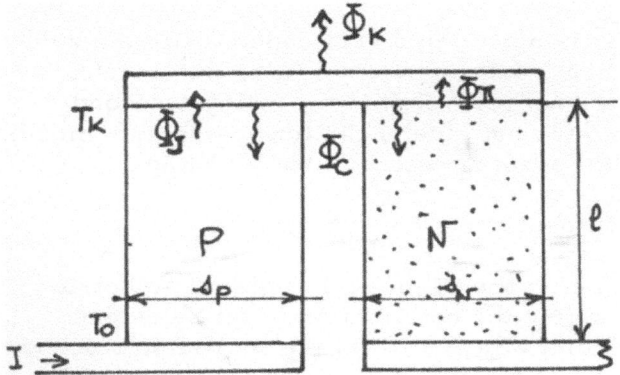

Fig. 23.   Energetic balance of thermocouple.

The efficiency of this system is limited.

9.   OTHER TYPES OF CYCLE THAT CAN BE USED IN GAS HEAT PUMPS

Numerous other types of cycle have been proposed, enabling wide differences of temperature to be obtained:

-    the Stirling cycle, as adapted by the engineers of the Philips Company[7];

-    the Gifford-MacMahon cycle[8];

-    the Gifford and Longsworth pulse tube[9], from which the Bertin thermal separator[10] is derived.

These cycles, which are basically intended for low-temperature refrigeration, can be used in heat pumps, if suitable technological modifications are made.

10.   THERMO-ELECTRIC HEAT PUMPS

10.1   These use the Peltier effect.   Let us consider a series of dissimilar conductors N, P, N, P, etc. "electrically" associated in series and connected by copper bridges Cu.   If a direct current I is made to flow in this circuit it will be found that, at the junctions with the same parity (in the case of fig. 21, the odd junctions 1, 3, etc.), heat Q_k is given off, while heat Q_0 is absorbed at the junctions of opposite parity (2, 4, etc.).   If these two different conductors are arranged like a chess-board (fig. 22) it can easily be seen that one face of the module thus constituted will become hot while the opposite face will become cold.

If a heat exchanger is placed in direct contact with the cold face, heat can be absorbed from the medium (air, water, etc.) which circulates in the exchanger.   Similarly, using a second exchanger in contact with the hot face of the termo-electric module, heat can be given up to the medium which is to be heated.

10.2   Theory of the thermo-electric heat pump

Let us consider an element NP of such a thermo-electric module (fig. 23), particularly the hot junction of this element. This junction PN is the point where certain thermal effects occur:

(1)   an increase in heat of thermo-electric origin (Peltier effect) proportional to the current I:

$$\Phi_\pi = (\pi_P - \pi_N) \, I = \pi_{PN} \, I \tag{43}$$

where π_{PN} is the Peltier coefficient of the junction PN. The laws of thermodynamics show that this coefficient is linked to the Seebeck thermo-electric factor for this junction:

$$\alpha_{PN} = \alpha_P - \alpha_N \tag{44}$$

by Kelvin's first law

$$\alpha_{PN} = \frac{\pi_{PN}}{T} \tag{45}$$

Hence, for our hot junction:

$$\Phi_\pi = \alpha_{PN} \, T_K I \tag{46}$$

(2)   an increase in heat due to the Joule effect. We shall assume, as does A. F. Joffe[10] that this heat increase is equally distributed between the hot and cold junctions. In his more complete theory, E. B. Penrod[11] has shown that this assumption is only approximate. However, we consider that it is adequate to illustrate our remarks.

$$\Phi_J = \frac{1}{2} \, RI^2 \tag{47}$$

where

$$R = R_P + R_N$$

is the total electrical resistance of the conductors P and N, neglecting the contact resistances between them and the copper bridges.

$$R = \rho_P \, \frac{1}{s_P} + \rho_N \, \frac{1}{s_N} \tag{48}$$

where ρ_P and ρ_N are the resistivities of the two conductors P and N, s_P and s_N their cross section and 1 their length.

(3)   a loss of heat by thermal conduction from the hot junction at T_k to the cold junction at T_O

$$\Phi_C = K (T_K - T_O) \qquad (49)$$

where K is the overall thermal conductance of the elements P and N taken together

$$K = \lambda_P \cdot \frac{s_P}{l} + \lambda_N \frac{s_N}{l} \qquad (50)$$

where λ_P and λ_N are the thermal conductivities of the conductors P and N;

(4)   an extraction of heat Φ_k which is transferred to the external medium. This constitutes the "useful effect" of the heat pump.

Assuming, as does Joffé, that the Seebeck coefficient is constant between temperatures T_O and T_k, which implies that the Thomson coefficient is $\tau = T\frac{d\alpha}{dT} = o$ the heat balance for the hot junction is as follows:

$$+ \Phi_\pi + \Phi_J - \Phi_C - \Phi_K = 0 \qquad (51)$$

$$\Phi_K = \Phi_\pi + \Phi_J - \Phi_C$$

$$\Phi_K = \alpha_{PN} T_K I + \frac{RI^2}{2} - K (T_K - T_O) \qquad (52)$$

This useful heating capacity is nullified for a current:

$$I_O = \frac{\alpha_{PN}}{R} \cdot T_K \left(\sqrt{1 + \frac{2}{Z} \frac{(T_K - T_O)}{T_K^2}} - 1 \right) \qquad (53)$$

taking

$$Z = \frac{\alpha^2_{PN}}{RK} \qquad (K^{-1}) \qquad (54)$$

this quantity being known as the <u>figure of merit of the thermocouple PN</u>.

It will be seen that, according to (53), for a given T_k, I_0 increases as does the difference in temperature $(T_k - T_0)$ between the junctions and as Z decreases, i. e. the thermo-element is less efficient (fig. 24)

The derivative of equation (52)

$$\frac{\delta \Phi_K}{\delta I} = \alpha_{PN} T_K + RI \qquad (55)$$

is always positive. This function $\Phi_k = f(I)$ is constantly increasing. On the other hand, we know that the refrigerating capacity of the module Φ_0 rises to a maximum depending on the current I.

For this heat pump to operate, the following electrical power must be supplied to the thermo-electric module:

$$W = \alpha_{PN} (T_K - T_O) \cdot I + RI^2 \qquad (56)$$

The first term corresponds to the thermo-electric counter-electromotive force $\alpha_{PN}(T_k - T_o)$ and the second to the Joule effect.

The heating coefficient of performance of the thermo-electric module can then be expressed as follows:

$$\varepsilon_C = \frac{\Phi_K}{W} = \frac{\alpha_{PN} T_K I + \frac{1}{2} RI^2 - K (T_K - T_O)}{I \{\alpha_{PN} (T_K - T_O) + RI\}} \qquad (57)$$

150

The derivative of this function, in relation to the current, will be:

$$\frac{\delta\epsilon_C}{\delta I} = \frac{-\alpha_{PN}\frac{RI^2}{2}(T_K+T_O) + 2KRI(T_K-T_O)^2 + K\alpha_{PN}(T_K-T_O)^2}{I^2\{\alpha_{PN}(T_K-T_O) + RI\}^2}$$

(58)

The coefficient of performance ϵ_C reaches a maximum for the value I_m of the current which annuls $\frac{\delta\epsilon_C}{\delta I}$:

$$I_M = \frac{\alpha_{PN}\cdot(T_K - T_O)}{\{\sqrt{1 + \overline{Z}\,T_m} - 1\}R}$$

(59)

The maximum value for ϵ_C corresponding to this current is then:

$$\epsilon_{C(max.)} = \frac{T_K}{T_K - T_O} \cdot \frac{\sqrt{1 + Z\,T_m} - \frac{T_O}{T_K}}{\sqrt{1 + Z\,T_m} + 1}$$

(60)

This maximum rises as Z increases (fig. 24).

If the values of Z could be sufficiently high for $\sqrt{1+Z\,T_m}$ to be large compared with unity, ϵ_C would tend to reach the maximum value:

$$\epsilon_{Cmax.} = \frac{T_K}{T_K-T_O}$$

Unfortunately this remark is purely theoretical, since the figures of merit for the available materials are so low that the actual coefficients of performance are well below this ideal value.

We saw that

$$Z = \frac{\alpha^2_{PN}}{RK} = \frac{(\alpha_P - \alpha_N)^2}{RK}$$

The two constituents P and N of the thermocouple are generally made from the same basic material, bismuth telluride, $Bi_2 Te_3$, which possess the optimum properties for such applications. By doping this material with suitable elements it is possible to obtain the two different conductors P and N. Conductor P is obtained by doping the bismuth telluride with atoms which will accept electrons and conductor N with atoms which give up electrons.

If we assume, as does Joffé, that:

-   the cross sections s_P and s_N of the two constituents of the thermo-electric element are the same, $s_P = s_N = s$;

-   the thermal conductivities of the materials P and N are identical $\lambda_P = \lambda_N = \lambda$;

-   their electrical resistivities are equal $\rho_P = \rho_N = \rho$;

-   their Seebeck coefficients are equal and of opposite sign

$$|\alpha_P| = |\alpha_N| = |\alpha|$$

we obtain from (48), (50) and (53):

$$Z = \frac{\alpha^2}{\rho \lambda} \tag{61}$$

The figures of merit of the elements at present available lie between two 2.10^{-3} and 3.10^{-3} (K^{-1}). Fig. 25 represents the real behaviour of a thermocouple for a thermo-electric module which is now on the market for a cold face temperature of $+ 15^{\circ}C$ (which makes it possible to absorb heat from the air at 20° or $25^{\circ}C$) and a hot face temperature of $45^{\circ}C$ (which enables heat to be transferred to the air at 35° or $40^{\circ}C$).

Study of the graphs in figs. 24 and 25 shows that:

-   the heating coefficients of performance ε_c remain very low and are generally well below 2. Thermo-electric heat pumps can scarcely compete with vapour heat pumps as regards efficiency;

152

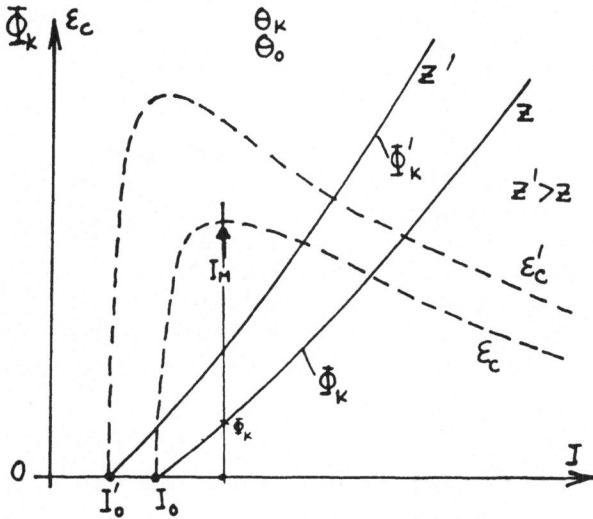

Fig. 24.   Performance curves of a thermo-electric heat pump.

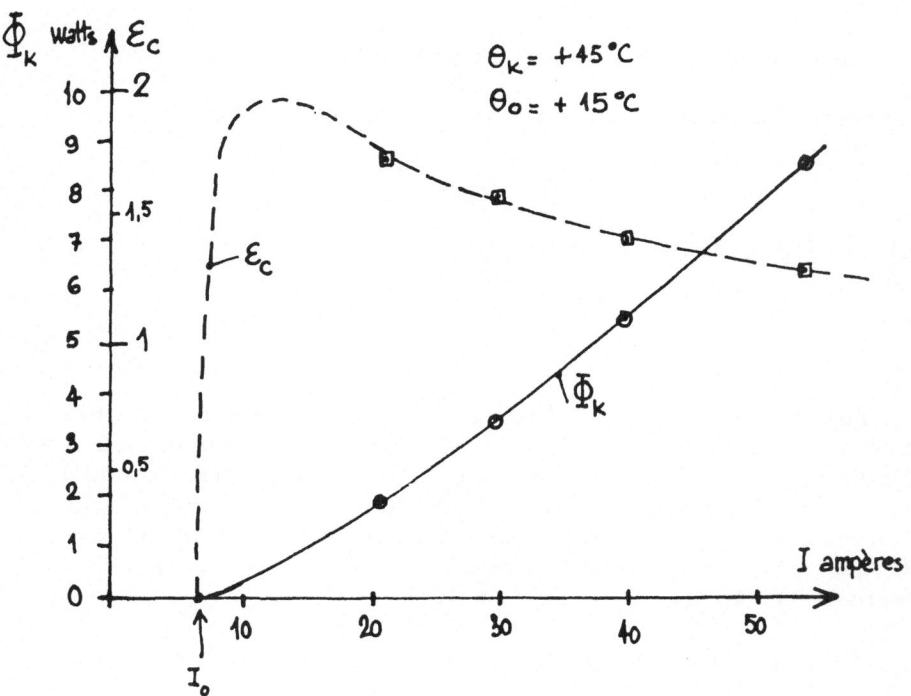

Fig. 25.   Performance test result of a thermo-electric heat
pump.

- thermo-electric heat pumps offer the maximum efficiency for very low heat outputs, which is a serious drawback.

If these systems are to produce considerable heat, a reduction in their efficiency must be accepted. It is generally agreed that although it is relatively inefficient, a thermo-electric module is a better proposition than a simple resistance. This is true as long as $\varepsilon_c > 1$, but we know that it is not true for a very low current, since part of the heat flows, by thermal conduction, to the cold face (figs. 24 and 25). Nor is it true for fairly strong currents, since in that case the cold face of the module ceases to absorb heat and part of the Joule effect heat is lost via this cold face. It might therefore be generally adopted for auxiliary heating, e.g. to supplement another type of heat pump, on cold days. This is obviously possible, but it must be realised that this type of module has certain disadvantages, as compared with a pure resistance, which cannot be overlooked:

(1)   For the same amount of heat produced, there is no comparison between the cost of a thermo-electric module and that of a resistance, even if the latter is made very large so that its surface temperature is low. The thermo-electric module is far more expensive.

(2)   It calls for the use of direct current supply equipment, generally a diode rectifier. Care must be taken that the ripple factor remains low if the thermo-electric efficiency is not to be decreased. The ripple factor must be less than 15% and if possible 10%. The cost of this supply equipment, which increases with the heat flow rate required, is by no means negligible and must be taken into account.

(3)   The currents absorbed by the thermo-electric modules are high (fig. 25) and the voltages low. This calls for fairly short, thick supply cables.

(4)   Unlike heating resistances, which can be installed any-where, arrangements must be made to supply heat to the cold source of the thermo-electric heat pump, which obviously restricts its flexibility of operation.

Despite this rather negative balance, it must be recognized that the thermo-electric module offers certain definite advantages:

- a total absence of noise, the plant being entirely static;

- complete safety;

- practically no maintenance;

- the possibility of very smooth adjustment.

At the present time the use of thermo-electric modules as heat pumps is extremely limited.

REFERENCES

1. Thermodynamic tables for R 11, R 12 and R 22, calculated by de Lepeleire, Institut International du Froid.
2. Properties and applications of Freon fluorocarbons, notice B 2, Du Pont.
3. Notice on Forane, Ugine Kuhlmann
4. Horvath, A. L. , ASHRAE Journal 7-1972.
5. Snider, K.L., ASHRAE Journal, 11-1967.
6. Eiseman B. J. , Jr. , Etudes d'actualités - Bulletin de l'Institut International du Froid no. 3, 1966.
7. Kohler, J. W. L. , and Jonkers, C. P. , Philips Techn. Rev. 16-1954 - 69-78.
8. Gifford, W. E. and MacMahon, H. O. , Report on 11th International Refrigeration Congress, Copenhagen, 1959, pp. 100-109, Advanced Cryogenic Engineering 1959, 5-368-372.
9. Gifford, W. E. and Longsworth, R. C. , ASME paper no. 63WA290, 1963.
10. Joffe, A. F. , Semi-conductor Thermo-elements and Thermo-electric cooling - Infosearch, London, 1957.
11. Penrod, E. B. , Extract from Bulletin de l'Institut International du Froid, XL no. 2 - 1960.

Duminil, M., Reports on seminars on the application of geothermics to domestic and industrial heating, Paris, 23rd-24th June, 1975.

DEVELOPMENT OF THERMAL PRIME MOVERS FOR HEAT PUMP DRIVE

G. Angelino

Istituto di Macchine - Piazza Leonardo Da Vinci, 32
20133 Milano, Italy.

SUMMARY

The energy savings connected with low grade heat generation by means of heat pumps powered by thermal engines whose waste heat is recovered are illustrated.

The performance of the system is compared with that of alternate methods of low temperature heat production.

The characteristics of thermal prime movers adequate for the proposed application which are either available or under development are reviewed.

Particular attention is devoted to organic working fluid cycles whose basic thermodynamics and main technical features are discussed.

Results are reported of an example of technical and economical analysis relating to the heating of a large building by means of a thermal engine-heat pump system.

If sea water is available as secondary heat source fuel savings larger than 50% and substantial financial gains appear possible.

INTRODUCTION

In modern central power stations one unit of fuel energy generates approximately 0.4 units work.

The efficiency of the process, defined as the ratio of actual to maximum work theoretically obtained can be shown to be around 40% (maximum work would be produced by carrying out oxidation in a reversible fuel cell, utilizing the heat developed during the reaction in an ideal thermal cycle and, finally, bringing the cool combustion products to equilibrium of concentration with the atmosphere by expanding each one of their chemical constituents isothermally and reversibly, resorting to semi-permeable membranes, until it reaches the same partial pressure it has in the atmosphere).

If the 0.4 units work obtained were used in an ideal heat pump, 5.9 units heat could be extracted from the ambient at $0^{\circ}C$ and delivered to indoor space at $20^{\circ}C$ for heating purposes. The fact that conventional heating systems deliver to the end user somewhat less than one unit heat for each one unit of fuel energy consumed gives an idea of how inefficient the process of low temperature heat generation is.

While in high temperature uses the largest single loss is, in general caused by irreversible combustion (around 30% of available energy in a central power station), in low temperature heat generation a second and prevailing loss is due to heat degradation from flame to indoor ambient temperature. Heat pumps, if developed to a sufficient level of efficiency, are in principle capable of reducing the energy cost of low temperature heat to levels much closer to the theoretical values than actually achieved.

Another effective way of reducing the cost of low grade heat is to make use of thermal discharges from prime movers such as the jacket cooling water of diesel engines or heat of condensation of steam power stations.

Dual purpose power and heat plants, however, although highly attractive from the theoretical point of view, have presently attained a limited application owing to difficulties encountered in matching power and heat generation and to the excessively high cost of distribution of low grade heat which is mostly produced in stations of great output while it is needed in a diffused mode over a wide area for space heating purposes.

Many of the advantages connected with the heat pump use and with waste heat exploitation can be obtained jointly by resorting to a system formed by a fossil fuel fired prime mover which drives directly a heat pump. The only useful output of the system is low grade heat, partially discharged by the power cycle and partly delivered by the heat pump.

The system capacity can be selected freely with reference only to the thermal needs of the consumer. In general small installations exhibit the best overall performance by minimizing the cost of heat distribution, both in terms of capital investments and, thermodynamically, in terms of a lower temperature level which is requested on account of the shorter and simpler distributing system.

SYSTEM PERFORMANCE

The thermal engine-heat pump system can be ideally represented as shown in Fig. la in which an amount Q_1^* of high temperature heat is delivered by a source at temperature T_1^* to the reversible engine M by which it is partly converted into work L while the residue Q_2^* is discharged at temperature T_1 at which is requested by the consumer. Work L actuates a heat pump, which lifts an amount of heat Q_2 from ambient temperature T_2 to end user's temperature T_1. The ideal coefficient of performance of the transformation, making reference to the cooling effect, is given by:

$$\mathcal{E}_{id} = \frac{Q_2}{L} = \frac{T_2}{T_1 - T_2} \tag{1}$$

or, with reference to the heating effect:

$$\mathcal{E}^*_{id} = \mathcal{E}_{id} + 1 = \frac{Q_1}{L} = \frac{T_1}{T_1 - T_2} \tag{2}$$

The overall process performed by the system is the conversion of high to low temperature heat according to the relationships

$$\left(\frac{Q_2^* + Q_1}{Q_1^*}\right)_{id} = \frac{T_1^* - T_2}{T_1^*} \cdot \frac{T_1}{T_1 - T_2} \tag{3}$$

158

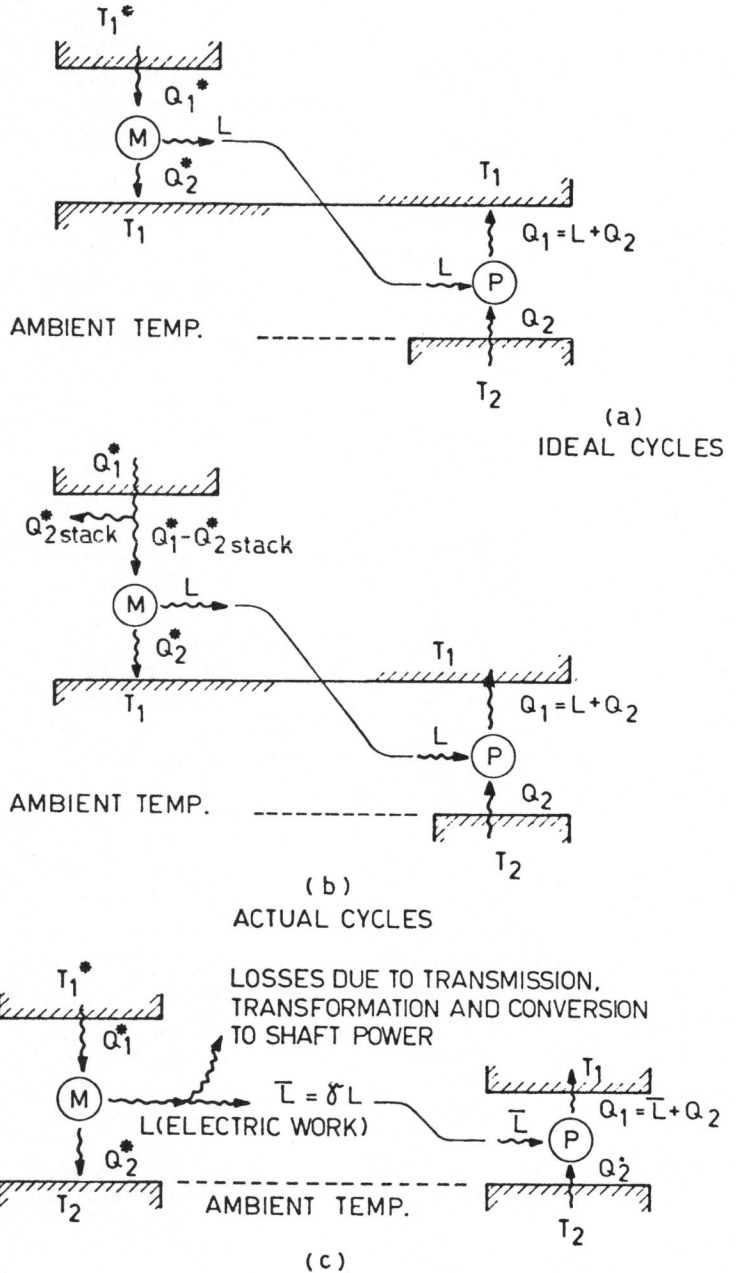

(a)
IDEAL CYCLES

(b)
ACTUAL CYCLES

(c)
ELECTRICALLY DRIVEN HEAT PUMP

Fig. 1. Schematic arrangements of possible thermodynamic
cycles for low grade heat generation.

which gives the heat multiplying effect of the cycles' arrangement under ideal conditions.

The behaviour of an actual thermal-engine heat pump system can be obtained from the ideal performance by introducing the following modifications (see Fig. 1b):

(1) the power cycle efficiency is a fraction of the ideal cycle efficiency:

$$\eta = \quad {}^{\alpha}\eta_{Carnot} = \quad \alpha \left(1 - \frac{T_1}{T_1^*}\right) \tag{4}$$

(2) the Coefficient of Performance of the heat pumps relating to the cooling effect is a fraction of the ideal COP:

$$\mathcal{E} = \quad \beta\mathcal{E}_{id} = \beta \; \frac{T_2}{T_1 - T_2} \tag{5}$$

(it could be shown that assumption (5) is better justified than a similar one relating to the COP $\mathcal{E}^*$ referring to the heating effect);

(3) not all the waste heat discharged by the prime mover can be utilized at temperature T_1, since a portion of it is lost through the stack.

A heater efficiency η_b can take into account the fraction of primary heat actually handled by the engine:

$$\eta_b = \frac{Q_1^* - Q_2^* \; Stack}{Q_1^*} \tag{6}$$

Under the afore stipulations the heat conversion ratio becomes:

$$\frac{Q_2^* + Q_1}{Q_1^*} = \eta_b \quad + \; \alpha\beta \; \frac{T_1^* - T_1}{T_1^*} \cdot \frac{T_2}{T_1 - T_2} \tag{7}$$

In order to compare the combined thermal engine-heat pump system with the more usual electrically driven heat pump, besides making use of coefficients α and β as previously illustrated a further loss must be evaluated accounting for transmission, and transformation to lower voltage and conversion to shaft power:

$$\bar{L} = \quad \gamma L \tag{8}$$

where $\overline{L}$ is the actual shaft work transmitted to the heat pump and L is the electric work delivered by the power station (Fig. 1c). The ratio of heat delivered to end user to primary heat consumed in the central station is, in this case:

$$\frac{Q_1}{Q_1^*} = \gamma\alpha \quad \frac{T_1^* - T_2}{T_1^*} \left(\beta \frac{T_2}{T_1 - T_2} + 1 \right) \tag{9}$$

Eq 3 giving the ideal performance of the combined system is represented in Fig. 2, from which it is seen, for instance, that heat at 600°C can be ideally converted into a 3.4 times larger amount of heat at 80°C.

In order that the behaviour of an actual system is correctly evaluated, coefficients α and β of Eqs 4 and 5 must be selected properly. On the base of the performance of various types of prime movers in a wide range of top cycle temperatures 0.60 seems a reasonable value for coefficient α. Similarly the performance of many existing heat pumps, in the upper output range, is predicted fairly well setting $\beta = 0.50$ (see Fig. 3); i.e. the coefficient of performance in the cooling mode is 50% lower than ideally possible (for heating water to 50°C using heat from the atmosphere at 0°C, for example, a cooling COP of 2.7 and a heating COP of 3.7 are predicted and can be actually attained with current, industrial type, equipment).

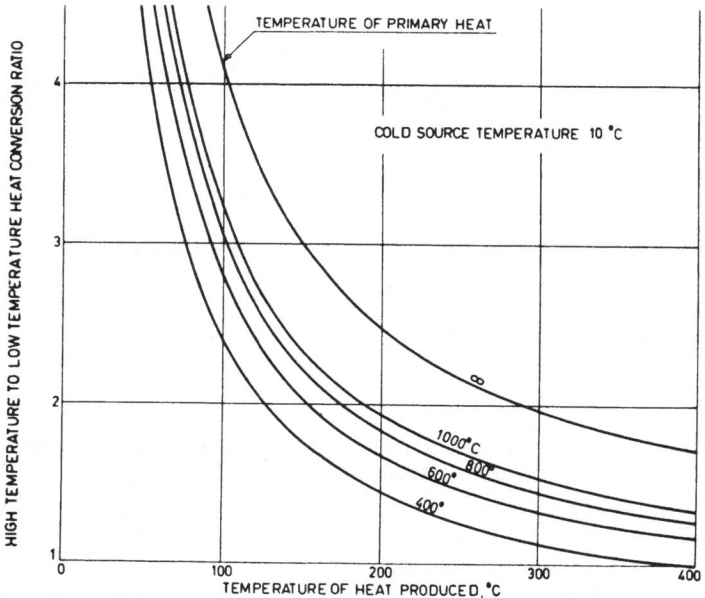

Fig. 2. Low temperature heat output, as a multiple of primary heat consumed, in an ideal thermal engine-heat pump system.

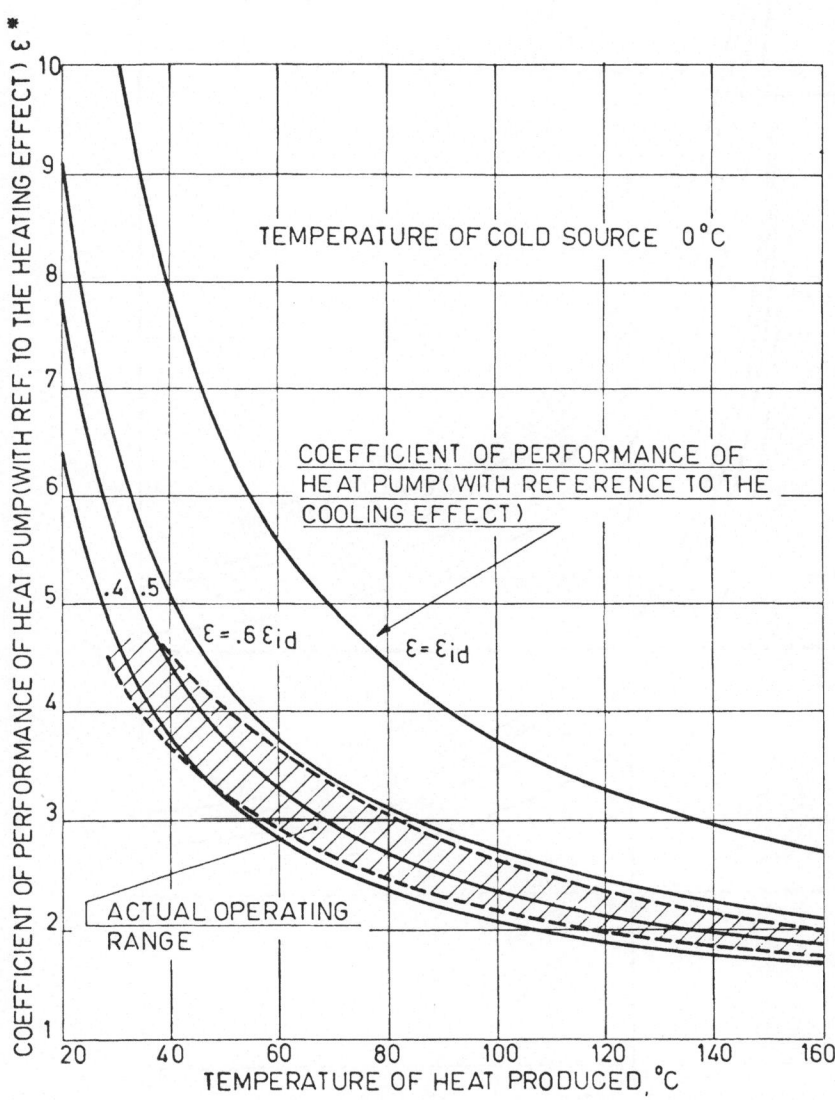

Fig. 3.   Ideal and actual performance of heat pumps as a
function of the temperature of the heat produced.

162

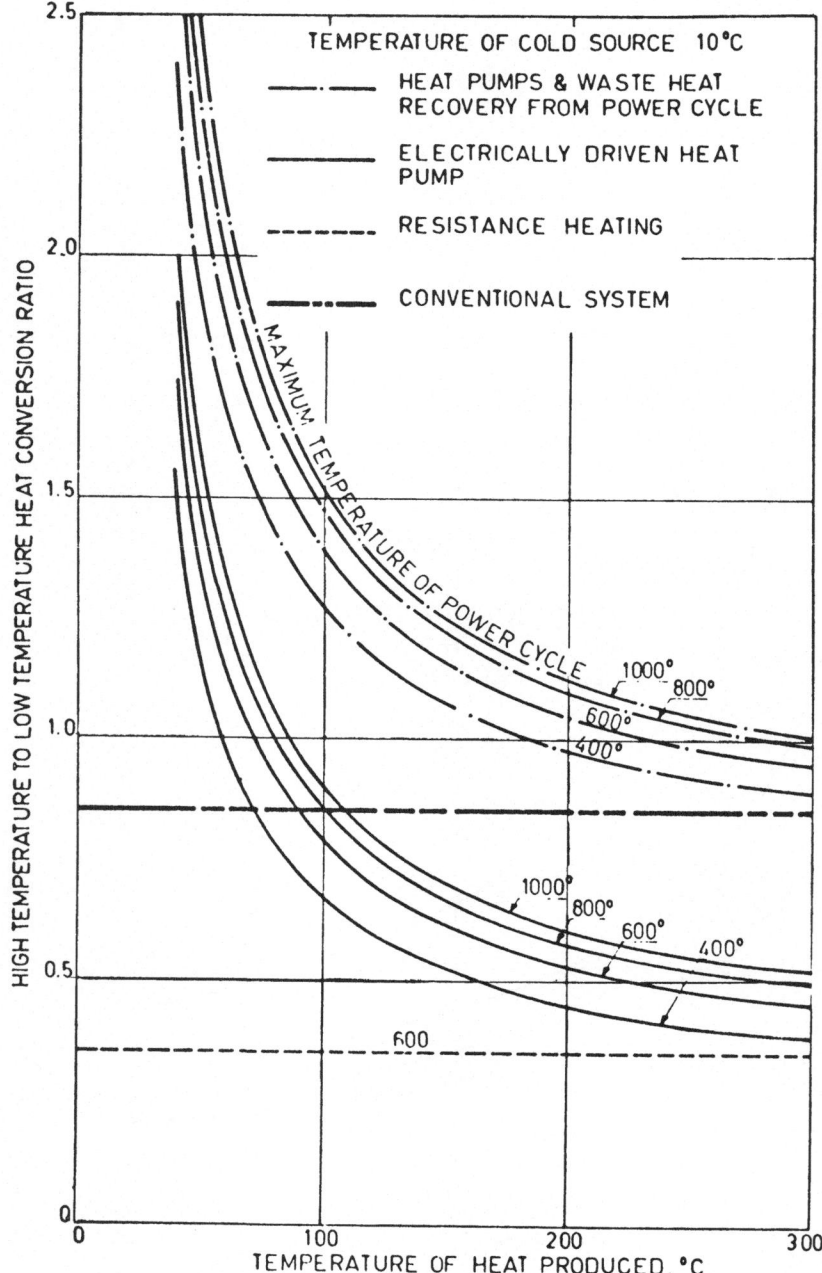

Fig. 4.   Low temperature heat output, as a multiple of
primary heat consumed, from an electrically driven
and from a thermal engine driven heat pump.

With respect to electrically powered heat pumps loss coefficient γ must be evaluated as the product of 3 efficiencies,
i. e. transmission and distribution, transformation to lower
voltage and conversion to shaft power. Depending on the power
level considered the last two efficiencies vary within a wide
range (for instance conversion to mechanical power is performed
with an efficiency lower than 75% in the kW power range and
better than 90% in the 100 kW power range). An overall value for
γ of 0.75 was assumed here as representative of average conditions in the power range of several kW.

In addition to the afore stipulations a cold source temperature of 10°C was assumed.

The performance of the electrically driven heat pump is compared with that of the system thermal engine-heat pump in Fig. 4,
obtained from Eqs 7 and 9, which gives the heat conversion ratio
as a function of the temperature of the heat produced. The in-
spection of Fig. 4 suggests the following observation:

(1) employing heat at a temperature as low as possible is vital
     in reducing primary energy consumption in heat pump
     systems of any type;

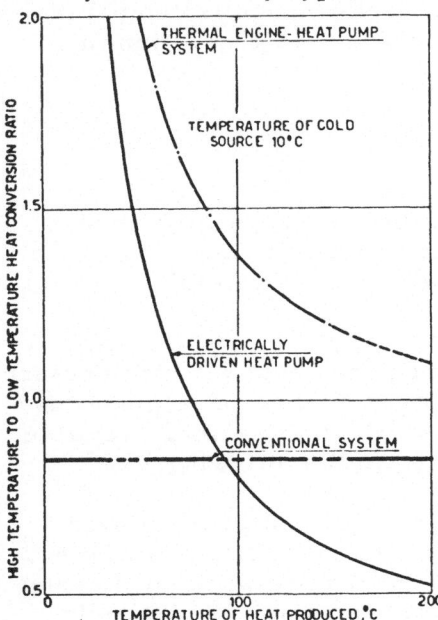

Fig. 5.   Low grade heat production, as a multiple of the heat
           consumed, for a thermal engine-heat pump system
           based on existing technology.

(2)   electrically driven heat pumps can consume less primary
       energy than a conventional system (boiler efficiency was
       assumed to be 0. 85) for temperatures of the heat produced
       lower than 80 to 90°C;

(3)   the thermal engine-heat pump system exhibits a definitely
       better performance than the electrically powered heat pump
       (at 80°C it delivers 1. 57 times the consumed energy while
       the electric system delivers only 0. 92 times the primary
       energy used). The difference in performance is significant
       over the whole temperature range considered and increases
       at increasing output temperatures.

At a temperature of 120°C, sufficient for many process heat
applications, a 1/3rd fuel economy is achievable by resorting to
the thermal engine-heat pump system.

Considering an actually available prime mover, i. e. the
diesel engine with an efficiency of 34%, the heat generating capability can be calculated similarly as done in the general case
previously described. The results are illustrated in Fig. 5 which
shows, for instance, that at a temperature of 60°C for the heat
produced the diesel engine-heat pump system can deliver 80%
more energy than it consumes. This performance is compared
with that of alternate methods of low grade heat generation in
Fig. 6.

PRIME MOVERS' AVAILABILITY

For most applications envisaged the thermal engine must
have a limited output (to cope with the problem of low grade heat
distribution) and must operate in urban areas, preferably without
the assistance of specialized personnel.

In order to compare the needed requirements with the present
day and with the perspective performance of thermal engines the
basic characteristics of prime movers which are either available
or are under development are reviewed in the following:

Diesel engine: is the most efficient, proved prime mover
suitable for total energy applications. Its durability is sufficient
under normal control, but may prove inadequate when considering
private-home unattended operation. It has some difficulties in
complying with noise and NO_x pollution requirements. Gas
engines are somewhat less efficient and have similar maintenance
characteristics.

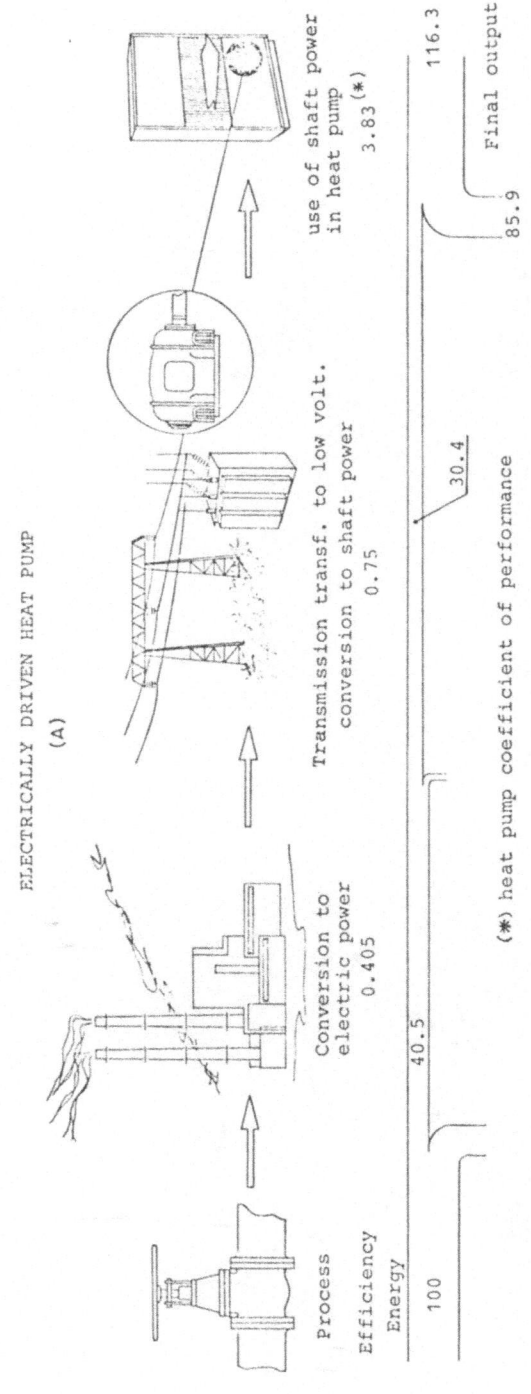

Fig. 6a. Energy balances for various heating systems - Electrically driven heat pump.

166

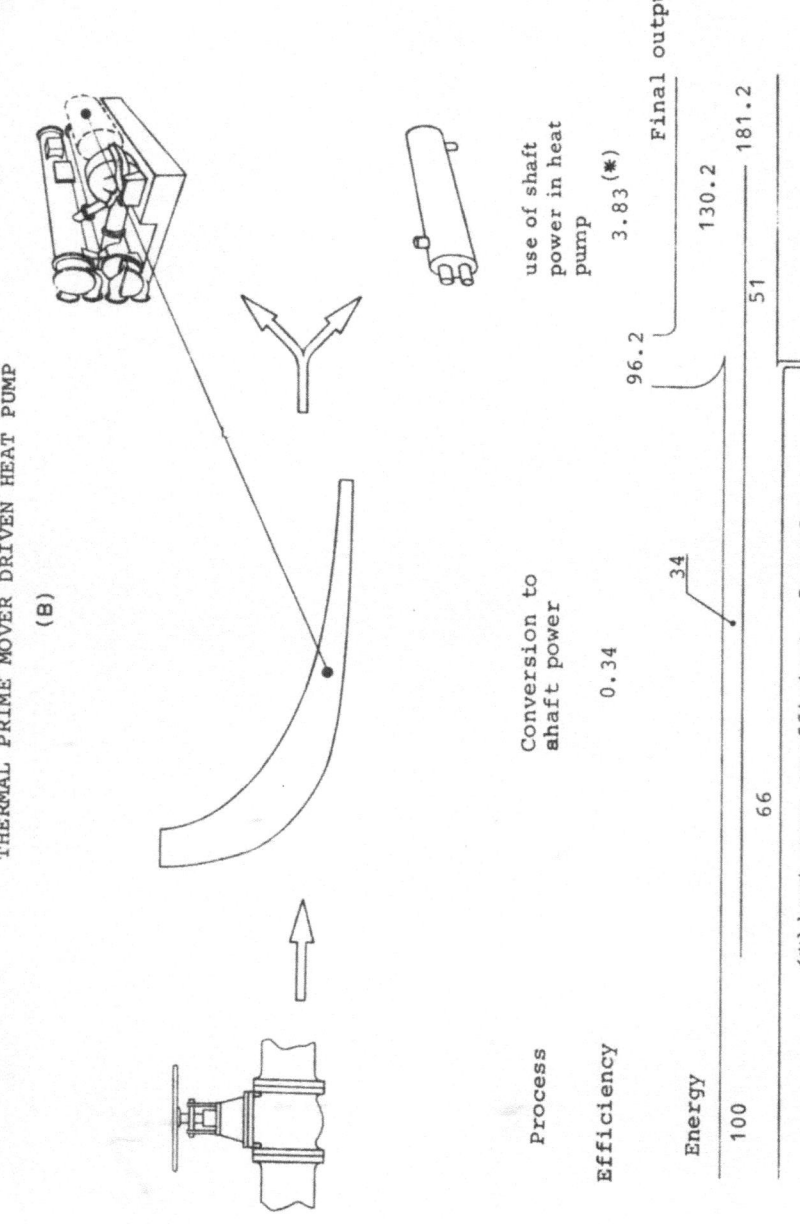

THERMAL PRIME MOVER DRIVEN HEAT PUMP

(B)

Process	Conversion to shaft power	use of shaft power in heat pump	Final output
Efficiency	0.34	3.83(*)	
Energy	66 34	96.2 51 130.2	181.2
100			

(*) heat pump coefficient of performance

Fig. 6b. Energy balances for various heating systems -
Thermal prime mover driven heat pump.

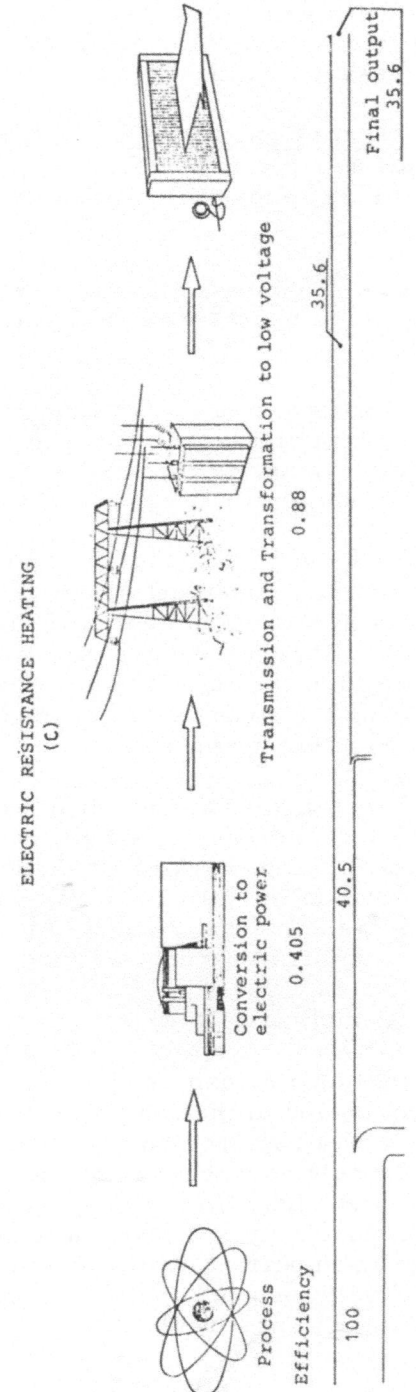

Fig. 6c. Energy balances for various heating systems -
Electric resistance heating.

Open cycle gas turbine (non-regenerative): can provide
quiet, unattended operation with potentially long time between
overhauls provided it is conservatively designed. Fuel economy,
however, is by far excessive, at least for units of small output.

Open cycle, regenerative gas turbine: in the low power
range most manufacturing experience for this type of engine is
connected with automotive applications adopting twin rotary heat
exchangers.

While a satisfactory efficiency is thus achieved, the life of
some critical components is probably insufficient for the proposed application.

The adoption of a fixed regenerator of adequate size could
improve reliability and durability at the cost of an increase in
volume and weight of the engine, which, however, are not critical characteristics for stationary use.

Closed cycle gas turbine: small units were developed for
space applications resorting to the most sophisticated techniques.
High efficiency was proved at turbine inlet temperatures which
require the use of super alloys for the primary heater (metal
temperatures around or in excess of 800°C); elevated first
cost and limited life being the consequence of the severe working
conditions of the main heat exchanger.

External combustion, assures a very clean exhaust. The
high degree of regeneration increases the temperature at which
external heat can be added to the working fluid; hence combustion
products cannot be cooled sufficiently in contact with the working
medium and must give up their sensible heat to a combustion air
pre-heater which adds complexity and cost to the power plant.

Stirling engine: it employs as working medium a permanent
gas which is, in principle, isothermally compressed, regeneratively heated at constant volume, expanded at constant temperature and regeneratively cooled to the compression temperature.
The cycle is performed by a reciprocating engine. As for the
closed cycle gas turbine only advanced materials and techniques
made it possible to achieve high efficiency, quiet operation and
comparatively long component life. In this case, too, combustion
products must undergo an important cooling outside the primary
heater. By virtue of a well designed continuous flow burner
emissions of pullutants are almost negligible.

Closed condensation cycles (Rankine): their most typical characteristic is high efficiency at moderate top temperatures. The use of steam is the most obvious choice; in the low power range, however, steam cycles are rather inefficient both from the point of view of the thermodynamics of the cycle and from that of the fluid-dynamics of the expander.

Organic working fluids offer a wide variety of thermophysical characteristics and yield, potentially, a much better performance. The low top temperature of the cycle minimizes material problems, while the moderate enthalpy change during the expansion, which is another peculiar character of heavy organic vapours, allows the use of low stress, low peripheral speed power turbines which are an ideal drive for centrifugal compressors.

Since thermal engines designed for heat pump drive are intended mainly for operation in urban areas, their pollution characteristics are of the greatest importance.

The diesel engine, owing to the transient nature of the combustion, to the high flame temperature and to the intensely cooled walls which confine the burning mixture tends to emit a noticeable amount of carbon monoxide and of nitrogen oxides (NO_x). The open cycle gas turbine of both the simple and the regenerative type, has difficulties in complying with new legislation with respect to NO_x emissions. Continuous flow burners equipping closed cycle systems experience pollution problems similar to those encountered with the gas turbine i. e. combustion can be easily brought to completion, thus eliminating carbon monoxide and unburned hydrocarbons, but reaction of atmospheric nitrogen with oxigen results in a certain irreversible generation of nitrogen oxides. The latter pollutant is the most difficult to fight since its formation is directly connected with high local combustion temperatures which are almost inevitably found in any burner. Lowering the temperature of the primary zone of combustion is the most effective way of reducing NO_x emissions. However, besides the technical difficulties that must be overcome, if temperature is lowered below a given range, combustion of hydrocarbons and of carbon monoxide cannot be brought to completion. This is illustrated schematically in Fig. 7a and b showing that there is only a window of definite amplitude in the temperature range of the primary zone which permits a simultaneously low level of emissions of all the basic pollutants. Another important parameter to control NO_x formation is residence time. Since NO_x concentration in the flame attains the equilibrium value only after an extremely long time, letting the combustion products

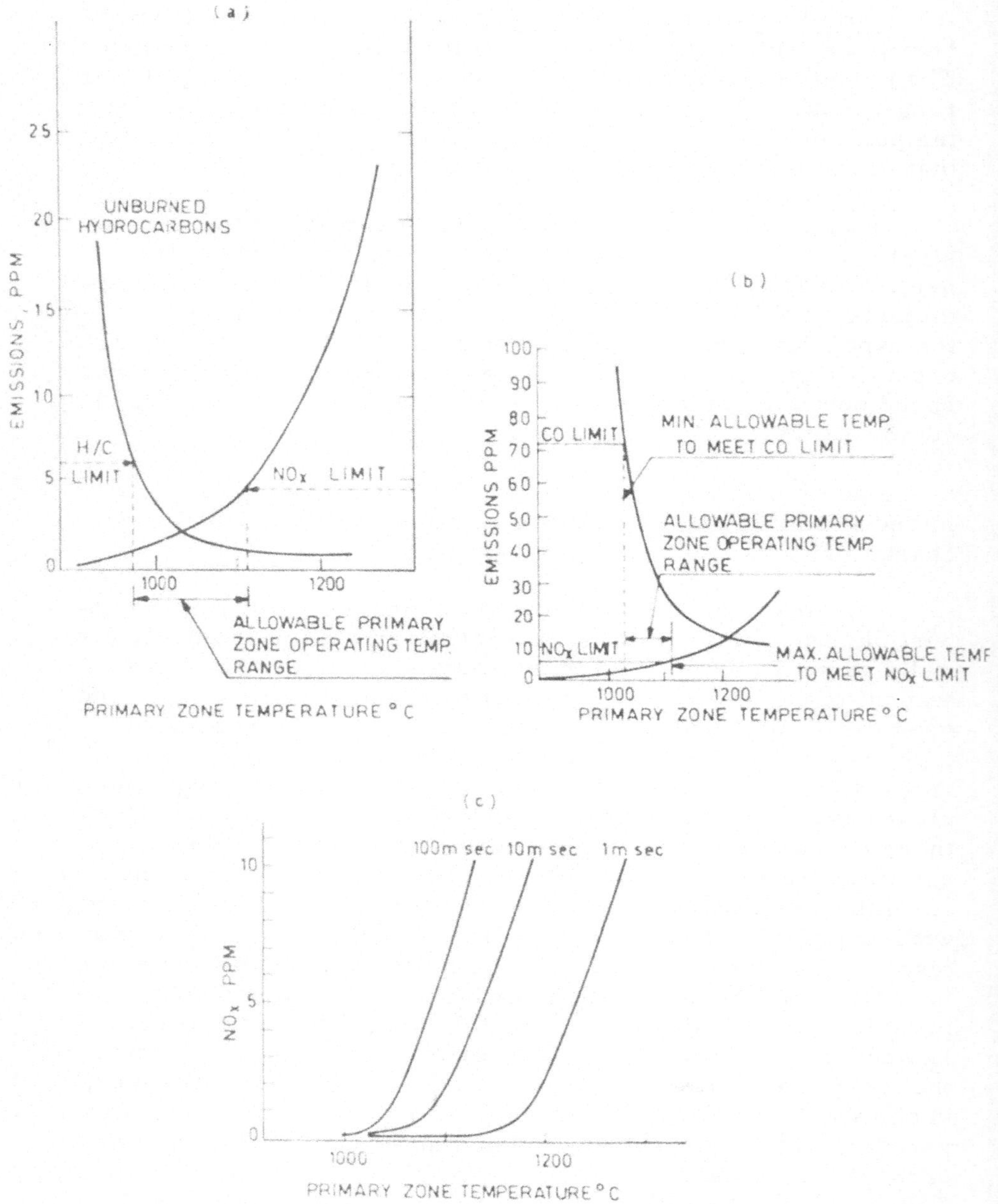

170

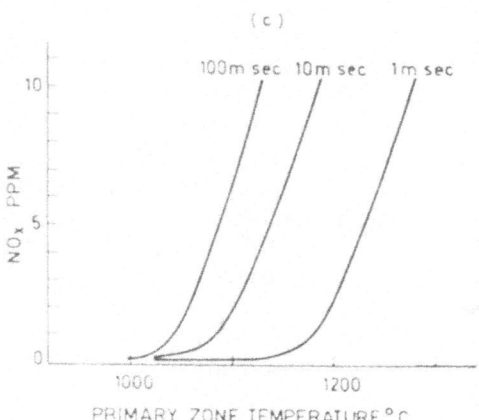

Fig. 7a,   Influence of flame temperature and of residence time
b and c.   on the emissions of unburned hydrocarbons, carbon
            monoxide and nitrogen oxides.

remain at the top temperature only for a very short time is effective in reducing nitrogen oxide emissions, as illustrated in Fig. 7c (temperature values given in Fig. 7 must be considered as relating to a particular burner and not as having a general validity

In continuous flow burners the following technical means can be employed separately or in a proper combination, to control NO_x generation [1-6]:

1 - pre-mix and, preferably, pre-vaporization of the liquid fuel. In this way a homogenous mixture can be obtained with sufficient excess air to reduce the adiabatic flame temperature to a desired level. If fuel is injected in the form of comparatively large droplets or is not well mixed to combustion air (even if in the gaseous form), combustion will take place locally in the vicinity of the fuel rich zones, with an energy release sufficient to attain almost stoichiometric flame temperatures, no matter how much overall excess air is present;

2 - radiation of a fraction of the heat of combustion from the flame to cooled walls. This effect is achieved by means of special techniques which make use of porous walls or of filamentary structures to support the non-adiabatic flame. Complete combustion with negligible NO_x formation has been demonstrated even for almost stoichiometric air-fuel ratios.

Power density, however, is much less than in conventional burners;

3 - exhaust gas recirculation. The inert nature of fully oxidized combustion products makes them suitable to be employed as diluting agents in order to reduce the specific chemical energy and oxygen content of the air-fuel mixture;

4 - staged combustion. By performing the combustion reaction in two stages, the first one leading to a partial oxidation of the fuel and the second one bringing the reaction to completion, chemical and thermodynamic conditions of the ambient where nitrogen oxides are formed can be controlled in a way to minimize NO_x formation;

5 - residence time control;

6 - combustion air inlet temperature control. At a given primary zone air-fuel ratio an elevated combustion air temperature results in high local temperatures and in a high NO_x

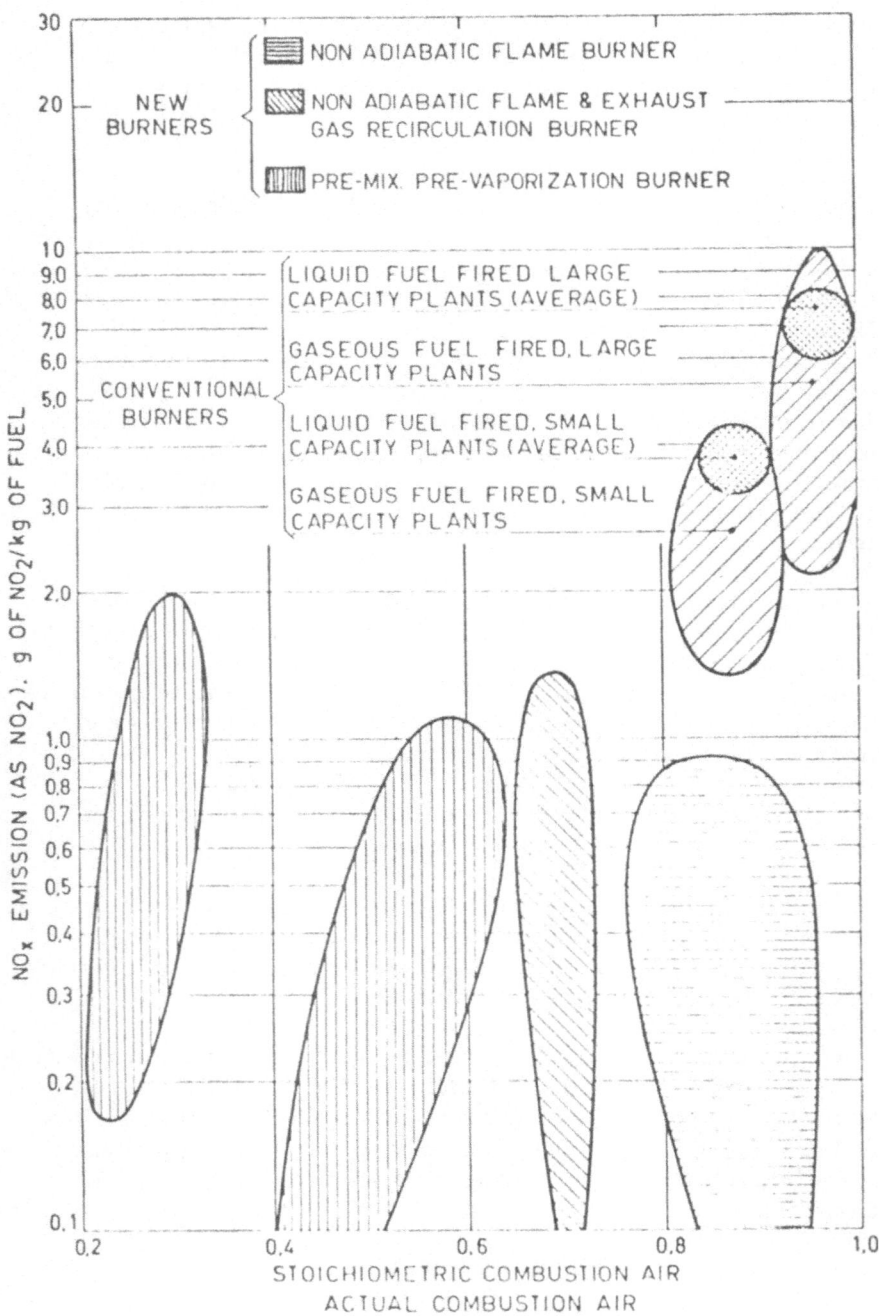

Fig. 8. Nitrogen oxide emissions from current and from advanced burners.

generation rate. If, on the contrary the average primary zone temperature is fixed, a higher air temperature helps in obtaining, (through a rapid fuel vaporization), a more homogenous mixture which reduces NO_x formation.

How effective the aforesaid measures are in limiting nitrogen oxides production can be seen with the help of Fig. 8 in which the performance of conventional and of advanced combustors is compared.

If a fully clean combustion has to be achieved, besides controlling the emission of unburned hydrocarbons, carbon monoxide and oxides formed by atmospheric nitrogen, the fuel should not contain either sulphur or nitrogen compounds which give rise, upon combustion, to the corresponding oxides.

THERMODYNAMIC ASPECTS OF ORGANIC FLUID CYCLES

It is known from the principle of corresponding states that the state diagrams of all fluids are similar if described in terms of reduced pressure, temperature and density. Referring to a typical fluid, as the one illustrated in Fig. 9, different portions of the state diagram can be employed to obtain power cycles having different characteristics.

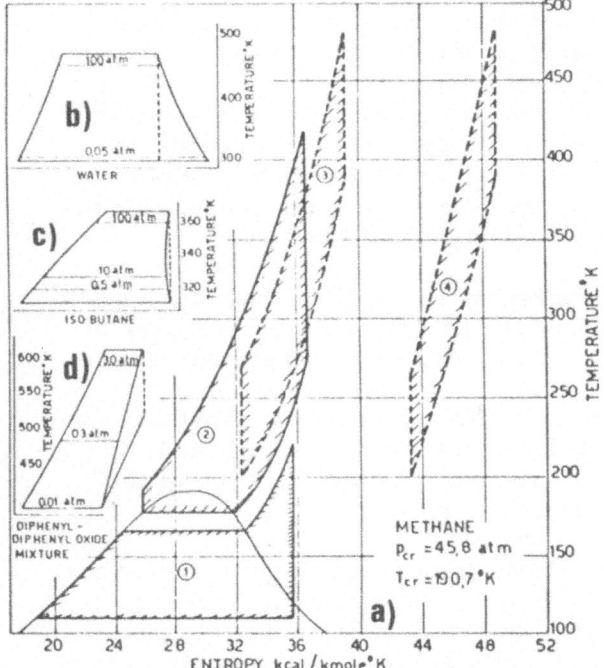

Fig. 9. Influence of the position of the power cycle on the state diagram of a typical fluid on cycle's configuration.

For instance cycle 1 of Fig. 9a has a low condensation pressure and a limited extension in the temperature field; cycle 2 while maintaining a substantially isothermal heat rejection and a liquid-phase compression has a much higher condensation pressure and receives heat at a continuously varying temperature; cycles 3 and 4 extend in the real gas and in the perfect gas region respectively. With reference to one particular fluid, not all the cycles which can be theoretically conceived are practically feasible due to heat sink limitations: in the case of air, for instance, the two phase region is inaccessible on account of its low temperature; in the case of water, gas phase compression at supercritical pressures and temperatures would involve heat rejection at a temperature by far too high. The possibility of adopting different fluids means then also the freedom of operating under normal temperature conditions within a desired region of the state diagram of the substance. Provided one particular region is selected, (for instance the two phase region including the saturation curve), the nature of the working medium plays still a very important role. As illustrated in Fig. 9b, c and d the shape of a vaporization-condensation cycle (Rankine) depends greatly on the thermal characteristics of the fluid considered. Isentropic expansion starting from saturated vapour conditions takes place with liquid condensation in steam, with superheating of the vapour in diphenyl and at a constant quality ($x = 1$) in iso-butane vapours[7]. Furthermore, the enthalpy drop of a power cycle, which has a determining influence on prime mover design, is strongly dependent on the molecular weight of the working medium, the heaviest substances giving the lowest work per unit mass. In summary the free selection of the operating fluid adds two degrees of freedom to the power cycle design: one concerning the portion of state diagram to be employed, the second one concerning the fluid characteristics within the chosen portion.

The variety of cycle configuration which is thus obtained is illustrated in Fig. 10a and 10b relating to two broad classes of fluids having a simple and a complex molecule respectively.

While in principle both inorganic and organic substances can supply working fluids of interest, only organic compounds are found in sufficient number to allow a true optimization of the fluid properties for each particular purpose.

Once the basic specifications of the closed cycle engine are given an optimum cycle and fluid can be selected. The most important single parameter in this choice is the power level involved: large engine output calls for high fluid density even at minimum pressure; small engine output requests low pressures

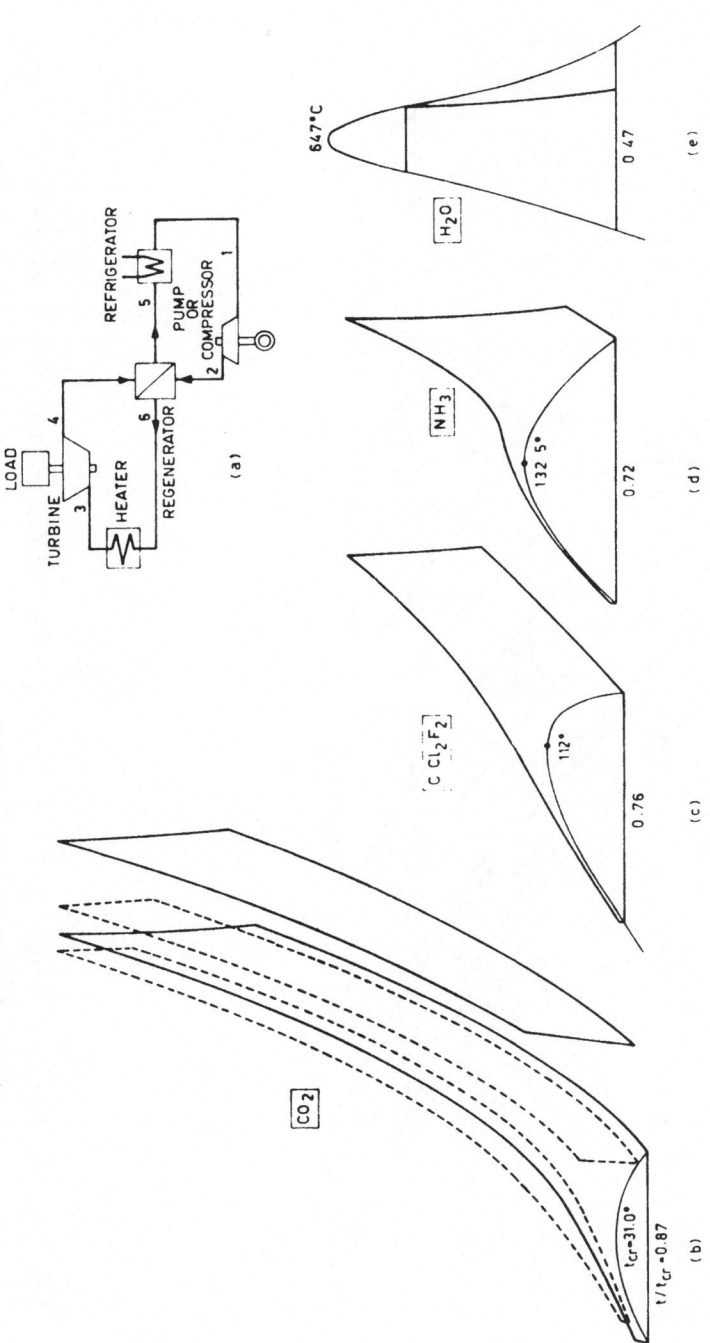

Fig. 10a.   Typical power cycle employing fluids of different critical temperatures for a low molecular complexity.

176

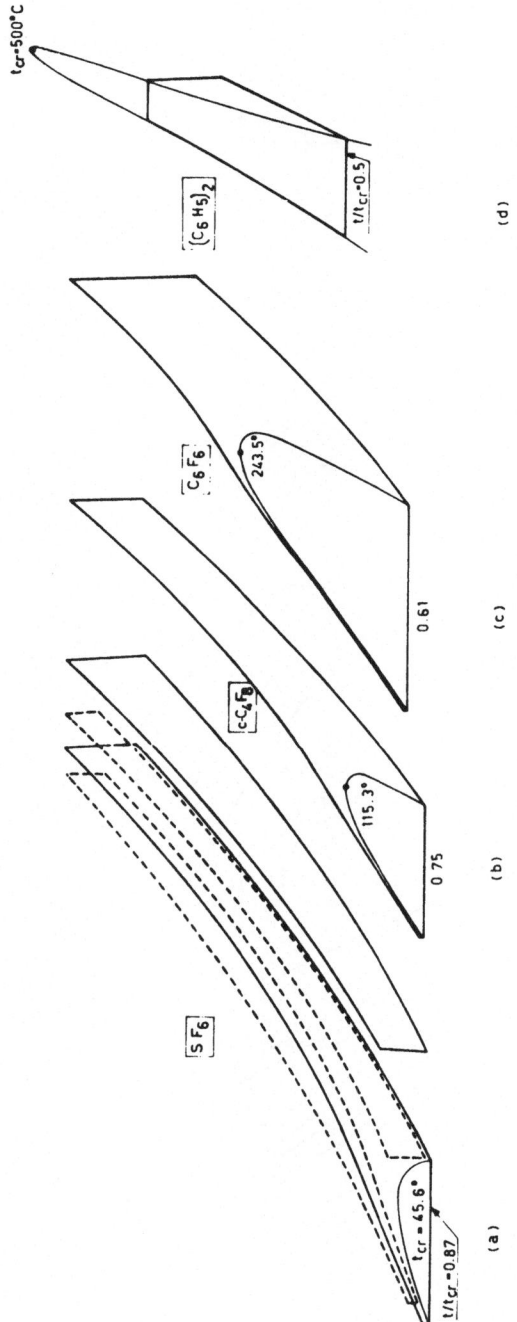

Fig. 10b.   Typical power cycle employing fluids of different critical temperatures for a high molecular complexity.

(and densities) to avoid the volume flow becoming too small.

Other technical details such as, for instance, the use of a turbine in place of a reciprocating expander or that of solar energy in place of fossil fuels can influence the selection of the working substance.

THE PROBLEM OF THERMAL STABILITY IN ORGANIC FLUIDS

From a strictly thermodynamic point of view there are literally hundreds of substances offering attractive power cycle performances.

Some of them are not sufficiently safe or are too costly. The strictest limitation to the number of fluids which can be taken into account, however, does not ensue neither from safety nor from availability reasons but from the limited thermal stability of all organic compounds.

Degradation at temperatures in excess of a given range is an irreversible process and is often accompanied by a corrosive action of the products of decomposition on building materials.

The simplest approach to the evaluation of the thermal stability of a given substance is to resort to static tests in which the pressure history within a vessel of a given material containing vapours of the substance to be evaluated is recorded as a function of time. Any degradation entails either the increase of the number of molecules in the vapour phase (due to formation of lighter fragments) or its reduction on account of the transition to the solid phase of the degraded portion of the substance (in the form of a solid deposit or as the result of a chemical attack of the walls). Any change in the number of molecules in the vapour phase results in a change in pressure. The fact that pressure increase and pressure decrease can, in some special cases, balance giving steady conditions, complicates sometimes the interpretation of the test.

A typical pressure diagram as a function of time is given in Fig. 11 relating to a test on hexafluorobenzene vapours in a stainless steel vessel. The initial pressure increase is due to variation in temperature which, at time zero had not yet reached the steady test value.

A search conducted by means of the pressure method[8] on high boiling organic substances gave the thermal stability limits

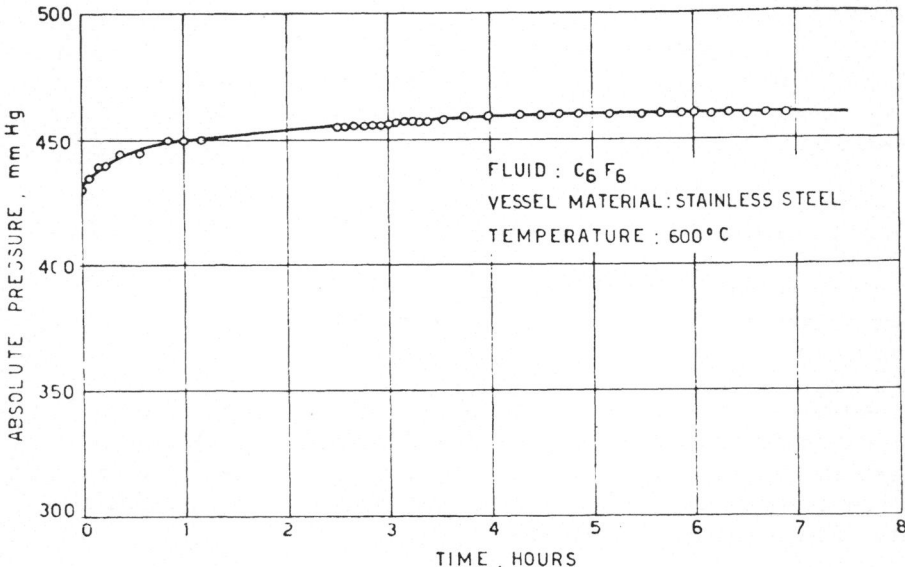

Fig. 11.   Typical pressure history within a constant volume
            vessel containing fluorocarbon vapours at elevated
            temperature.

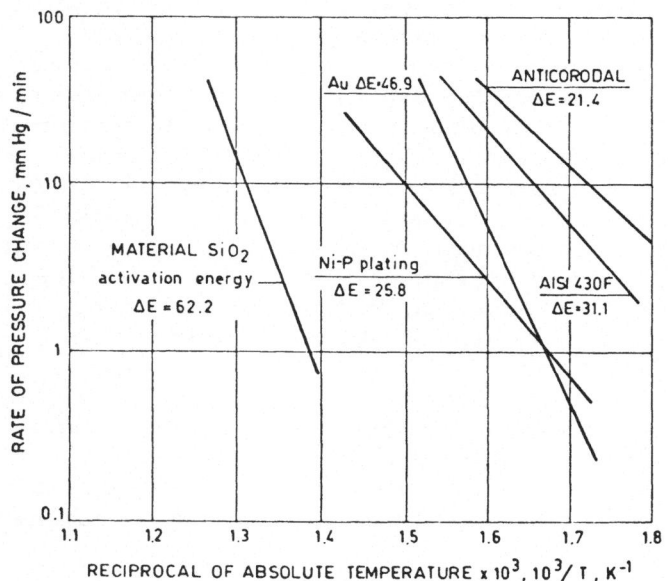

Fig. 12.   Influence of materials in contact with an organic
            vapour on decomposition rates.

TABLE 1

Conventional decomposition temperature (1% per hour decomposition) for some classes of stable organic fluids

Fluorine Compd. (a)	Decompn. Temp. °C
Hexafluorobenzene	650
Perfluorocyclobutane-perfluoroisobutylene	670
Tetrakis(perfluoromethyl)pyrazine	650
Perfluorocyclohexane	620–650
Cyanuric fluoride	530
2,4,6-Tris(perfluoromethyl)-s-triazine	480
Benzotrifluoride	430
Perfluorodiethylcyclohexane	430
Nonfluorine Compd. (b)	
Quinoline	650
2,2'-Bipyridine	620–650
Pyridine	620–650
Imidazole	620–650
Pyrimidine	620–650
Naphthalene	620–650
Benzene	600
Thiophene	590–620
2-Phenylimidazole (c)	600
Benzonitrile	570–600
Toluene	565–590
Benzothiazole	550–565
Thiazole	540
Biphenyl	510–540
Benzimidazole	400–430
1,2,4-Triazole	Under 430
Indazole	Under 480

(a) All tests in Monel tube.
(b) Vycor tube.
(c) From 1-phenylimidazole by thermal rearrangement at 510-540°C. during test.

reported in Table 1 (decomposition temperature is defined as the temperature at which the degradation rate is 1% per hour).

Completely fluorinated substances are the most stable compounds, but also many hydrocarbons exhibit an exceptionally good stability, with decomposition temperatures in excess of 600°C.

The question arises, however, whether the reported thermal stability limit gives also, at least as a first approximation, the upper operating temperature. The answer to this question is not straightforward. A closer examination of the thermal degradation of organic substances reveals a rather complicated picture which cannot be summarized in a simple decomposition temperature.

Among the phenomena which must be considered are the following:

1 - influence of the nature of the walls in contact with the fluid. There is evidence that decomposition is strongly influenced by the catalytic action of the walls. A clear example of the importance of this parameter is given in Fig. 12 in which the velocity of decomposition and the resulting activation energy are reported for methylene chloride vapours in presence of various materials (quartz, nickel-phosphorus plated steel, gold, ferritic stainless steel, aluminium alloy).

Even quartz, which is in general considered as one of the most inert materials, has a rather strong action on decomposition as shown in Fig. 13 which gives the pressure increase within a vessel containing CH_2Cl_2 vapours before and after treating the surface with an inhibiting substance.

Not only the rate of degradation is considerably lower within the inhibited vessel, but also the total amount of vapour decomposed before the system reaches a steady chemical state (horizontal asymptote of curves) is much less.

2 - Corrosion. Sometimes decomposition rate in certain operating situations is so low that the working fluid remains practically unaffected.

Nevertheless, the materials of the testing loop can experience a corrosive action which could endanger the integrity of the system. In this situation thermal stability and corrosion must be considered as two aspects of a single phenomenon.

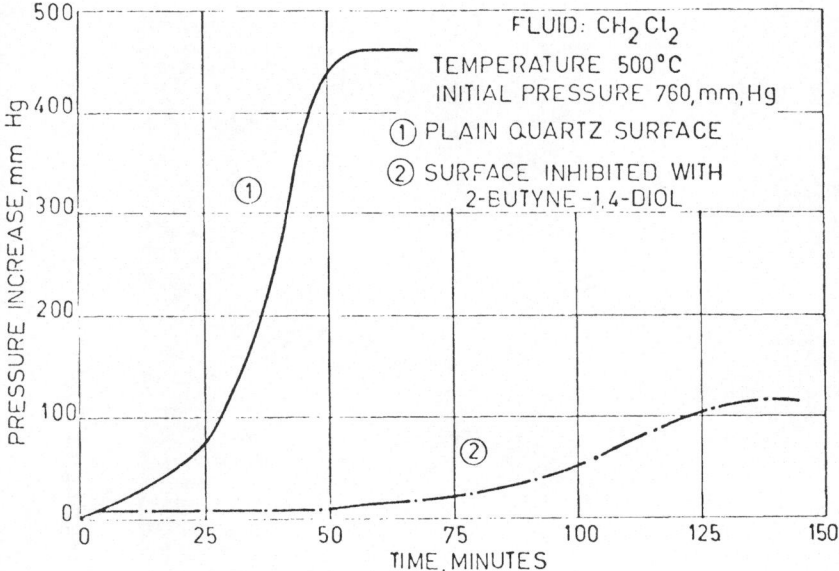

Fig. 13.   Effect of an inhibiting agent applied on a quartz
surface on the decomposition rate of CH_2Cl_2 vapours.

If the products of decomposition are highly aggressive,
corrosion is an inevitable consequence of fluid degradation and
corrosion tests can become more useful than other types of
stability measurements.

As an example of search for the maximum allowable temperature for two fluids, i. e. $C Cl_3 F$ and CH_2Cl_2, having a
limited intrinsic stability, let us consider the results of a corrosion test at 300ºC on a nickel net. The selection of the net as
metal sample was dictated by two reasons: 1) - owing to the well
defined geometry of the net and to the small diameter of the
threads (0. 15 mm) even a small corrosion would have been
visually evidenced as a loss of roundness of the thread; 2) -
since fluid degradation is usually associated with formation of
a carbonaceous deposit, any tendency of metallic surfaces to
undergo fouling would cause the filling of the small voids between
any two threads.   Tests were performed in flowing vapours and
lasted 320 hours for CH_2Cl_2, 160 hours for $C Cl_3F$.

Fig. 14a and b show the negligible corrosion suffered by
nickel under the action of CH_2Cl_2 (in other tests chemically
plated carbon steel was found to perform as well as solid nickel);

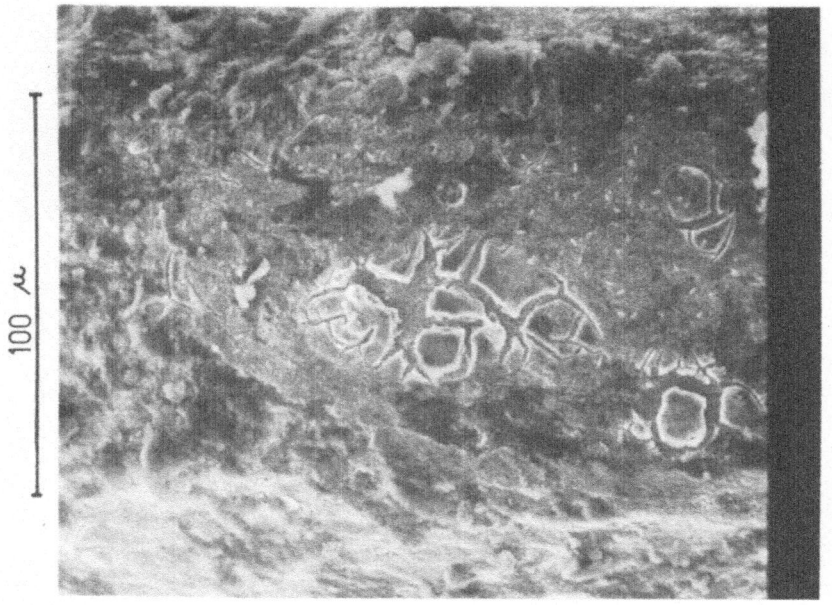

Fig. 14a and b.   0.15 mm nickel thread after 320 hours testing at 300°C in CH_2Cl_2 vapours.

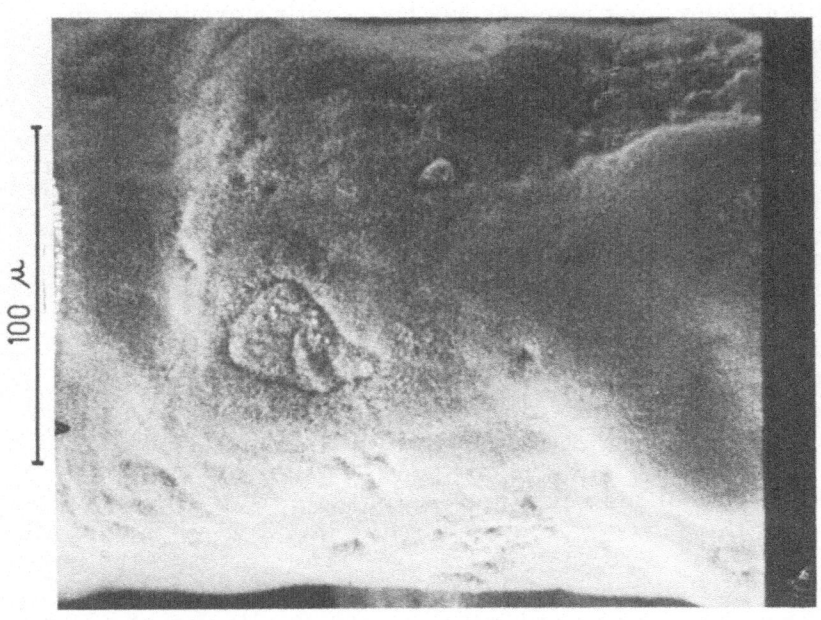

Fig. 14d. 0.15 mm carbon steel thread after 240 hours testing at 450°C in air.

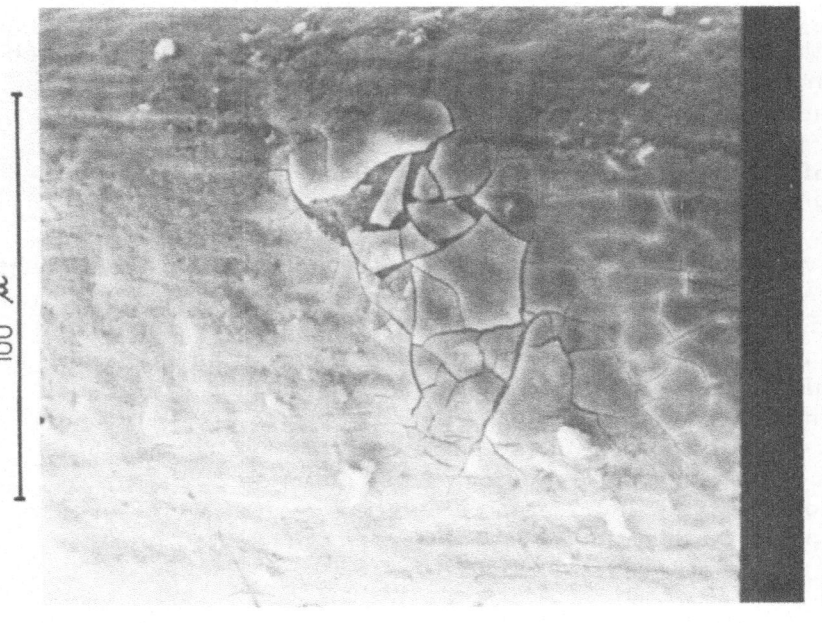

Fig. 14c. 0.15 mm nickel thread after 160 hours testing at 300°C in C Cl₃ F vapours.

Fig. 14c records a very small corrosion due to the action of C Cl$_3$ F; Fig. 14d illustrates the attack of air on carbon steel at 450°C (240 hours test) and shows how easily corrosion can be visually evaluated on a thin thread sample.

A more stable, completely fluorinated fluid, the perfluoro-1. 3-dimethyclyclohexane was tested for 400 hours at 400°C with a 20 hours overheating to 600°C. A cross section of the hottest end of the heated coil is shown in Fig. 15a (tube before test) and 15b (tube after test). The upper edge of the picture, showing the inner surface of the tube in contact with the working fluid, exhibited no noticeable sign of corrosion. The outside tube surface, on the contrary, suffered a severe attack from atmospheric air (lower edge of picture). A thin, black, adherent deposit was found in the hot portion of coil (Fig. 15c) and a white deposit in the cold end (Fig. 15d). No clogging of filters in the circuit was observed.

3 - Extension of the results of static tests to flowing systems. On account of the active role played by the metallic surface of the containing vessel, important differences are to be expected between static and flow tests. In static pressure tests which are the most usual screening procedure, a very small amount of low pressure fluid is kept in contact with a wide surface and an unfavourable balance is established between the number of molecules in the vapour system and the number of active decomposition centres on the metallic surface.

Furthermore, in static tests an induction period during which the decomposition rate is very low even at high temperatures is often observed (see Fig. 13). In flow systems, on the contrary, the equivalent ratio of the active metal surface of the hottest portion of the coil to the number of molecules passed through the heater is much more favourable.

Active decomposition centres are, in proportion to the amount of fluid tested, by far less numerous. Whenever more reliable information is needed on the suitability of a fluid to be employed as working medium in a power plant, flow tests of adequate duration should be run. In order to avoid any perturbating influence of foreign substances, which have been proved to have a strong effect on decomposition modes and rates, flow systems should be designed either without moving parts, actuated by natural circulation or with unlubricated circulating devices which prevent any oil contamination (photographic records presented in Figs. 14 and 15 relate to a natural circulating loop).

185

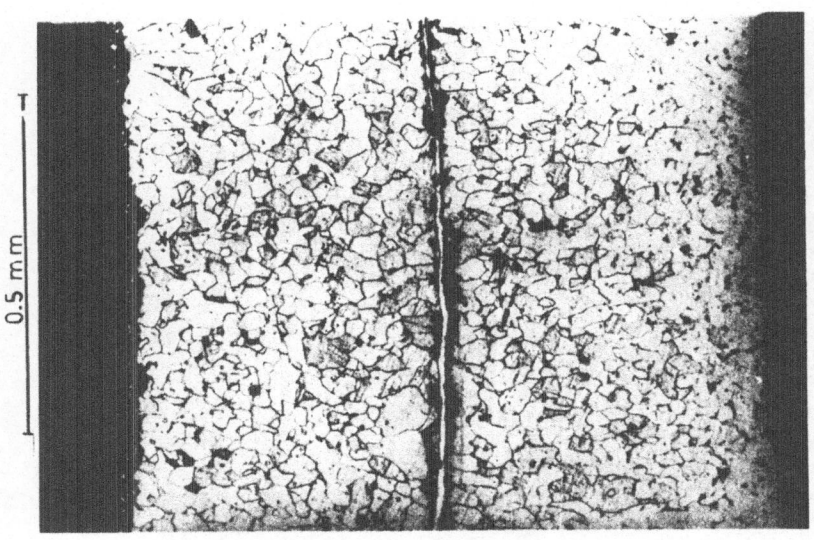

Fig. 15b. cross section of the hot end of the heating coil after 400 hours testing with C_8F_{16} vapour at temperatures between 400 and 600°C.

0.5 mm

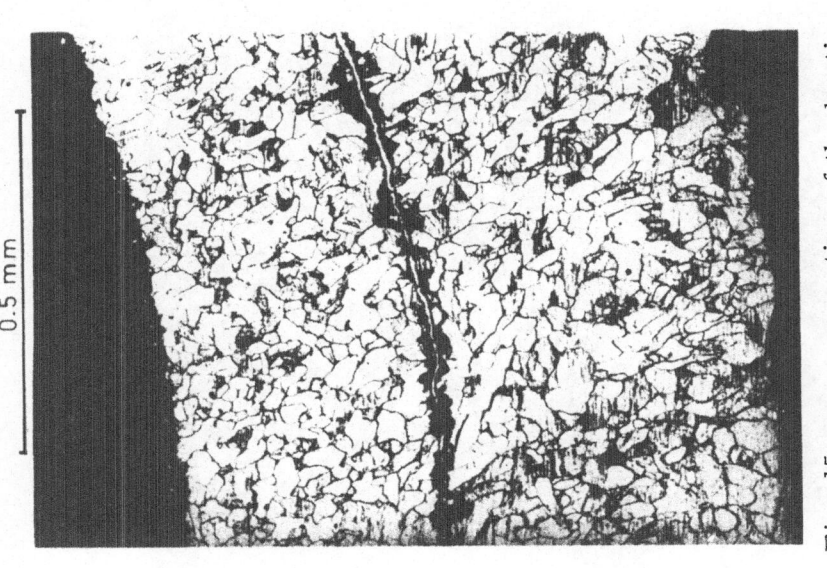

Fig. 15a. cross section of the heating coil before testing.

0.5 mm

186

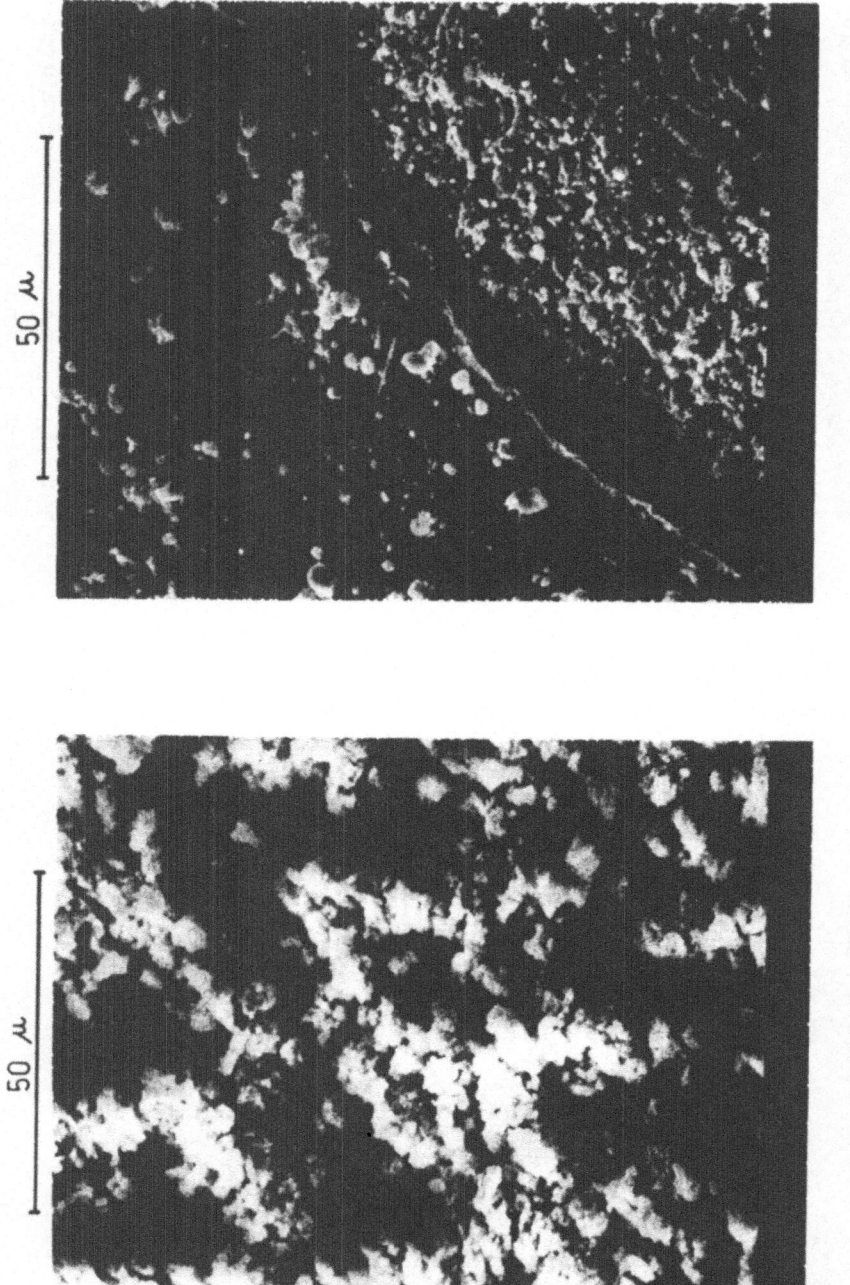

50 μ

Fig. 15c.   structure of deposits in the
hot end of the coil.

50 μ

Fig. 15d.   structure of deposits in the
cold end of the coil.

Besides the data previously reported, the capability of organic substances to withstand thermal conditions which are typical of the operation of power plants was demonstrated in ref. 8 in which one working fluid formed by a mixture of C_6HF_5 and C_6F_6 was successfully tested for more than 1,000 hours at 380°C without any decomposition or corrosive action on materials.

By combining the results of various types of tests (static tests in presence of different materials, flow tests in natural and in forced circulation loops, corrosion tests) with data reported in the literature and relating mainly to the work done in view of the automotive application of closed cycles[9-10], it can be stated confidently that a large number of working media are available for operation up to at least 400°C top cycle temperature.

ORGANIC FLUID ENGINE DEVELOPMENT

When designing an organic fluid power cycle for maximum reliability and minimum maintenance, turbine expanders should be preferred. The thermodynamic analysis of a large number of power cycles based on thermally stable fluids (chiefly completely fluorinated hydrocarbons) showed that a large expansion ratio is needed for optimum cycle efficiency. Reciprocating expanders exhibit a limited capability of handling such large expansions, while turbines, if properly designed, can efficiently perform the requested task.

It is known that turbines have an intrinsically high capacity of handling large volume flows.

In the low and medium power range (10 to 100 kW) working fluids should be selected in such a way that the exhaust volume flow be sufficiently large to lead to turbine diameters and blade heights which are compatible with ordinary fabricating techniques. Two fluids, among those which exhibited a sufficient thermal stability, were found adequate also with respect to turbine design, i. e. hexafluorobenzene (C_6F_6) and perfluoro-1.3-dimethylcyclo-hexane (C_8F_{16}).

On account of the lower condensation pressure, the latter, fluid is more suitable for low power engines.

With reference to low temperature heating systems, as those using radiating ceiling panels, exhaust heat from the power cycle is required at temperatures in the range 40 to 60°C. At these heat discharge levels and for a safe top cycle temperature

TABLE 2 - THERMAL PRIME MOVERS' CHARACTERISTICS

Characteristic	Prime mover (100 kW output class)							
	Diesel engine	Gas engine	Open cycle gas turbine (non-regenerative)	Open cycle gas turbine (regenerative)	Closed cycle gas turbine	Stirling engine	Steam turbine	Organic fluid turbine
Type of motion	reciprocating	reciprocating	uniform rotation	uniform rotation	uniform rotation	reciprocating	uniform rotation	uniform rotation
Efficiency	high	medium	low	medium	medium	high	low	medium
First cost	low	low	potentially low	potentially medium	high	high	potentially low	potentially low
Content of high cost materials	low	low	high	high	high	high	low	low
Maximum temperature of stressed components, °C (approximate values)	600	600	800	800	800	750	450	400
Maximum internal pressure, bar	60	60	4 to 10	4 to 7	10 to 20	70 to 150	10 to 40	10 to 40
Life of the most critical basic component	limited	limited	limited	limited	limited	limited	unlimited	unlimited
Maintenance and servicing characteristics	satisfactory	satisfactory	potentially excellent	potentially good	potentially excellent	potentially good	potentially excellent	potentially excellent
Multi-fuel capability	limited	limited	limited	limited	wide	wide	wide	wide
Heat rejection modes	cooling water (cylinder jacket) and exhaust gases	cooling water (cylinder jacket) and exhaust gases	exhaust gases	exhaust gases	cooling water (pre-cooler)	cooling water	cooling water (condenser)	cooling water (condenser)
Environmental characteristics	marginal	acceptable	acceptable	acceptable	good	good	good	good

(380 to 450°C) both fluids exhibit cycle efficiencies better than 30% (Fig. 16 and 17), which appears as a remarkable performance if compared with alternate prime movers.

Furthermore, the organic fluid turbine is almost ideally suited for driving the centrifugal compressor of the heat pump cycle (as for expanders, also for compressors maximum reliability and simplicity is obtained resorting to continuous flow machines).

To gain a better understanding of the possibilities offered by organic fluids as power plants' working media, the detailed design of two units was developed, for two different power levels (i. e. 145 and 42 mechanical kW) employing C_6F_6 and C_8F_{16} as working fluids respectively, following the guidelines of maximum reliability and easiest maintenance. The configuration of the two turbo-compressors is illustrated in Fig. 18a and b: both adopt a one stage reaction turbine of the axial type which drives a two stage centrifugal compressor. Typical features of the units are simplicity (very limited number of moving parts), small dimensions, low stresses (both mechanical and thermal) and use of ordinary construction materials.

A comparison of the main characteristics of the organic fluid (turbine) engine with those of other prime movers is given in table 2.

To illustrate the potential gain in simplicity resulting by the use of the organic fluid engine the turbocompressor design of Fig. 18 can be compared with a conventional solution envisaging the adoption of a 4 cylinder diesel engine driving a 12 cylinder reciprocating compressor. The diesel engine has a total of 5 main bearings plus 27 secondary bearings (connecting rod and wrist pin, cam shaft bearing etc.); furthermore it has 16 valves (8 admission and 8 exhaust), 4 sets of piston rings and 2 main oil seals. The reciprocating compressor besides the cylinders' valves, piston rings and main seals has 5 main and 24 secondary bearings.

All these components are liable to wearing or to malfunctioning. The organic fluid unit illustrated has only 2 main bearing and 2 main seals. It has the potential of yielding a silent, unattended operation with long component life and minimum servicing.

In organic fluid engines as far as heat exchangers are concerned, no particular design or operating problem should arise, provided a working fluid is selected which does not present any

190

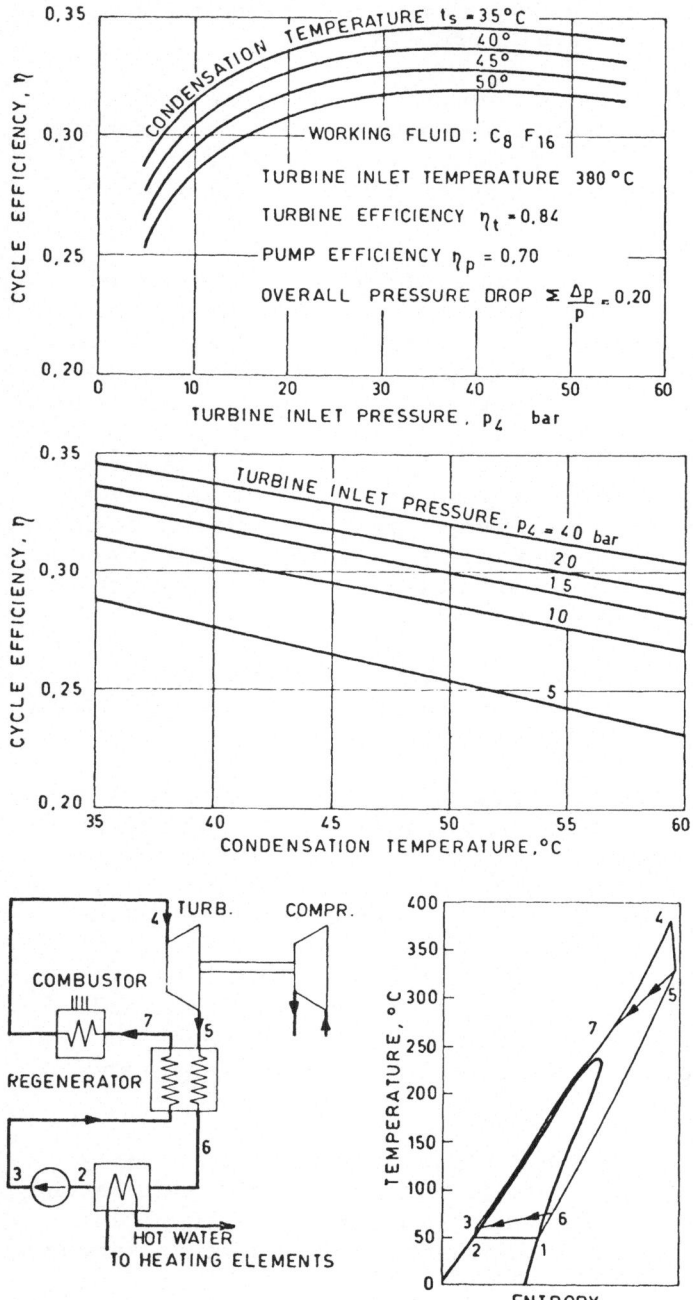

Fig. 16.   Configuration and performance of C_8F_{16} power cycles.

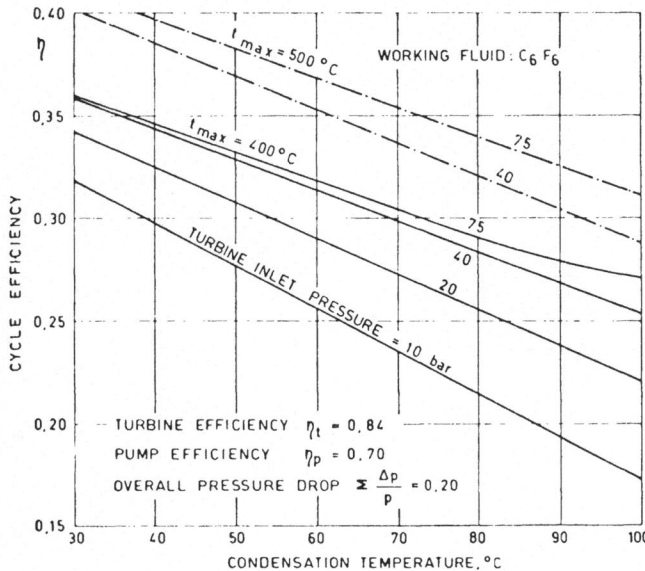

Fig. 17.   Performance of C_6F_6 power cycles.

corrosive action neither cause any fouling of heat transfer surfaces (the availability of such fluids has been demonstrated).

A DETAILED STUDY ON THE APPLICATION OF AN ORGANIC FLUID ENGINE-HEAT PUMP SYSTEM

In order to evaluate the energy saving capability and the
economic viability of thermal engine-heat pump systems, a
detailed example was worked out and compared with a conventional solution[11].

The main choices for the system characteristics and the
results of the analysis are discussed in the following.

Power level: the selected thermal power rating (at 0°C
ambient temperature) was 1,000 kW.   A similar power was
known to be required to heat a large single building taken as
reference.

Confining the heat supply to one building, allows the direct
circulation of water from the power cycle and the heat pump
condensers to the final heat exchangers, thus avoiding the

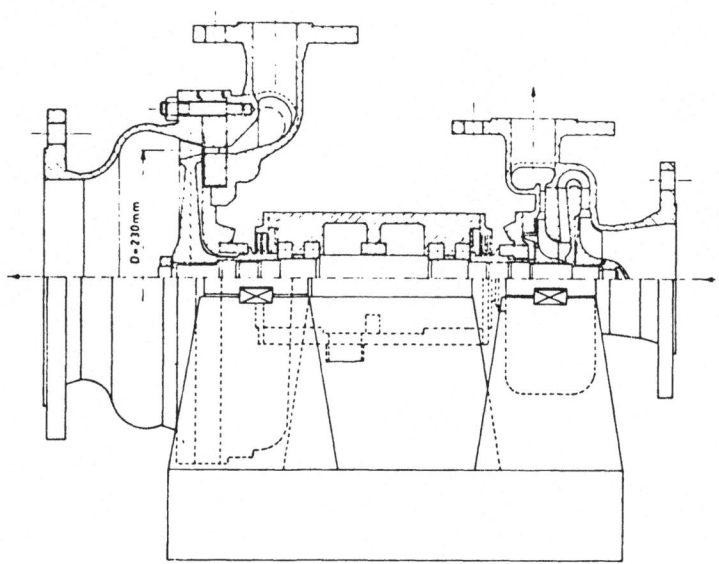

Fig. 18a. Organic fluid turbo-compressors for direct heat pump
drive: C_6F_6 - R12, 145 kW, 25,000 rpm unit.

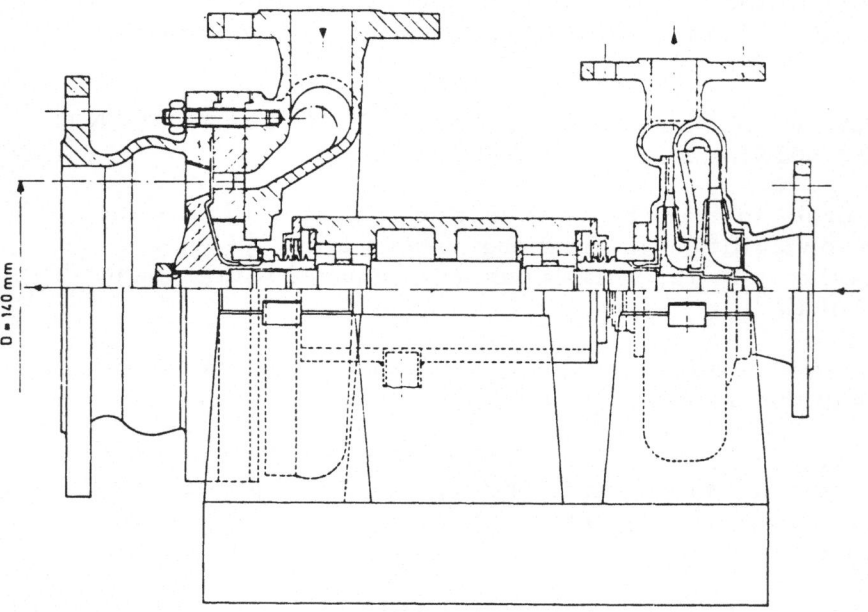

Fig. 18b. Organic fluid turbo-compressors for direct heat pump
drive: C_8F_{16} - R12, 42 kW, 30,000 rpm unit.

thermodynamic and financial losses connected with an intermediate heat carrying loop. The mechanical power which
corresponds to the quoted thermal power is around 150 kW which
permits a proper design of a high efficiency turbine driven centrifugal compressor.

Heat source and heat sink: the heating station considered
was intended for location on the border of the sea, namely in
the city of Venice. At the selected power level, air-heated
evaporators would pose very difficult problems, not only technical, and would not consent the achievement of the best fuel
economy. Radiating ceiling panels were assumed as final heat
exchangers. In order not to limit the validity of the analysis to
this heating mode, a preliminary study was conducted to inves-
tigate the possibility of employing, in place of ceiling panels,
low temperature thermoconvectors. By making use of properly
arranged extended heat transfer surfaces, similar to those
employed in vehicular radiators, it seems possible to design a
natural circulation convector for water temperatures similar to
those used in ceiling panels. The relative sizes of a conventional
high temperature radiator, of a forced circulation and of a natural
circulation convector of advanced design are shown in Fig. 19.
The ability of complying with the requested task at the minimum
temperature of the heat carrier is of great help in improving
system performance. .

Type of engine and control mode: the selected prime mover
was a C_8F_{16} organic fluid engine working at 380°C top cycle
temperature.

A fully subsonic, multi-stage turbine was preferred to the
simpler supersonic, single stage solution since it was decided
to make use, as far as possible, of proven components having a
well predictable performance. The turbine drives directly,
through a common shaft, a two stage R11 centrifugal compressor.

At ambient temperatures higher than the design value the
decreased heat load calls for a reduction in the mechanical power
of the turbine which is obtained by a similar reduction in mass
flow and in inlet pressure.

At the same time the engine condenser is cooled more
efficiently by the colder water coming from the ceiling panels
and temperature differences between condensing organic fluid
and water become smaller on account of the overabundant exchange surface available at reduced loads. As a result, the
condensing temperature and pressure decrease, which partly

194

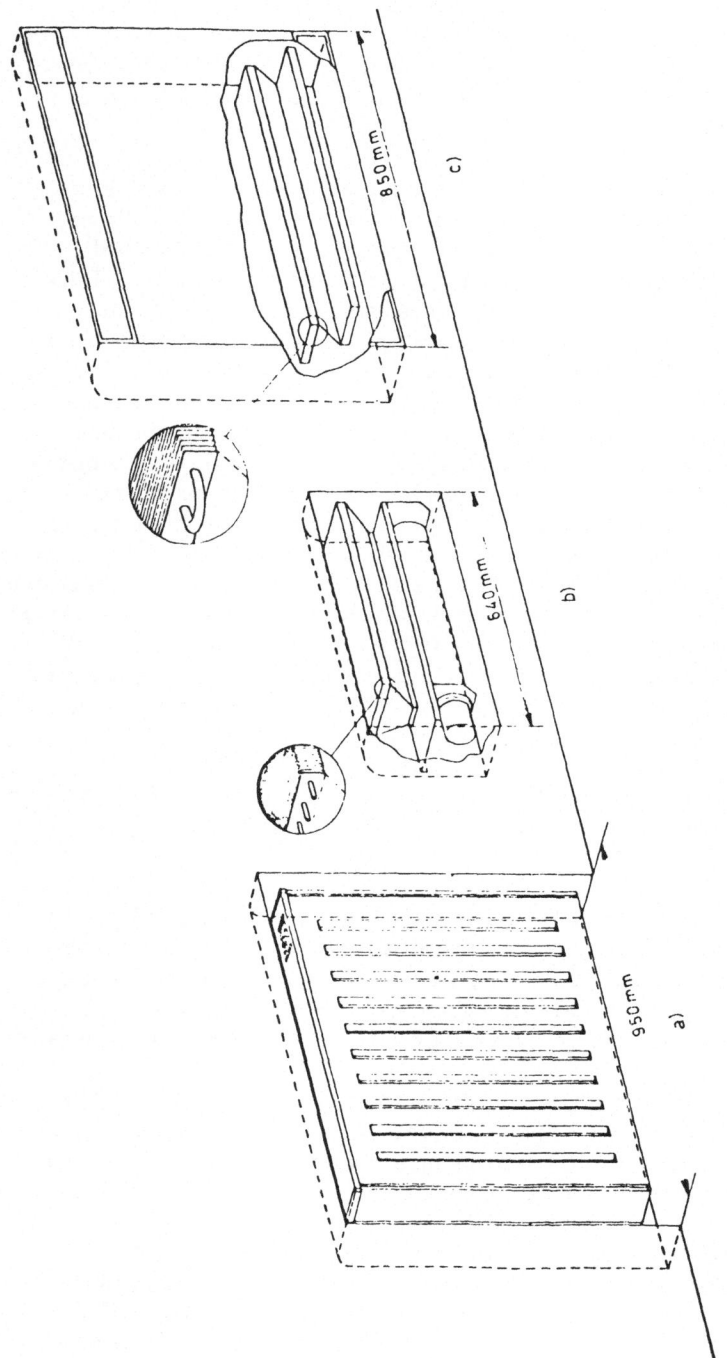

Fig. 19a, b and c. Size and configuration comparison of conventional high temperature radiator with low temperature forced and natural circulation convectors.

offsets the reduction in expansion ratio produced by the reduction in inlet pressure. At high ambient temperatures the compressor also is requested to work at a reduced capacity and head, which is done partly by decreasing the rotating speed of the turbo-compressor shaft. The combined effect of condenser's pressure reduction and rotating speed control yields an excellent part load performance.

Energy balance: At 0°C ambient temperature the energy balance of the system is as follows.

A thermal power of 403 kW is transferred from the fossil fuel to prime mover's working fluid and generates 124 kW mechanical power while discharging 279 kW waste heat in the power cycle's condenser.

An additional thermal power of 579 kW is "lifted" from the low temperature level of the cold heat source (sea-water at 7°C) to the level required by the heating system (around 43°C). An overall thermal power of 1,000 kW is delivered to the consumer which is 2.5 times that transferred to the working fluid by fossil fuel.

At higher ambient temperatures the good part load efficiency of the engine and the improved performance of the heat pump lead to a still greater energy saving (see Fig. 20).

The overall fuel consumption extended to the whole heating season is 2.76 times lower than that of a conventional system.

Economic analysis: In order to evaluate the capital and operating costs of the heat pump system a detailed design of all major components was carried out (plant layout obtained in this phase is given in Fig. 21 where it is compared with that of a conventional station). Costs were then estimated by selected manufacturers. The greatest contribution to the final cost is due to conventional equipment closely related to that employed in the refrigerating practice (about 50% of the total cost accounts for the sea-water heated evaporator). The cost of non-conventional components such as the turbo-compressor or the power plant regenerator is limited to about 10% of the total cost.

The capital cost of the installed kW (evaluated at an ambient temperature of -5°C) is It. L.7,000 for the heat pump version and It. L 5,130 for a conventional solution. The large fuel savings allowed by the heat pump plant lead to an energy cost of It.L 6.9/ kW-h which must be compared with a cost of It. L 10.7/kW-h for

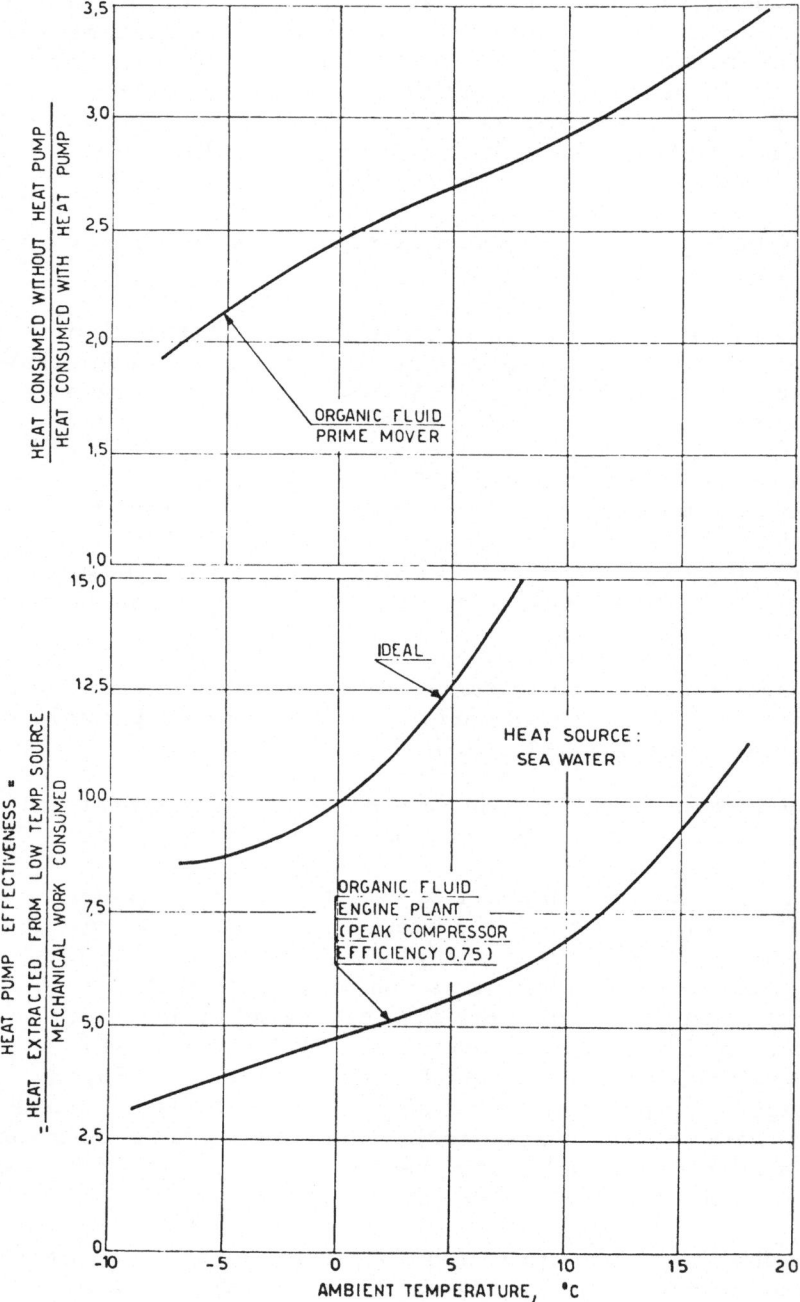

Fig. 20.   Heat pump and thermal engine-heat pump system
performance as a function of ambient temperature.

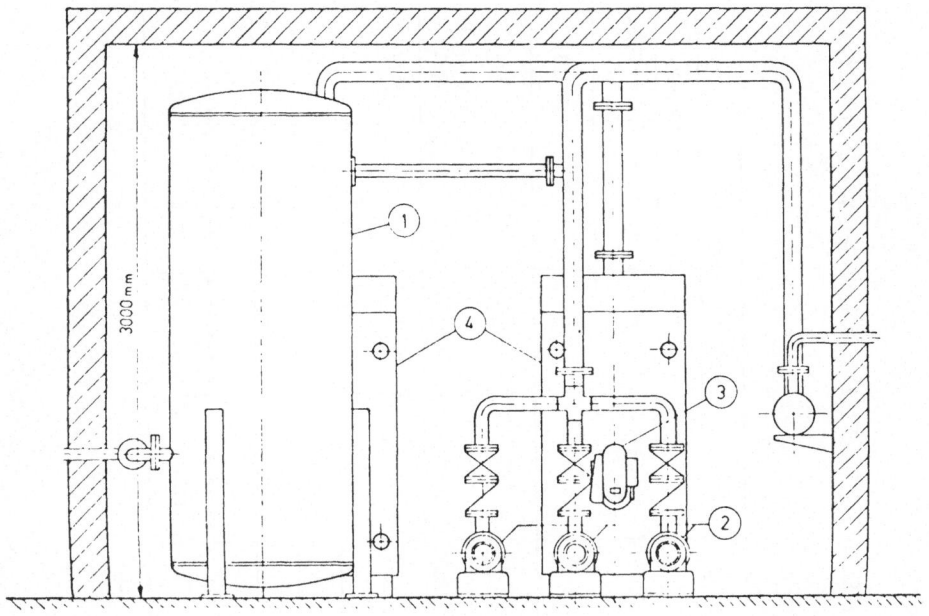

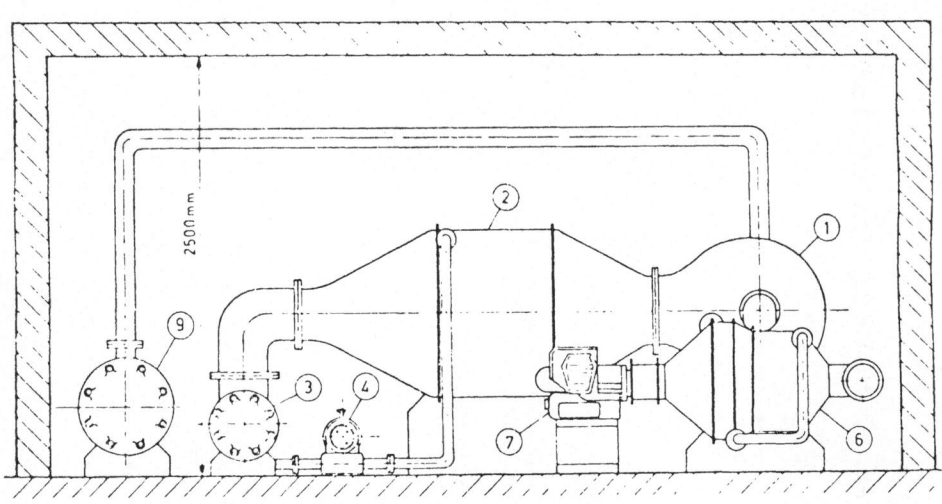

Fig. 21a   Plant layout of a conventional and of a thermal engine-
and b.   heat pump system for 1,000 thermal kW output.

TABLE 3

CAPITAL AND OPERATING COSTS OF HEATING PLANTS IN THE 100 KW POWER RANGE

Item	Installation cost				Operating cost (*)			
	Conventional plant		Heat pump plant		Conventional plant		Heat pump plant	
	absolute It. L /KW	fractional %	absolute It. L /KW	fractional %	absolute It. L /(KW-h)	fractional %	absolute It. L /(KW-h)	fractional %
Heat generating station with reserve units	1930	37.6	3800	54.2	0.75	7.0	1.48	21.3
Distributing network and heating system inside the buildings	3200	61.4	3200	45.8	1.25	11.6	1.25	17.9
Total capital investments	5130	100	7000	100	2.0	18.6	1.73	39.2
Fuel (gas oil)					7.43	69.3	2.49	35.8
Personnel					0.32	3.0	0.40	5.8
Maintenance					0.56	5.2	0.86	12.3
Electric energy consumption					0.35	3.3	0.4	6.2
Administration, Insurance etc.					0.05	0.5	0.05	0.8
Total operating costs					10.7	100	7.0	100

(*) assumed plant life 20 years, interest for capital investments 9% per year.

the conventional plant (assumed plant life 20 years; interest on capital investments 9%; cost of fuel (gas oil) It.L 67/kg). Even doubling the capital costs of the heat-pump plant would result in an energy cost lower than that of the conventional station.

Further details on cost breakdown are given in table 3.

CONCLUSION

On the base of the results of a continuing research pro-gramme which was partly reported here, the following conclusions can be drawn:

(1) combined thermal engine-heat pump plants offer an extremely attractive potential performance in low temperature heat generation. Fuel savings in excess of 50% with respect to a conventional station seem possible even at the present state of the art.

Space heating is the most direct application which can be envisaged, but also low grade process heat production could benefit from a substantial fuel economy;

(2) the key component of the combined system is the thermal engine which must comply with a number of technical and environmental requirements. In particular it should allow a silent unattended operation with a minimum of maintenance and servicing; it should be capable of burning a variety of fuels and should have a clean exhaust. A prime mover truly adequate for the proposed application is not found among existing thermal engines. On the contrary new types of engines, presently under development, and, in particular, those based on organic fluid condensation cycles could re-present an ideal heat pump drive.

With respect to environmental implications, external com-bustion engines have already demonstrated their capability of reducing pollutants' emissions to almost negligible levels;

(3) as in other heat pump applications the effectiveness of the heat generation process is very sensitive to the thermal level at which heat is produced. An effort should be made to reduce heat temperatures to the lowest acceptable value, renouncing to unnecessary margins.

For instance in space heating, where the end use of thermal energy is around 20°C, new methods of heat delivery should be developed, allowing the use of heat at temperatures in the range 30-40°C whereas, at present, heat is employed in the range 40-80°C.

REFERENCES

1. Zwick, E. B., Mills, T. R. and FioRito, R., "Evaluation of a Low NO$_x$ Burner", Paxve Report USG-1, Newport Beach, USA, 1971.
2. Pietsch, A. and Rackley, R. A., "Low Pollution Closed Brayton Cycle Engine", Intersociety Energy Conversion Engineering Conference, Boston, 1971.
3. US Environmental Protection Agency, "Low Pollution Power System Development Program", Document N. 1, Ann Arbor, Michigan, 1973.
4. Duffy, T. E. et al., "Low Emission Burner for Rankine Cycle Engines for Automobiles", Environmental Protection Agency, Document APTD-0707, Ann Arbor, Michigan, 1971.
5. Hazard, H. R., Fisher, R. D. and McComis, C., "Low Emission Burners for Automotive Rankine Cycle Engines", Environmental Protection Agency, Document APTD-1516, Ann Arbor, Michigan, 1973.
6. Hazard, R. H., "NO$_x$ Emission from Experimental Compact Combustors", ASME paper N. 72-GT-108, 1972.
7. Casci, C. and Angelino, G., "Some Thermodynamic Aspects of Special Fluid Power Plants", Problems of Fluid-Flow Machines, Warszawa, 1968 p. 167.
8. Johns, I. B., McElhill, E. A. and Smith, J. O., "Thermal Stability of Some Organic Compounds", Journal of Chemical and Engineering Data, Vol. 7 n. 2, 1962 p. 277.
9. Miller, R. D., Null, H. R. and Thompson, Q. E., "Optimum Working Fluids for Automotive Rankine Engines - Vol. II" US Environmental Protection Agency, Document APTD-1564, Ann Arbor, Michigan, 1973.
10. Technical Information from the ISC Chemicals Limited, Avonmouth, England, 1974.
11. Angelino, G. et al., "A Proposal for a Low Pollution Central Heating System for the City of Venice", Proceedings of the 6th International Congress of Climatistics, Paper IV-04.1.0, Milano, 1975.

INDUSTRIAL APPLICATIONS OF HEAT PUMPS

P. Kolbusz, Dipl. Ing.

The Electricity Council Research Centre,
Capenhurst, UK.

1. INTRODUCTION

The importance of efficient energy utilization in industry has been recognised for some time, and it is not surprising that the use of heat pumps is being increasingly recommended. Some manufacturers are actively engaged in the marketing of heat pumps for domestic space heating and cooling. A few specialised heat pump applications have also been reported: heating of swimming pools, greenhouses and domestic water. While these heating methods are of great significance it has been recognised that there are possibilities for industrial applications as well.

The object of this paper is to review some of the methods that show promise for industrial heat recovery by heat pumping methods.

2. THE OPERATION AND PERFORMANCE OF HEAT PUMPS

The heat pump, like the mechanical refrigerator, is a simple thermodynamic machine: consisting of a compressor, a condenser, an expansion valve and an evaporator (Fig. 1). The significant difference between the two machines is in the application: the refrigeration plant is installed with the object of cooling, i. e. the removal of heat from a space or a process, the heat pump, on the other hand, is built either for heating or for both heating and cooling.

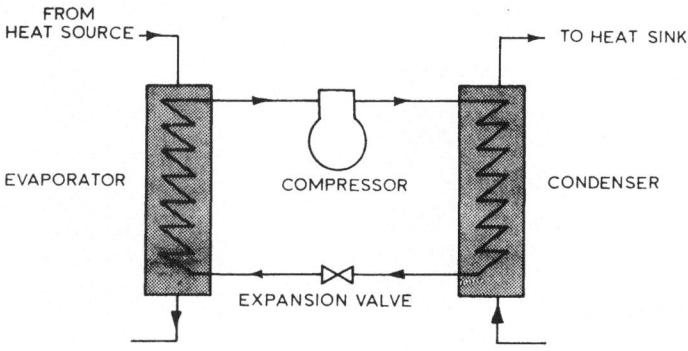

Fig. 1. Heat pump: simplified flow diagram.

To carry out a heat pumping duty the temperature of the
heating surface (i. e. condenser) has to be above that at which
heat is to be used and the temperature of the heat absorbing
surface (i. e. evaporator) has to be below that of the heat source.
If all losses are neglected, the heat output is equal to the thermal
energy absorbed from the heat source plus the thermal equivalent
of the mechanical energy which is required to bring about the
change of temperature levels. The ideal heat pump is a reversed
heat engine operating on the Carnot cycle (Fig. 2). The cycle is
usually shown on the temperature-entropy (T-s) diagram, where
the heat equivalent of the mechanical energy input (W) is proportional to the area 1234, the heat absorbed from the heat source
(Q_e) to 6145, and the heat transferred at higher temperature level
(Q_c) to 6235. It is customary to express the performance of the
heat pump as the ratio of the heat supplied to the heat equivalent
of the mechanical work:

$$\epsilon_h = \frac{Q_c}{W}$$

(The subscript h signifies the coefficient of performance of
heating as opposed to that of cooling: $\epsilon_c = \frac{Q_e}{W}$).

The coefficient of performance of a Carnot-cycle operation
can also be expressed by the ratio of the absolute temperature
of heat rejection to the difference between the temperatures of
heat rejection and heat source:

$$\epsilon_h = \frac{T_c}{T_c - T_e}$$

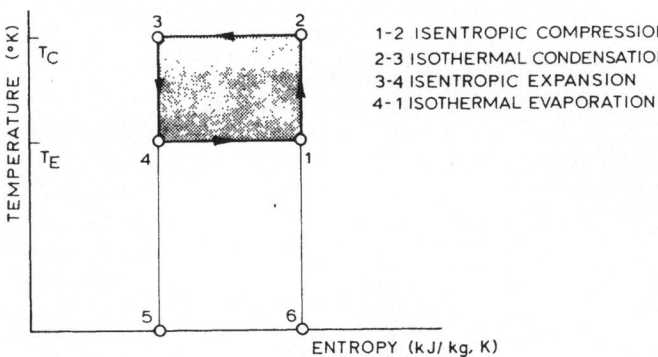

Fig. 2. Carnot-cycle heat pump.

While it is useful to obtain the coefficient of performance for the Carnot-cycle, it should only be used as a guide. The Carnot-cycle cannot be realised for obvious thermodynamic reasons and the real coefficient of performance of an actual heat pump is always considerably less than what the Carnot efficiency indicates (Fig. 3).

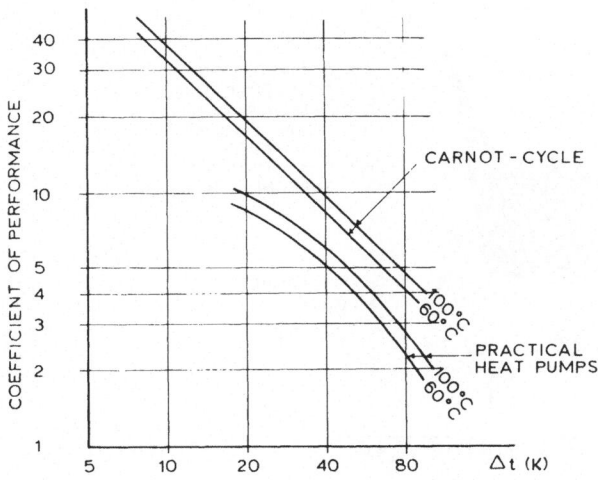

Fig. 3. Coefficient of performance of heat pumps.

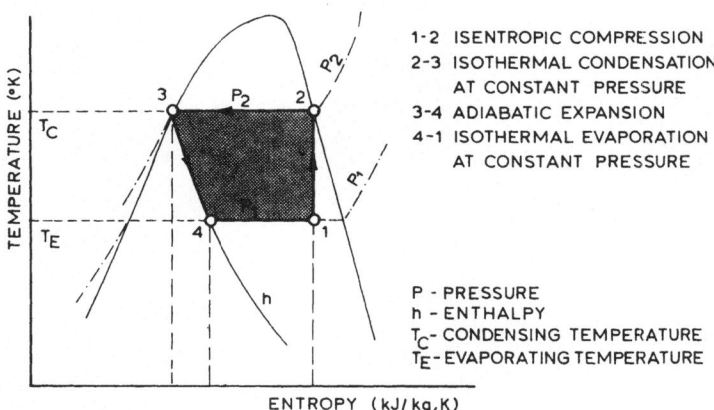

1-2 ISENTROPIC COMPRESSION
2-3 ISOTHERMAL CONDENSATION
     AT CONSTANT PRESSURE
3-4 ADIABATIC EXPANSION
4-1 ISOTHERMAL EVAPORATION
     AT CONSTANT PRESSURE

P - PRESSURE
h - ENTHALPY
T_C- CONDENSING TEMPERATURE
T_E- EVAPORATING TEMPERATURE

Fig. 4.　Theoretical Rankine-cycle heat pump.

　　Practical heat pumps operate with refrigerant fluids according to the reversed Rankine-cycle. The nearest to the Carnot-cycle operation is the cycle in which a partially saturated vapour is isentropically compressed to a fully saturated condition (Fig. 4). However, for practical reasons, the vapour on entry to the compressor should be saturated, or slightly super-heated, to avoid damage to the compressor (Fig. 5).

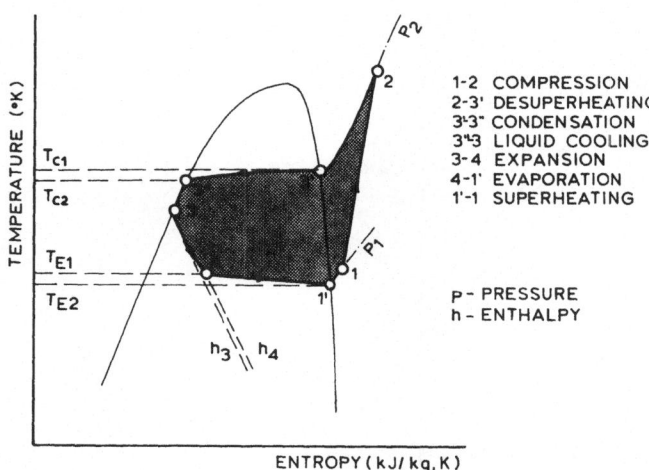

1-2 COMPRESSION
2-3' DESUPERHEATING
3'-3" CONDENSATION
3"-3 LIQUID COOLING
3-4 EXPANSION
4-1' EVAPORATION
1'-1 SUPERHEATING

P - PRESSURE
h - ENTHALPY

Fig. 5.　Practical Rankine-cycle heat pump.

The coefficient of performance of a real heat pump is lower than that of an ideal heat pump for a number of reasons: the non-isentropic nature of compression; heat transfer at variable temperature (because of super-heated vapour and pressure drop in the condenser and in the evaporator; non-reversible, adiabatic expansion; and the necessary final temperature difference in the heat exchangers. The use of auxilliary equipment (fans, pumps) also reduces the actual coefficient of performance.

3. HEAT SOURCES

In view of the high capital cost of heat pump plants, it is important that the operating cost is comparable to or lower than that of a fuel fired heating system of the same heating capacity. For an electrically powered heat pump to be energetically more efficient than a fuel fired system the coefficient of performance (ϵ_h) must generally exceed η_f / η_e, where η_f is the nett efficiency of the fuel fired system and η_e is that of the electricity generation and distribution. As an example, if $\eta_f = 80\%$ and $\eta_e = 27\%$ the coefficient of performance must be in excess of 3. In most instances it is difficult to achieve a coefficient of performance as high as that for an industrial heating application by using naturally occurring heat sources such as ambient air, river or lake water or ground. While it is possible to show economies of heat pumping for domestic space heating, industrial applications generally require higher temperature heat sources. Fortunately there are many low-grade thermal energy sources in industry at temperatures well above those of natural sources: cooling processes can be considered as the most important source of low-grade heat. Certain industrial effluents are also suitable for heat pumping applications.

Another promising source of heat is the warm, moist air of drying processes. This method is discussed later on in the paper.

An often quoted source of heat is that of the cooling water of power stations. There are several problems concerning the application of this source. The main objection is that pass-out steam turbines can provide a more economic method of thermal energy supply than heat pumps. A suitable balance between thermal and electrical energy demand is essential for the successful operation of a district-heating power station and it is outside the scope of this paper to deal with that problem.

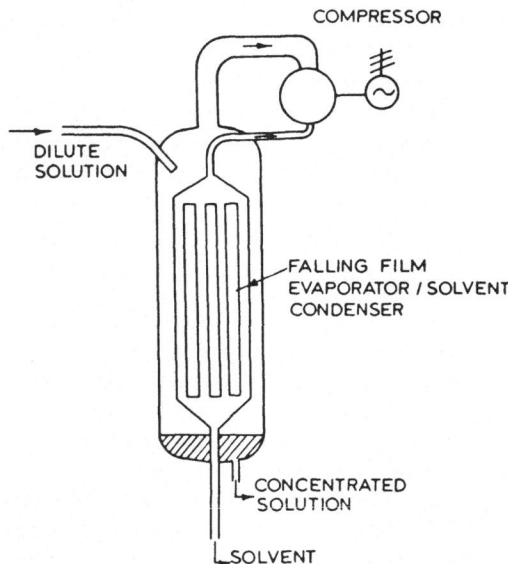

Fig. 6. Vapour-recompression evaporator.

Very high coefficients of performance can be achieved in vapour-recompression type evaporators (Fig. 6). The medium on which the compression is carried out is the evaporating liquid and since the same heat exchanger serves as evaporator on one side and as condenser on the other, the temperature difference between the evaporating and condensing liquid can be very small.

4.   CHOICE OF WORKING FLUID

The working fluid of a heat pump installation is similar to that used in refrigerating plants and is referred to as a refrigerant. The selection of a suitable refrigerant is made according to thermodynamic and lubricating properties:

(a)   vapour pressure at condensing and evaporating temperatures;

(b)   latent heat of condensation and evaporation;

(c)   specific volume;

(d)   critical temperature and pressure;

Table 1. Comparative data of selected thermodynamic Properties [1] and Heat Pumping Performance [2] of refrigerants [3] (in feet).

Refrigerant	Boiling point at 1 atm. t_b (°F)	Critical Temperature t_{cr} (°F)	Critical Pressure P_{cr} (psia)	Evaporating Pressure P_{ev} (psia)	Condensing Pressure P_{cond} (psia)	Compression ratio r	Enthalpy of sat. suction vapour h_1 (BTU/lb)	Enthalpy of superheated vapour h_2 (BTU/lb)	Enthalpy of saturated liquid h_3 (BTU/lb)	Σ_h (4)	Specific volume of saturated suction vapour v_s (ft³/lb)	Volume flow in suction line V (ft³/min, ton of refrigeration) (5)
11	74.9	388.4	639.5	25.16	95.12	3.78	104.4	115.5	48.8	6.01	1.652	5.948
12	-21.6	233.6	596.9	139.33	408.63	2.93	87.4	97.0	57.4	4.10	0.291	1.946
21	48.0	353.3	750.0	43.00	155.0	3.60	131.7	145.6	60.1	6.18	1.280	3.572
22	-41.4	204.8	721.9	222.4	644.3	2.90	112.3	123.0	76.2	4.36	0.242	1.343
113	117.6	417.4	495.0	11.35	50.0	4.40	97.0	106.0	51.0	6.11	2.95	12.83
114	38.4	294.3	474.0	48.9	166.5	3.40	90.0	97.6	56.6	5.39	0.70	4.183
142b	15.0			77.0	255.0	3.31	117.5	128.0	65.0	6.02	0.67	2.549
152a	-12.0			134.0	420.5	3.14	164.5	182.0	96.9	4.86	0.566	1.673
216	97.0	356.0	399.5	17.4	71.2	4.10	91.0	98.2	56.6	5.63	1.62	9.419
C318	21.0	239.5	401.0	73.0	246.3	3.37	87.5	94.0	67.0	4.15	0.382	3.727
500	-28.3	221.9	641.9	164.8	485.6	2.95	103.9	114.0	70.5	4.32	0.293	1.752

NOTES: (1) Evaporation at 104°F (40°C); Condensation at 194°F (90°C). Σ carnot, $h = 7.26$

(2) Isentropic compression is generally assumed, for refrigerants 113,114,216 and C318 the isentropic compression is assumed to begin after the saturated suction vapour is superheated by the condensate to the point from where after isentropic compression saturated vapour state is reached.

(3) Some refrigerants (142b, 152a, 216 and 318) are not commercially available.

(4) $\Sigma_h = \dfrac{h_2 - h_3}{h_2 - h_1}$, the coefficient of performance at isentropic compression.

(5) 1 ton of refrigeration = 12000 BTU = 3.516 kWh.

Table 1. Comparative data of selected thermodynamic properties[1] and heat pumping performance[2] of refrigerants[3] (in metres).

Refrigerant	Boiling point at 1 atm.	Critical Temperature	Critical Pressure	Evaporating Pressure	Condensing Pressure	Compression Ratio	Enthalpy of sat. Suction Vapour	Enthalpy of super heated Vapour	Enthalpy of Saturated Liquid	Coefficient of Performance[4]	Specific Volume of Saturated Suction Vapour	Compressor Displacement[5]
	t_b (°C)	t_{cr} (°C)	P_{cr} (bar)	P_{er} (bar)	P_{cond} (bar)	r	h_1 (kJ/kg)	h_2 (kJ/kg)	h_3 (kJ/kg)	ε_n	v_s (m³/kg)	$V(10^6 m^3/kJ)$
11	23.83	198.0	44.1	1.73	6.56	3.78	242.6	268.4	113.4	6.01	0.1031	798.3
12	-29.78	112.0	44.1	9.61	28.2	2.93	203.1	225.4	133.4	4.10	0.0181	261.2
21	8.89	178.5	51.7	2.96	10.7	3.60	306.1	338.4	139.7	6.18	0.0799	479.4
22	-40.78	96.0	49.8	15.3	44.4	2.90	261.0	285.8	177.1	4.36	0.0151	180.2
113	47.55	214.1	34.1	0.782	3.45	4.40	225.4	246.3	118.5	6.11	0.1841	1722.0
114	3.55	145.7	32.7	3.37	11.5	3.40	209.2	226.8	131.5	5.39	0.0437	561.4
142 b	-9.44			5.31	17.6	3.31	273.1	297.5	151.1	6.02	0.0418	342.1
152 a	-24.44			9.24	29.0	3.14	382.3	423.0	225.2	4.86	0.0353	224.5
216	71.67	180.0	27.5	1.20	4.91	4.10	211.5	228.2	131.5	5.63	0.1011	1264.2
C 318	-6.11	115.3	27.6	5.03	17.0	3.37	203.3	218.4	155.7	4.15	0.0238	500.2
500	-46.07	105.5	44.2	11.4	33.5	2.95	241.5	264.9	163.8	4.32	0.0183	235.1

NOTES : (1) Evaporation at 40° C. Condensation at 90° C. $\varepsilon_{carnot, h}$ = 7.26

(2) Isentropic compression is generally assumed; for refrigerants 113, 114, 216 and C318 the isentropic compression is assumed to begin after the saturated suction vapour is superheated by the condensate to the point from where after isentropic compression saturated vapour state is reached.

(3) Some refrigerants (142 b, 152 a, 216 and 318) are not commercially available.

(4) $\varepsilon_h = \dfrac{h_2 - h_3}{h_2 - h_1}$, the coefficient of performance at isentropic compression.

(5) Theoretical compressor displacement per unit of cooling.

(e) miscibility with the lubricating oil.

There are also some safety factors to consider; the refrigerant should not be toxic, inflammable, corrosive or chemically unstable with the lubricating oil. Some refrigerants decompose above certain temperatures which can be below the thermodynamic critical temperature.

A list of some refrigerants with thermodynamic properties and some comments as to their application is shown in Table 1

To develop heat pumps for high temperature use (i. e. condensing above 80°C) research on new refrigerants will have to be carried out.

5. REFRIGERANT COMPRESSORS

5.1 Reciprocating piston compressor

The reciprocating piston compressors are among the most frequently used in heat pump applications. These compressors can operate at high pressure and low rates of volume flow. There are three distinct types of construction: hermetic, semi-hermetic and open.

Both the hermetic and the semi-hermetic compressors have the common feature of enclosed compressor and motor construction thus eliminating the need for rotary seal which is necessary for the open-type construction.

The hermetic compressor was originally developed for refrigeration duty, it is cheap to produce but it cannot be easily repaired. Most hermetic compressors are unsuitable for heat pump application because of the high evaporating temperature demands on heat pumps. The semi-hermetic design is generally produced in sizes up to about 150 kW absorbed power and is the most widely used in heat pumping systems.

The open-type compressors have the advantage of being independent of the drive (i. e. can be driven by a number of different prime-movers) and are less sensitive to high evaporating temperatures.

Among the most expensive open-type compressors are the cross-head construction units. These have been made for specialized refrigeration and there are no reports of these being used in heat pumping systems.

5.2 Turbo compressors

Large refrigeration and air conditioning plants are generally operated with turbo (or centrifugal) compressors. The absorbed power is normally above 150 kW and the compressors are supplied as water-cooled condensing units. It is understood that Sulzers are developing a high temperature condensing turbo-compressor unit for heat pumping application.

6. HEAT PUMP APPLICATIONS

6.1 Recovery of sensible heat

The industrial use of heat pumps (apart from vapour-recompression evaporators) is a relatively recent development and there are only a few examples to quote. The common feature of these heat recovery systems is that the cooling as well as the heating requirements are simultaneous. Chilled water is generally used to increase the rate of production of injection moulding plants. At a recently commissioned plant in England the manufacturers installed a refrigerated water chiller system with a dual condensing arrangement: roof mounted air-cooled condensers are used in the summer, but when space heating is required ceiling mounted condensers can be brought into use by a simple change-over valve. A moderately small extra capital expenditure thus enables the full utilization of otherwise wasted thermal energy. A similar heat recovery system has been reported by the Westinghouse Electrical Company [1,2] in connection with a welding plant. The company forecasts considerable market in this respect and have already coined the word "Templifier" for this type of industrial heat recovery.

Proposals by Electricité de France [3] suggest future heat pump developments with heating temperatures in excess of 120°C.

6.2 Drying by dehumidification

A very promising application, hitherto in limited use, is the heat pump dehumidifier for industrial drying.

The efficiency of drying in a single pass dryer (Fig. 7) is subject to the temperature of the air at inlet, after heating, and at the point of exhaust. The process of humidification of the air in the dryer can be shown on a psychrometric diagram (Fig. 8). It is assumed that drying takes place at constant wet bulb

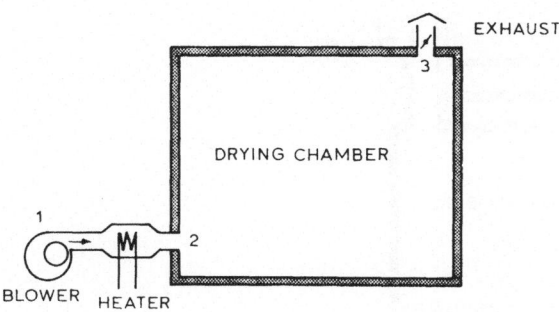

Fig. 7. Single-pass dryer.

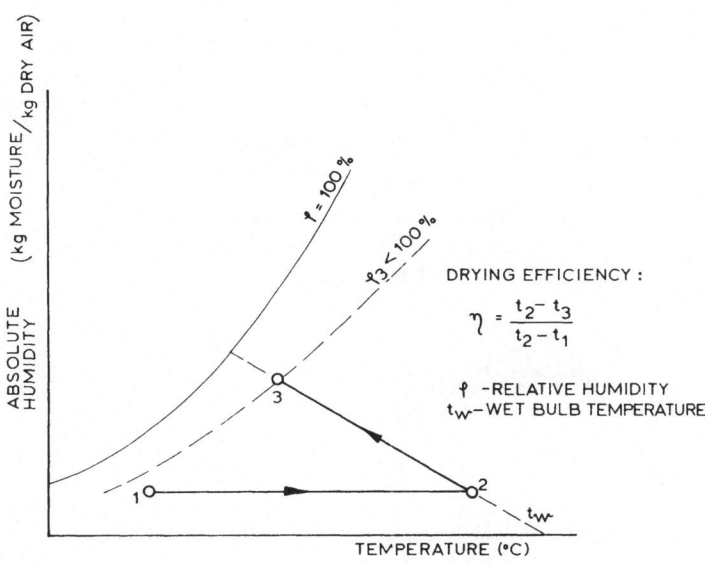

Fig. 8. Conditions of air in a single-pass dryer.

212

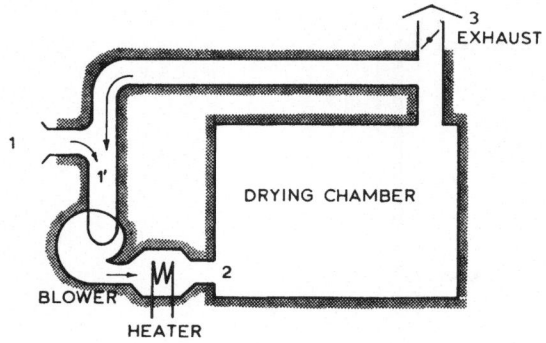

Fig. 9.　Dryer with recirculation.

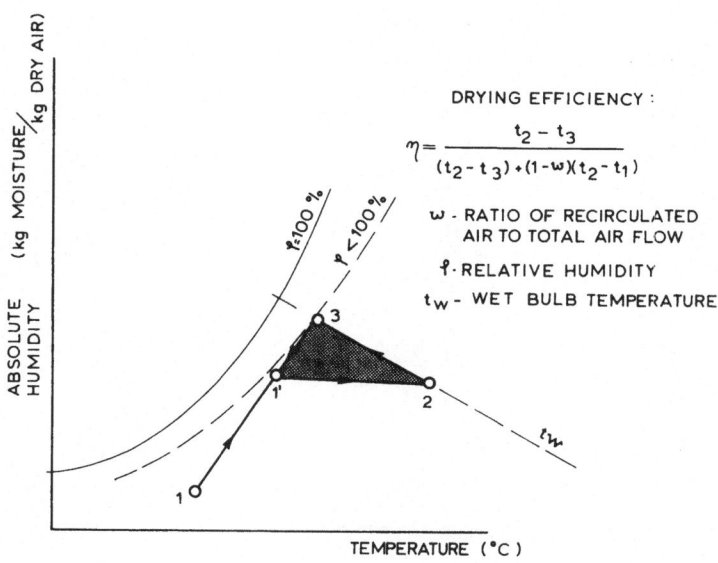

Fig. 10.　Conditions of air in dryer with recirculation.

temperature. There are two methods for increasing the efficiency of drying in a "heat-and-vent dryer:" (a) by raising the temperature of the exhaust air as near to saturation as possible and (b) by recirculating a proportion of the air within the dryer (Figs. 9 and 10). With increased recirculation the efficiency of drying approaches 100% (Fig. 11). In practice, however, full recirculation can only be achieved in a closed kiln from which the moisture is removed by a dehumidifier (Fig. 12).

The dehumidifier is essentially a heat pump. Drying of timber has been practised for some years by this method and it is expected that other applications will follow.

The evaporator of the dehumidifier (Fig. 13) cools the air below the dew-point and the heat of cooling (sensible and latent) and also the heat equivalent of the work of compression and fan power is transferred to the dehumidified air. The full cycle of moisture removal and subsequent humidification of the recirculated air is shown in a psychrometric diagram (Fig. 14). Following the cycle, the air is first cooled by the evaporator from condition 1 to 2 and some of its moisture, ΔX, condenses. The condensate is being continuously drained from the kiln and the rate of flow is an indication of the rate of drying. Although the efficiency of drying is theoretically high (100%, provided there are no air leaks from the kiln) the overall efficiency is subject to the coefficient of performance of the heat pump. The coefficient improves with decreasing difference between the temperatures of the condensing and evaporating refrigerant. The temperature of condensation can be reduced by an increased rate of air flow over the condenser. This is achieved by mixing chilled and saturated air with unchilled air in a plenum chamber before the air is passed over the condenser.

Subject to heat loss through the fabric of the kiln, the overall efficiency of drying improves with increasing kiln temperature (Fig. 15). Tests on a dehumidifier in a test room indicate that the rate of specific moisture extraction improves with increasing kiln temperature (Fig. 16).

7. CONCLUSIONS

The use of heat pumps in industry has, so far, been limited to a few heat recovery applications and to drying by dehumidifiers. Efforts are being made by various organizations to extend the range and field of application of heat pumps. Work on the development of suitable high temperature refrigerants and compressors will have to be completed before the heat pump can fulfill its industrial role.

214

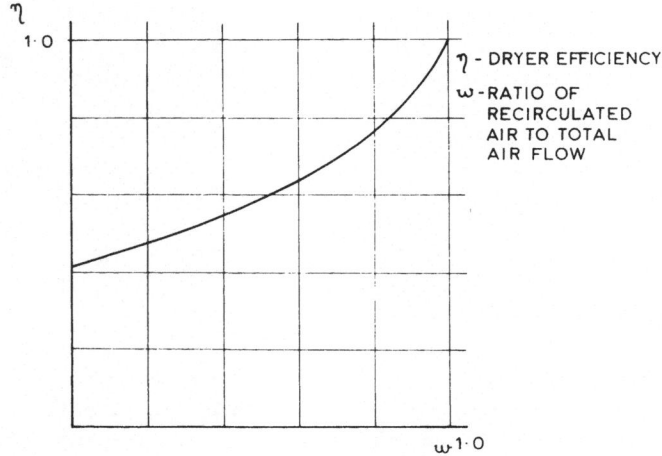

Fig. 11.   Effect of recirculation on dryer efficiency.

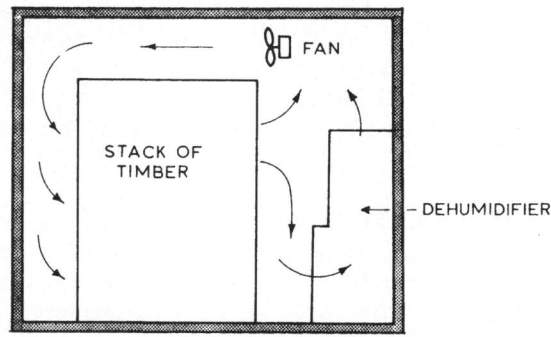

Fig. 12.   Dehumidified kiln.

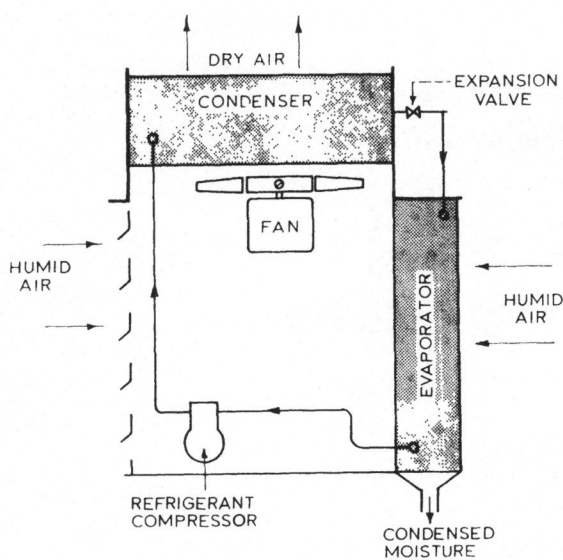

Fig. 13.   Heat pump dehumidifier.

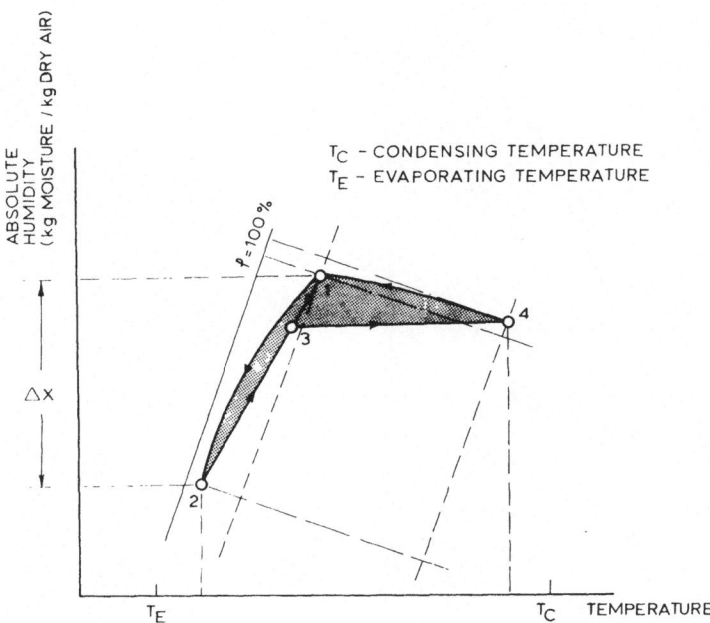

Fig. 14.   Cycle of moisture by dehumidification and subsequent absorption.

216

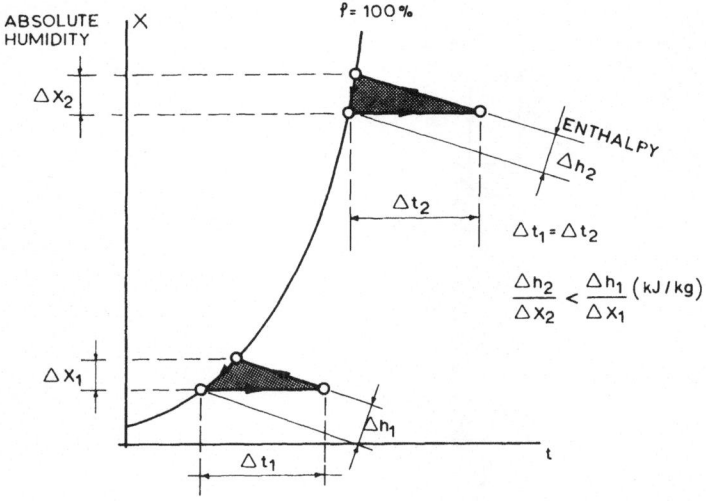

Fig. 15.   Dehumidification at high and low temperatures.

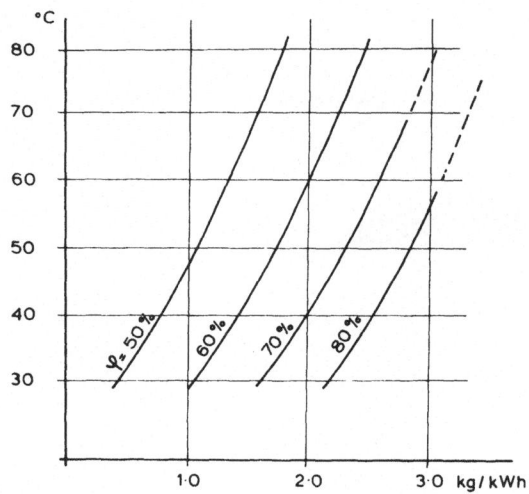

Fig. 16.   Rate of specific moisture removal by dehumidification.

REFERENCES

1. Trumbower, S.A. , The Westinghouse Templifier Steam or
    Hot Water Generator (October 1974), Westinghouse Electric
    Corporation, 700 Braddock Avenue, East Pittsburgh, PA.

2. Ross, P.N. , The Templifier for Process Heat, EEI
    Conservation & Energy Management Division Annual
    Conference, Atlanta, Georgia, (16th-18th March, 1975),
    Westinghouse Electric Corporation.

3. L'Hermitte, J. , and Douay, D. , La Pompe à Chaleur dans
    ses Applications Industrielles , E.D.F. Journées
    d'Informations Electro-Industrielles, Paris, November 1974.

AIR DRYING BY HEAT PUMPS WITH SPECIAL REFERENCE TO TIMBER DRYING

B. Geeraert, Engineer.

Head of the Electro-heat Department of Laborelec, Belgium.

SUMMARY

Since about ten years, heat pumps are being used for drying purposes. With these equipments the solvent is no longer evacuated in the vapour phase with the extraction air, as it is with classical dryers; but it is condensed onto the cold evaporator surface and evacuated in the liquid phase. In such a manner the heat of condensation can be recovered. When heat pumps are being used correctly a reduction of primary energy use is possible as far as the dryers are seriously conceived. The aim of this report is to show that there are optimal working conditions for heat pumps being used as air dryers. Several mounting schemes are compared with each other and the primary energy reduction, by replacing the classical ventilation dryers by heat pumps, is evaluated. The examples are dealing with timber drying, but the same technique can be used in several other fields. The only difference lies in the fact that the desorption isotherms are not the same as those for timber.

Manufacturers and research centres do permanent efforts to increase the heat pump performances and efficiency modifying the internal components or the circuit used. These modifications are directly profitable for heat pumps used for heating or cooling purposes. For air dehumidification the problem is somewhat different. With the heat pump a high COP can be realized and nevertheless the dehumidification efficiency can be bad. In the following text the dehumidification efficiency is characterized by the energy, per kilo of condensed water, taken by the compressor motor.

1. CONVENTIONAL DRYERS

Conventional dryers proceed through ventilation. The wet air is extracted from the dryer and replaced by an equivalent fresh air volume preheated in order to increase its humidity absorption capacity. Two different designs are in use, depending on air-reheating schemes.

1.1 Single pass system

If air reheating only occurs at the air inlet, the dryer is called "adiabatic" or "single pass". The reheated air being in contact with the wet product is adiabatically saturated during its humidification. In figure 1 the path the representative air point is passing through is represented on the wet air diagram: the trajectory ABD. The exhaust state of the air (D) should be more and more dryer when the drying process of a hygroscopic product progresses. In order to maintain the drying rate at a sufficient high value, a constant gradient has to be maintained between air humidity at the dryer output and the wet product equilibrium humidity. This equilibrium humidity may be deduced from the moisture isotherms of the wet product. A typical form of such a moisture isotherm is outlined at figure 2.

For hard wood such as oak, the air humidity/equilibrium humidity relation is maintained, f. i. at 1.8 ... 2; for soft wood this relation may range up to 4. Energy consumption per kilogram of evacuated water is given by: $(h_D - h_A)/\Delta X_1$; $(h_D - h_A)$ being the enthalpy difference between the outlet and inlet air. The adiabatic dryer is rather scarcely used for hard wood drying.

1.2 In the second variant, there is an air recycling inside the dryer; thus the air is reheated many times before it is evacuated. The dryer is maintained at a constant temperature and is said to work "isothermally". At its limit the representative air point passing through the system is following the trajectory ABC (fig. 1). Energy consumption per kilogram of evacuated water is expressed by: $(h_C - h_A)/\Delta X_2$. The change of energy consumption as a function of relative humidity and temperature to be maintained within the dryer, is plotted at figure 3 (curves a - without heat-exchanger). The calculations are based on saturated fresh inlet air at +10°C (average temperature in Brussels over the whole year). From these curves, the energy consumption for drying of a given product may be calculated, provided the moisture desorption isotherm is known.

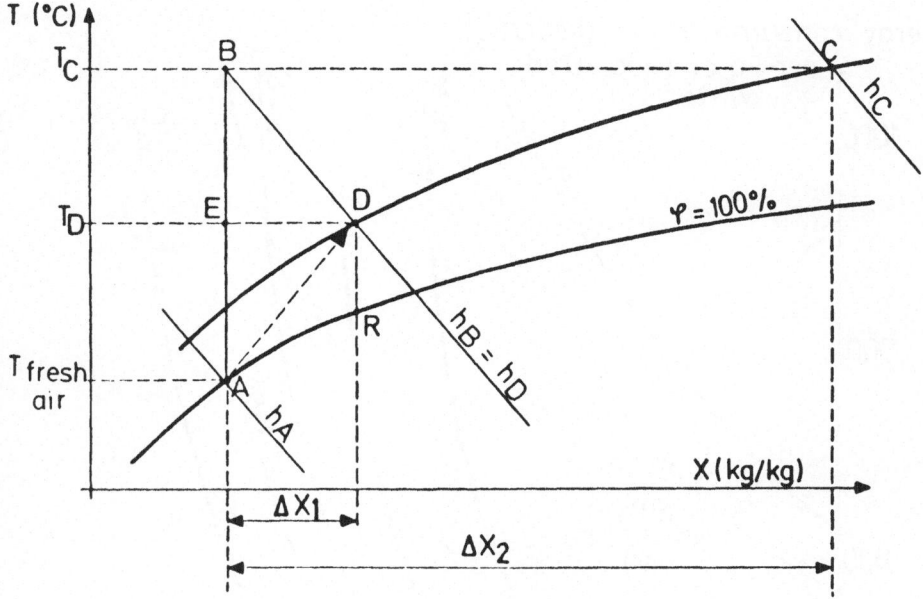

Fig. 1.   Paths followed by representative air point on the wet
air diagram.   ABD in a single-pass dryer; ABC in an
"isothermal" dryer.

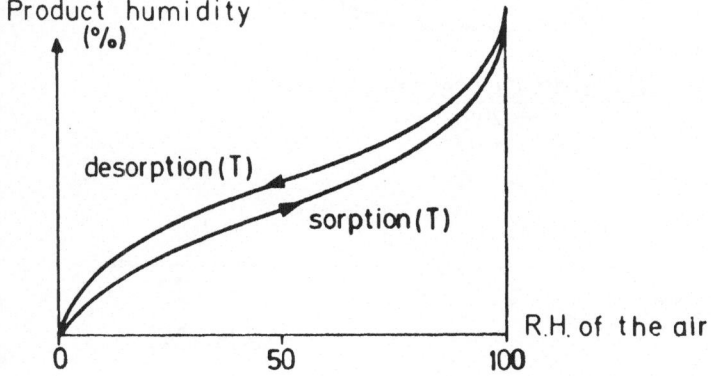

Fig. 2.   Desorption and sorption isotherm of wet hygroscopic
product.

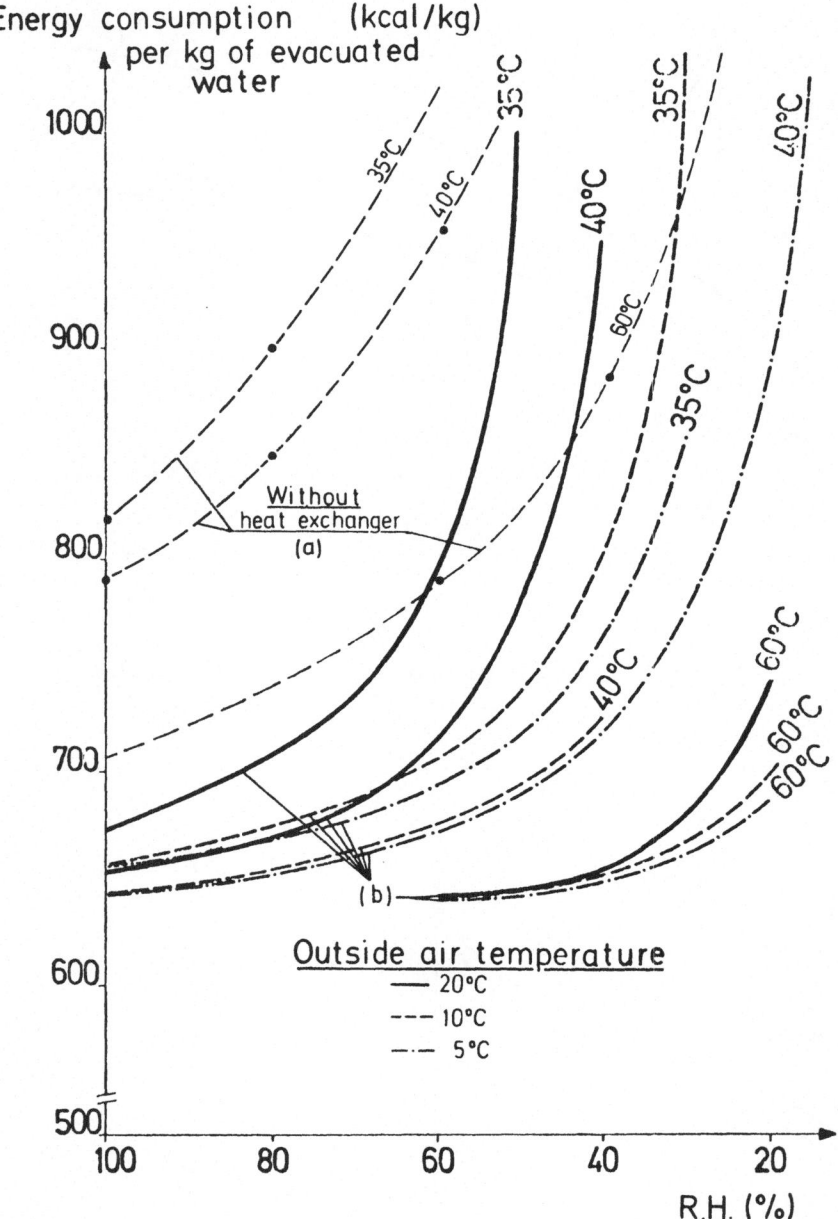

Fig. 3.   Energy consumption with and without heat exchanger
for air treatment in isothermal ventilation dryer.

Table I gives an example of such a calculation for oak sawed into 25 mm thick planks and dried from 40% down to 9% moisture content.

These are the meanings of the symbols used:

H timber: timber moisture content on dry basis (%)
R. H. air: relative air humidity within the dryer (%)
θ air: dry air temperature within the dryer (°C)
$h_{air, o}$: air enthalpy within the dryer (kcal/kg)
$\Delta h = h_o - h_{inlet}$ = air enthalpy variation between the inlet
(fresh air) and the outlet of the dryer
X_o: absolute air humidity of the dryer (g/kg)
$\Delta X = X_o - X_{inlet}$: variation of the absolute air humidity between
inlet and outlet
E: water amount to be evacuated per m^3 of timber, at each
stage of the drying process (kg/m^3)
E/ΔX: mass of the carrier air absolutely necessary for water
evacuation (kg/m^3)
Consumption = E . $\Delta h/\Delta X$: energy consumption for reheating
and temperature maintaining of the air during
its dehumidification, per drying process stage
and per m^3 of timber.

A similar calculation on Sipo or Afrormosia - also dried at 40°C from 40% down to 9% moisture content, but maintaining H timber/H equilibrium = 2.6 - leads to a total energy consumption for the air treatment of 304 kWh/m^3.

2. CONVENTIONAL DRYER WITH STATIC HEAT EXCHANGER IN THE EXTRACTION CONDUCT.

In the isothermal dryer enthalpy variation for each kilogram of air, between inlet and outlet corresponds to ($h_C - h_A$). Using an ideal static heat exchanger it would be possible to reheat the fresh air, by consuming energy taken from the exhausted air, up to the temperature existing within the dryer, and to recover ($h_B - h_A$) (fig. 1). In practice this cannot be achieved by real heat exchanger. Supposing the static heat exchanger reheats the fresh air up to ($T_B - 5$), energy consumptions for the air treatment in the isothermal dryer have been plotted again on figure 3 (curve bundle b). Calculations have been carried out for various meteorological conditions within the dryer (relative humidity and temperature) and for 3 meteorological conditions of the fresh air, taken in from the outside (saturated air at temperatures of respectively 5, 10 and 20°C). The profit

TABLE I: Ventilation drying at 40/60°C

H timber (%)	R.H. air (%)	θ air (°C)	hair,o (kcal/kg)	Δh $h_o - 4,4$	x_o (g/kg)	Δx $x_o - 0,53$	E to be evacuated water (kg/m3)	$\Xi/\Delta x$ (kg/m3)	Energy consumption (kcal/m3 timber)
40/28	82	40	39,3	34,9		48,8	78,3	1 604,5	55 997
28/25	82	40	39,3	34,9		48,8	19,56	400,8	13 989
25/23	80	60	88	83,6	117	116,5	13,04	112	9 363
23/21	78	60	85	80,6	113	112,5	13,04	115,9	9 342
21/19	74	60	82	77,6	108	107,5	13,04	121,3	9 413
19/17	70	60	78	73,6	102	101,5	13,04	128,5	9 456
17/15	62	60	68	63,6	88	87,5	13,04	149	9 478
15/13	55	60	62,5	58,1	77	76,5	13,04	170,5	9 904
13/11	47	60	53,5	49,1	62,5	62	13,04	210,3	10 327
11/ 9	37	60	44,2	39,8	48	47,5	13,04	274,5	10 926

total : 148 195
i.e. 172,3 kWh/m3

resulting from the installation of a static heat exchanger may be directly deduced from figure 3. This profit ranges up to nearly 25% but the energy consumption per kilogram of evacuated water remains higher than 600 kcal/kg, anyway. Energy consumption tends towards the infinite when the absolute humidity of the fresh air tends towards the absolute air humidity within the dryer. Energy consumptions mentioned on figure 3 are theoretical values, implying - at each stage of the drying process - the air flow rate should exactly correspond with the strictly necessary flow rate; this requires an automatic flow rate control at hygrometric conditions. Experience has denoted that in practice, the energy consumption for air treatment is from 20 to 100% higher than the theoretical value, depending on the flow rate control accuracy.

3. THERMODYNAMIC RECUPERATOR

Installing the evaporator of a heat pump in the exhaust channel and the condenser at the fresh air inlet (fig. 4) it becomes possible to recover the sensible heat of the exhaust air <u>and</u> the latent heat of condensation of the water vapour it contains. This mounting is called a "recuperator heat pump".

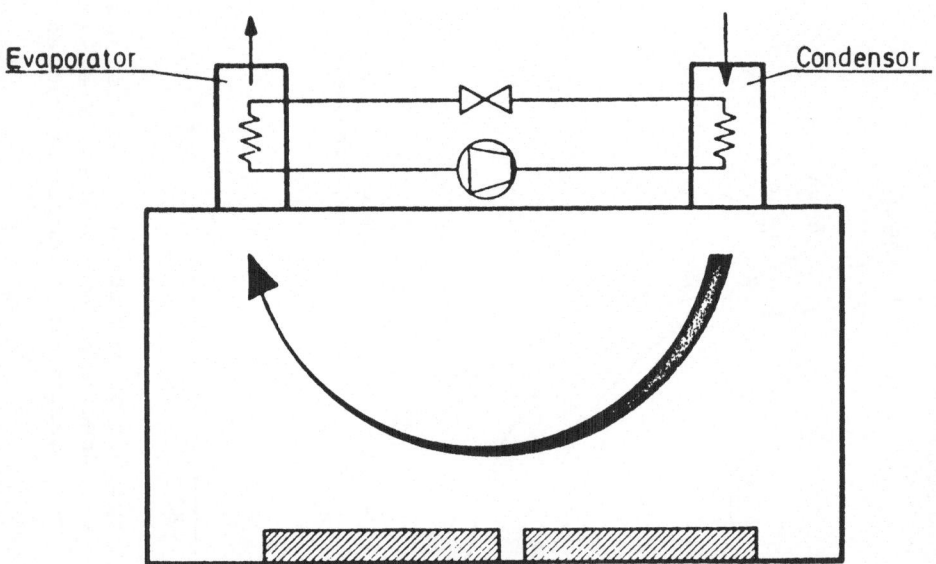

Fig. 4.   Schematic representation of a heat pump mounted as recuperator in a drying kiln.

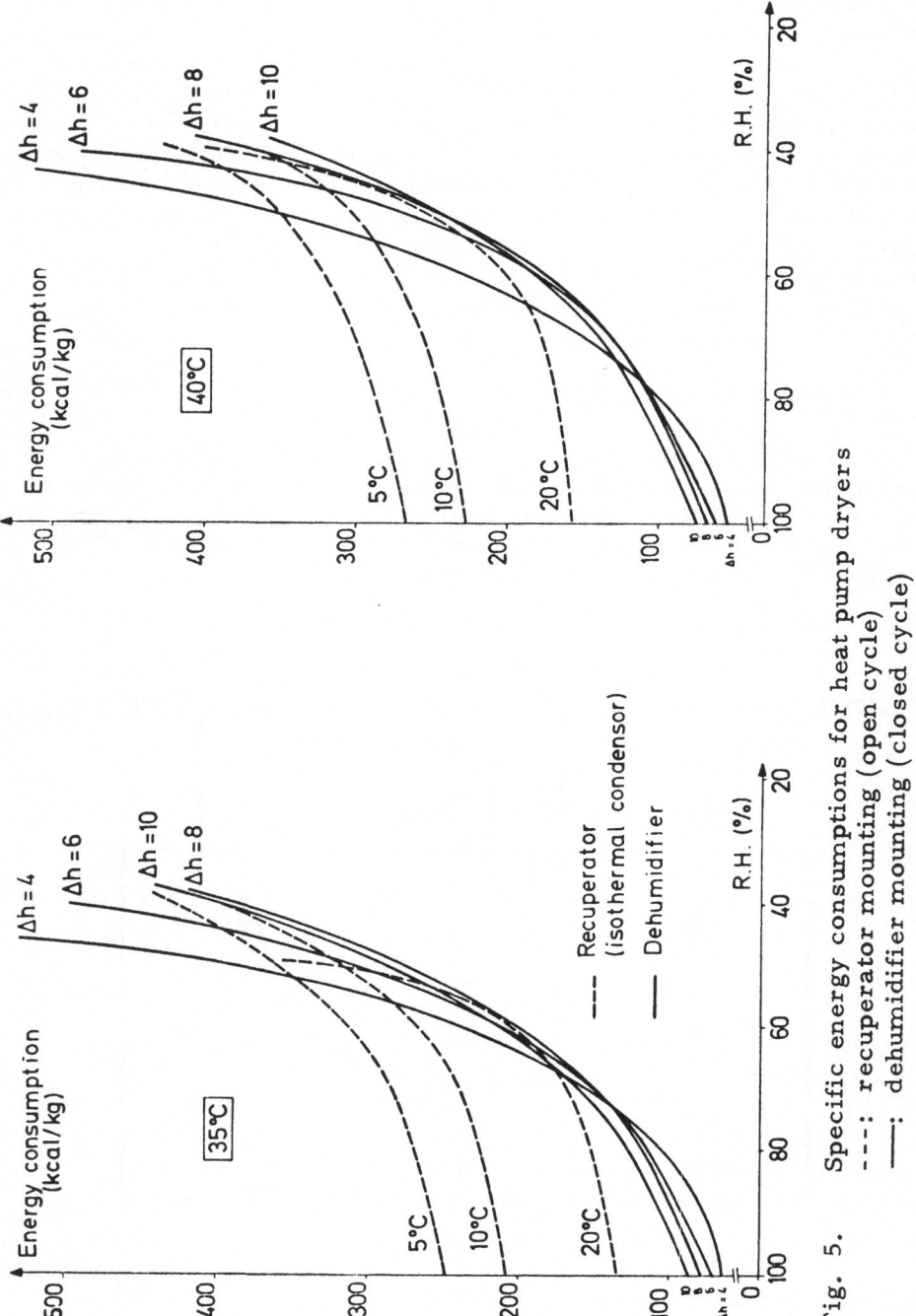

Fig. 5. Specific energy consumptions for heat pump dryers
--- : recuperator mounting (open cycle)
——— : dehumidifier mounting (closed cycle)

Placing the condensor within the dryer itself and circulating a much higher air flow rate on the condenser than on the evaporator may be shown to be more advantageous. At the limit, if the mass flow at the condenser becomes infinite it will be maintained isothermal at the dryer temperature. Assuming the extraction air is repelled into the atmosphere after it has been cooled down to the fresh air temperature (the heat pump permits to cool it more, but the optimum discharge temperature can be proved to be near the fresh air temperature); the energy consumption has been calculated per kilogram of water condensed on the evaporator. The results, for two different dryer temperatures, respectively 35°C and 40°C are shown on figure 5 (dotted line curves) as a function of the relative fresh air humidity at three different temperatures (5, 10 and 20°C).

The representative points of the drying air are indicated on figure 1 in the wet air diagram. At the evaporator, the air is cooled from D down to R (dew point). Along the whole RA trajectory, there is humidity condensing, which is evacuated in liquid form. The heat liberated at the evaporator ($h_D - h_A$) is transferred to the condenser by means of the transfer fluid. Moreover, at the condenser appears the power consumed by the driving motor: ($h_D - h_A$)/COP. Knowing the extreme cold and warm air temperatures and taking into account a temperature drop of 5 K on each heat exchanger, the coefficient of performance is calculated according to (1).

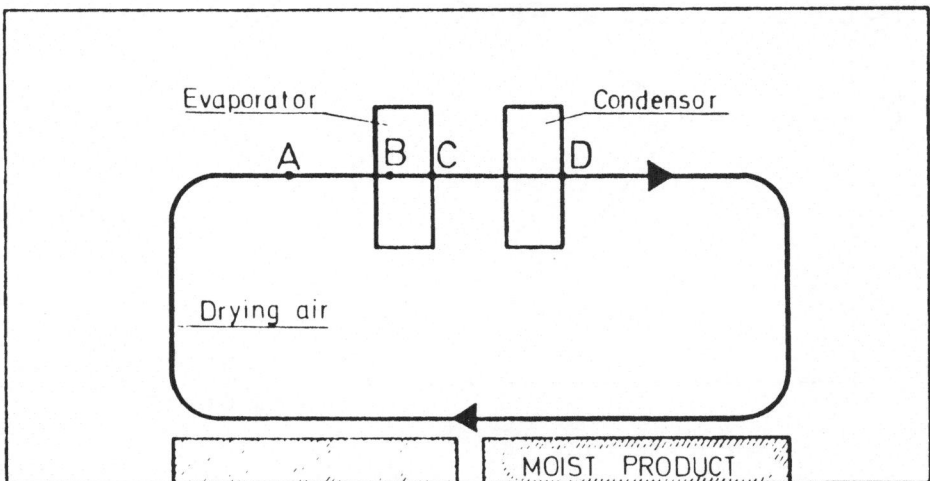

Fig. 6. Heat pump mounted as "dehumidifier" with the same air flow on the evaporator and condenser.

228

$$(COP)_{\text{cooling group}} = K_e \cdot \frac{T_A - 5}{T_D + 5 - (T_A - 5)} \qquad (1)$$

K_e: coefficient which is a function of the exergetic efficiency of the heat pump under given utilization circumstances (mean value of K_e = 0.5; K_e may vary between 0.4 and 0.7).

The energy consumption is nearly three times lower as for a ventilation dryer provided with a static exchanger and depends on the outside weather conditions.

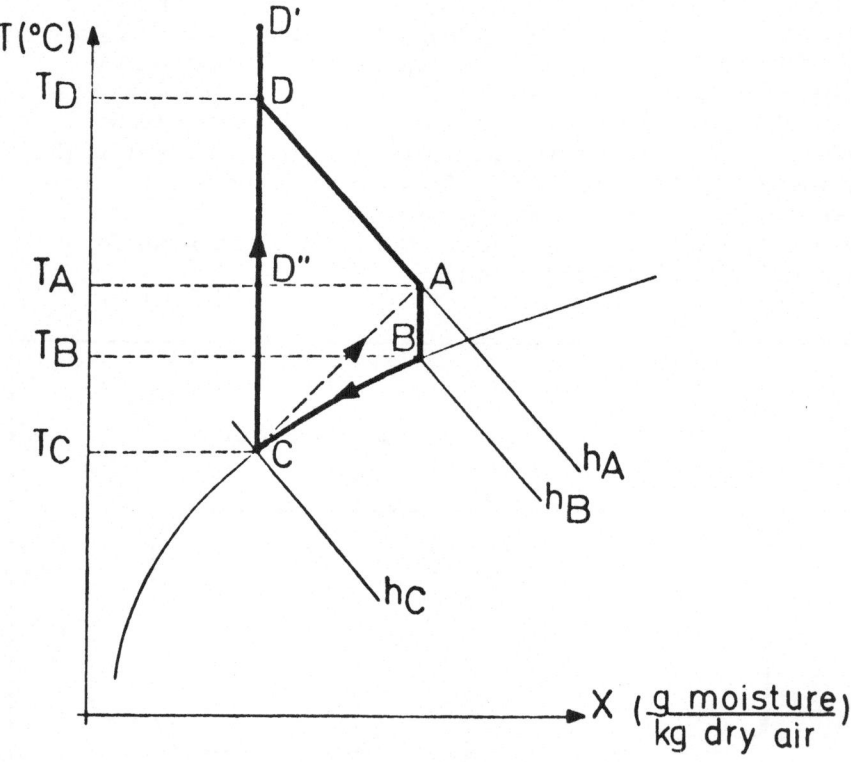

Fig. 7.   Paths followed by the representative wet air point plotted on a wet air diagram for dehumidifiers.

4. AIR DEHUMIDIFIER

4.1 Another mounting consists of installing the heat pump for air dehumidification within the dryer. This mounting is outlined in figure 6 and called "dehumidifier mounting". The drying air circulates in closed cycle on the evaporator, the condenser and the wet product. The representative points on the wet air diagram are plotted at figure 7. Because of the motor power, the representative point of the air after passing over the condenser is not located at D, but at D′ (higher than D). Let us start as a first approach from the assumption that this excess power is just sufficient to compensate transmission losses of the dryer. The air then follows the closed cycle ABCD. The energy consumption for each kilogram condensed water is expressed by: $(h_A - h_C)/$ $[COP (X_A - X_C)]$. This consumption represented at figure 8 is calculated as a function of the enthalpy change the air passing on the evaporator is submitted to. All the curves present a very sharp minimum and the left branch tends towards the infinite (the air is not cooled down to the dew point). The minimum itself shows a displacement to higher Δh values with decreasing relative humidity.

At the present time all manufacturers provide operating in the area of low Δh values (2 ... 4 kcal/kg) and without Δh adjustment as a function of dryer humidity. In this way they realise a high COP but when humidity decreases, energy consumption for each kilogram condensed water becomes enormous. These installations should be modified in order to enable their economical operation. Unfortunately the minimum is a critical one.

4.2 The following alternative - not yet used - allows even lower energy consumptions and transforms the sharp minimum into a less critical one. The mounting design is illustrated at figure 9.

The condensor is maintained nearly isothermal by passing over it a much higher flow rate of drying air (f.i. the total circulating air flow rate). The representative wet air points are again plotted at figure 7. Two boundary cases are conceivable and reality probably locates between both.

4.2.1 The air cooled on the evaporator (point C) no longer passes over the condenser, but is mixed with the dryer air circulating at infinite flow rate over the condenser, thus maintaining its representative point at A (CA is the mixing line).

230

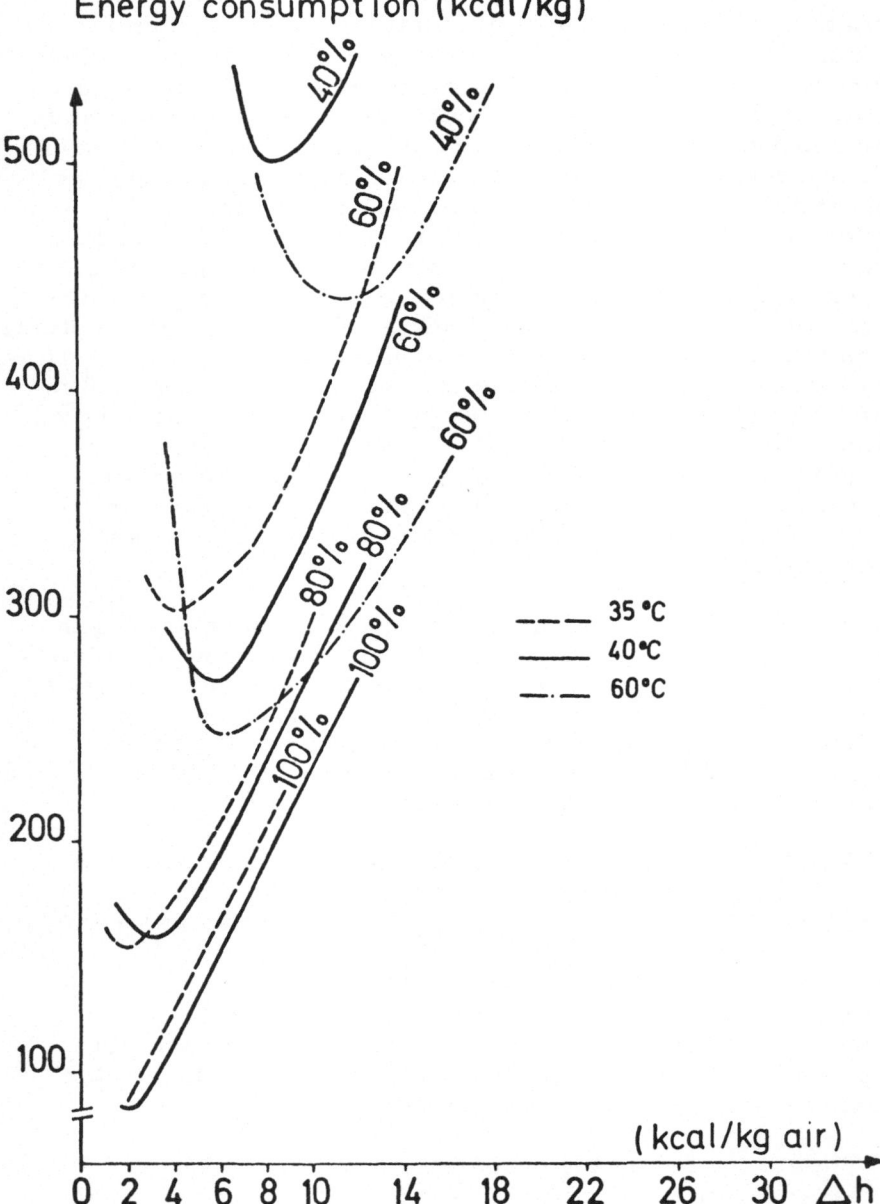

Fig. 8.   Specific energy consumption for the dehumidifier
           mounting as a function of the air conditions in the
           kiln.

4.2.2 The air cooled on the evaporator (point C) passes over the condenser where it will be reheated (following CD″)without mixing during reheating. Then mixing occurs in the dryer according to the trajectory D″A.

This mixture, however, enables the maximum temperature to be lowered at the condenser from T_D down to $T_{D″}$, increasing thus the coefficient of performance. But there is a much more significant advantage. Energy consumption curves charted in figure 10 no longer present this sharp-built minimum but, on the opposite, a rather flattened minimum area. The minimum energy consumption area is no longer critical and, moreover, it is no longer absolutely necessary with decreasing relative dryer humidity to control automatically the Δh values of the air passing over the evaporator.

Δh = 10 kcal/kg appears to be an economical value convenient for a relative humidity between 100% and 40% (and even 30%). Energy consumptions as a function of relative humidity have been superposed at figure 5 (full lines, each curve corresponds to a fixed Δh value). For humidity values above 60% (at 40°C), "dehumidifier" mounting (cf. 4) is obviously better than "recuperator" mounting (cf. 3) from the point of view of energy

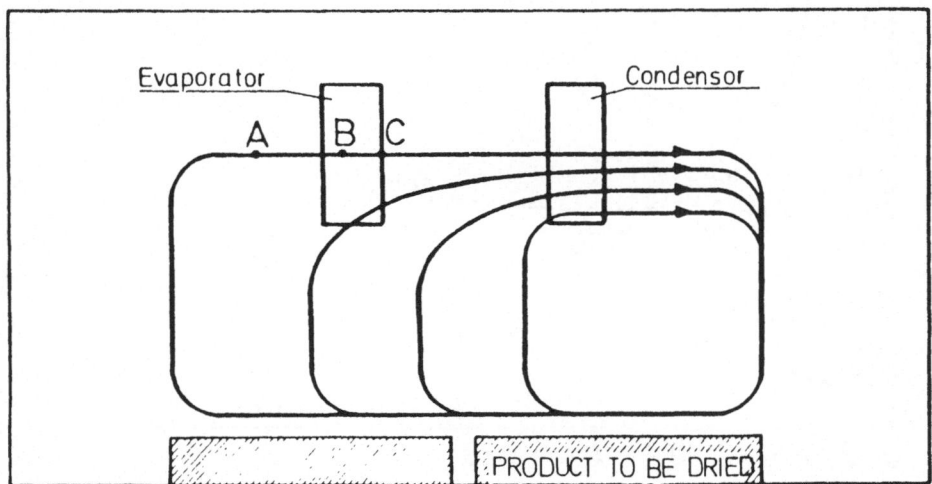

Fig. 9.   Schematic representation of a heat pump mounted as "dehumidifier" with an increased airflow on the condenser.

232

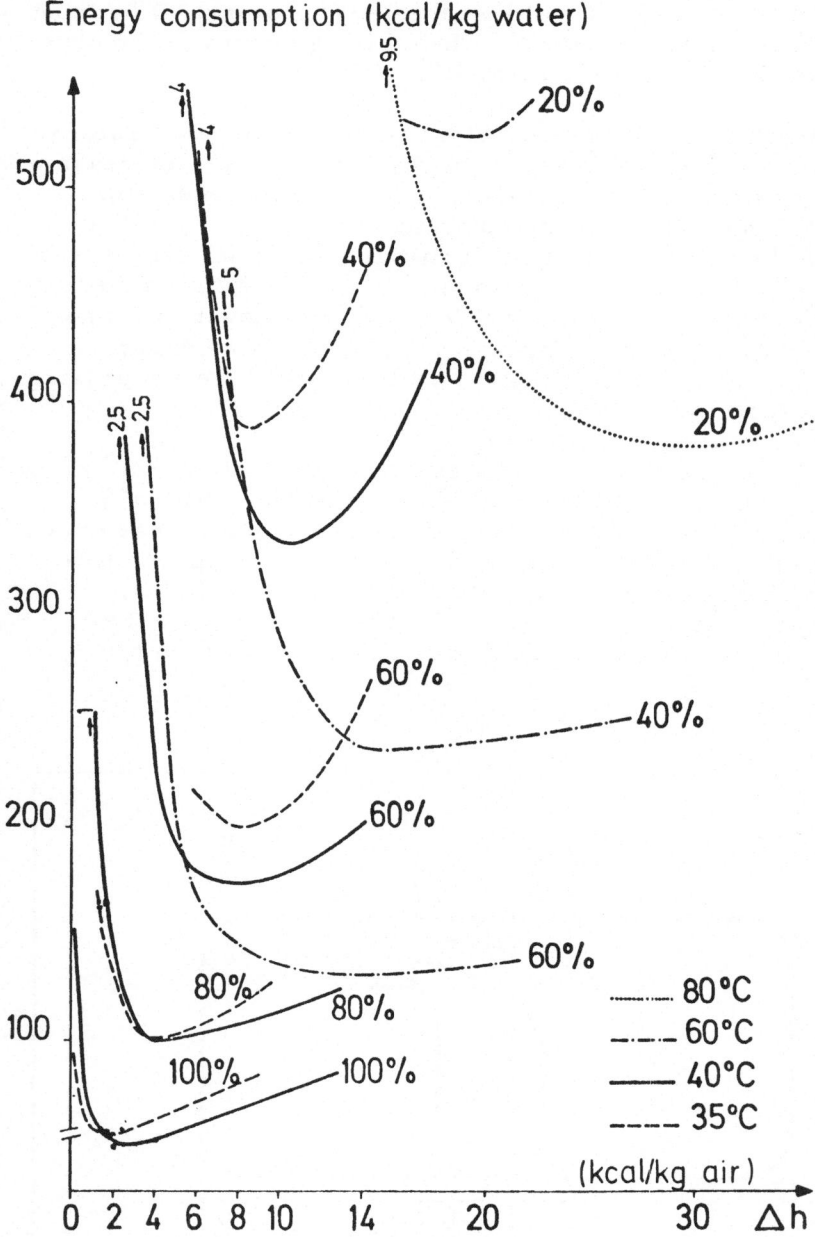

Energy consumption (kcal/kg water)

Fig. 10. Specific energy consumption of a dehumidifier with nearly isothermal condenser.

consumption. Between 60 and 40% relative dryer humidity, occurs a crossing of the energy consumption curves at smaller Δh values. As soon as Δh = 8 ... 10 kcal/kg, the energy consumption of the "dehumidifier" mounting becomes smaller than the one for the "recuperator". Ideally Δh should be adjusted as a function of the relative humidity that is to be maintained within the dryer. So the energy consumption curve becomes the envelope of the consumption curves at constant Δh, minimising in such a way the consumption for each climatic condition in the dryer (this requires a low Δh at high humidity values and Δh progressively increasing with decreasing air humidity).

5. SUMMARY OF DRYER ENERGY CONSUMPTIONS AT 40°C

(without taking into account transmission losses and the power required for heating the wet product up to dryer temperature) cf. Table II.

6. PRIMARY ENERGY CONSUMPTION AND INFLUENCE OF DRYER TEMPERATURE

6.1 Heat pumps used until now are rather exclusively of the compression type; they are fed by electric power mainly generated in thermal power plants (for Belgium). Heat/electric power conversion efficiency, including network distribution losses, nearly ranges between 33 and 35%. So primary energy consumption of the heat pump dryers will be three times as large as the values indicated in Table II. Despite this, their primary energy consumption will always remain smaller as is the case for conventional fuel-heated ventilation dryers, even if the latter ones are provided with a static heat exchanger. Figure 11 summarises the minimum primary energy consumptions for the various dryer systems.

For the ventilation dryer, normally fossil fuel heated, efficiency is supposed to be the same as the combustion efficiency (i.e. 80%) of the boiler operating at full power. In practice the performance is much lower and varies, for hard timber dryers, between 25 and 60% (taking into account the heat required for heating, the transmission losses, the distribution losses, ...). The static heat exchanger is supposed to operate with 5 K temperature difference between both fluids (the one cooled and the other heated) at the outlet. Energy consumption of ventilation dryers depends on outside temperature. The calculations are based on an outside temperature of 5°C and saturated fresh air.

234

TABLE II: summary of energy consumptions of several types of dryers working at 40°C.

Nr	Relative humidity	Energy consumption (kcal/kg evacuated water)		
		100%	80%	60%
1	Dryer without recuperation	790	850	960
2	+ static heat exchanger (t_{out} = 10°C)	640	650	670
3	+ recuperator heat pump isothermal condenser (t_{out} = + 10°C)	210	225	250
4	+ dehumidifier heat pump isothermal condenser (Δh = 8)	70	105	180

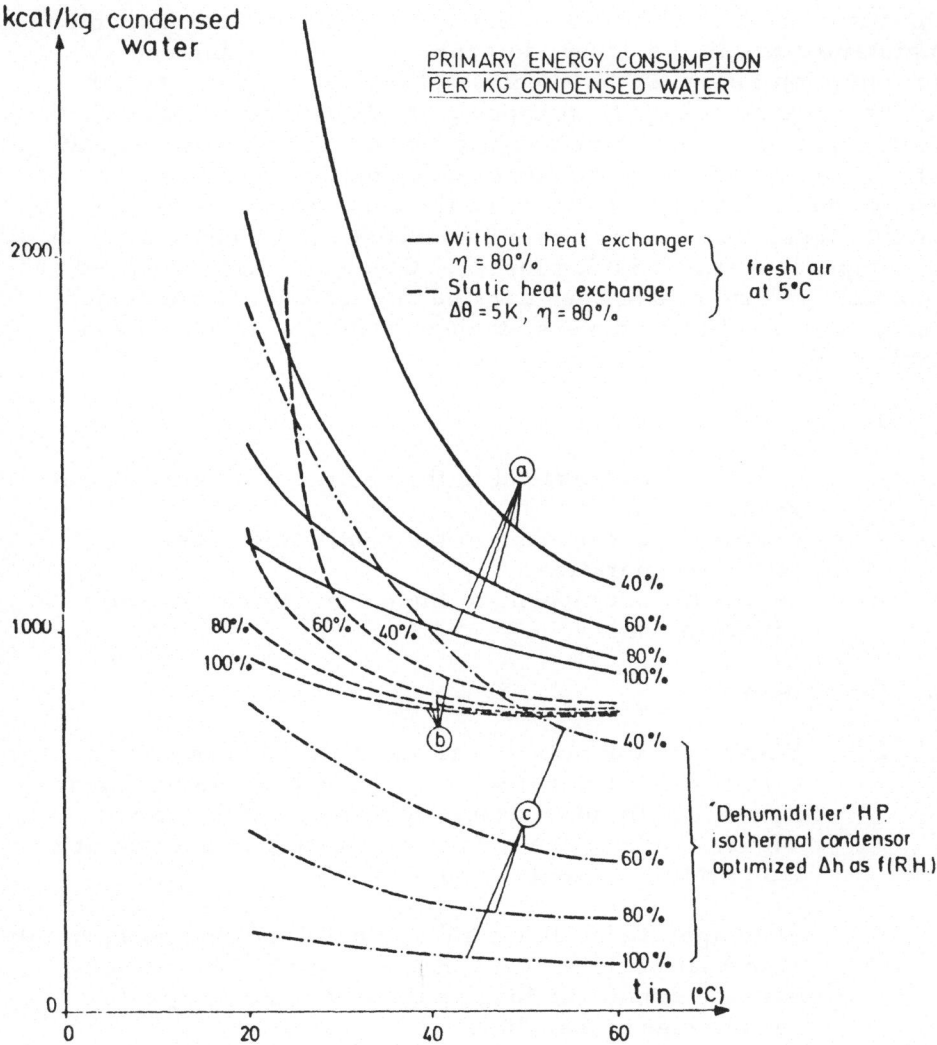

Fig. 11. Primary energy consumption per kg condensed water as a function of the air conditions in the kiln for three different systems:
- simple isothermal ventilation dryer;
- isothermal ventilation dryer with static heat exchanger;
- optimised heat pump air dehumidifier.

Heat pump energy consumption is based on the "dehumidifier" mounting with nearly isothermal condenser (variant 4.2) and continuous control of Δh as a function of the relative humidity to be maintained within the dryer, in order to minimize the energy consumption following the envelope of the curves indicated at figure 5. A closed cycle heat pump air dryer recuperates all the heat required for moisture evaporation and carrier-air reheating. The power absorbed by the compressor motor appears as an excess power which tends to raise the oven temperature. In the calculations, this excess power is supposed to be just sufficient to compensate transmission losses. Other calculations based on an adiabatic dryer have been carried out but are not mentioned here. For ventilation dryers, transmission losses have been neglected.

In figure 11:

-      curves (a) correspond to the ventilation dryer without recuperation;
-      curves (b) correspond to the ventilation dryer with static recuperation;
-      curves (c) correspond to the dehumidification dryer with heat pump.

6.2 Conclusions

6.2.1   Energy consumptions decrease with increasing temperature. If compatible with the product to be dried, it is certainly profitable, from the point of view of energy consumption for air treatment, to operate at a temperature as high as possible.

6.2.2   At temperatures above 45°C, obviously the appropriately used heat pump is more interesting than the conventional systems, as long as the dryer humidity may remain higher than 40%.

6.2.3   At 30°C temperature and 50% relative humidity (climatic conditions in swimming pools), the heat pump is only slightly more interesting than the fossil-fuel heated ventilation system equipped with a static heat exchanger, provided the efficiency of the conventional system, distribution losses included, reaches at least 80%.

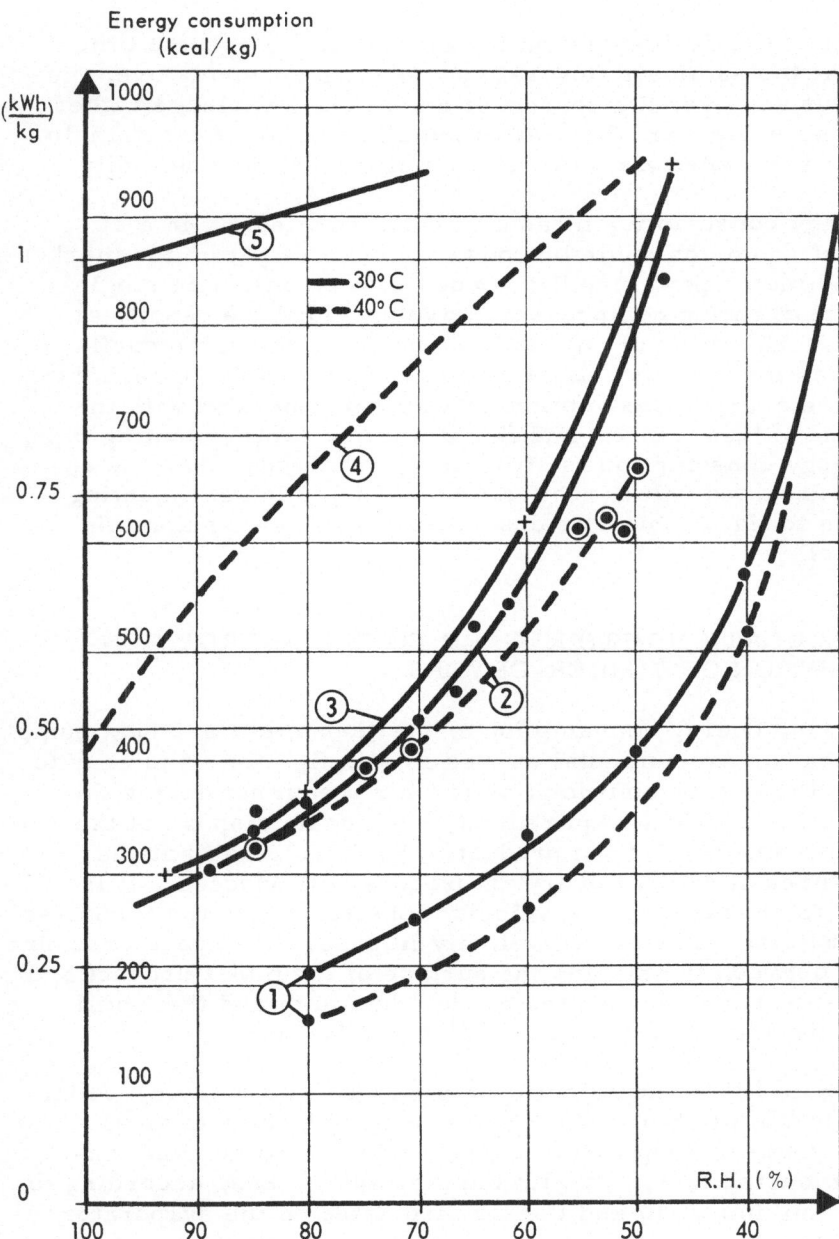

Fig. 12.   Measured specific energy consumption per kilo
of condensed water as a function of the air conditions
in the kiln for five different equipments.

7. MEASURED ENERGY CONSUMPTIONS OF DEHUMIDIFIERS

Most of the devices put on the market at the present time operate at the same air flow rate passing successively over the evaporator and over the condenser and at our knowledge rather exceptional equipments have an automatic control of the enthalpy change of the cooled air as a function of the relative humidity.

Energy consumption measurements, carried out in a climatic test room, on some low powered devices put on the market (a few kW motor power), all equipped with an hermetic single-stage reciprocating compressor, have provided the results at figure 12. The trend of curves 1, 2 and 3 has the same configuration as for the theoretical curves. For the best apparatus tested, the energy consumption exactly corresponded with the forecasted values (cf. 4.1 and fig. 8) for the same mounting type. This energy consumption is still nearly 60% higher than the optimum forecasted value (realisable on assumption of mounting according to fig. 9, such as described in 4.2, with automatic Δh control).

8. CALCULATION WITH REGARD TO THE ENERGY CONSUMPTION OF TIMBER-DRYERS

Knowing energy consumption of the "dehumidifier" heat pump (theoretical or experimental curves as f.i. fig. 12), it is easy to calculate the energy consumption for air treatment during the drying process. As a requirement, one should dispose of the moisture isotherms (of the product to be dried), valid at the temperatures at which the drying process will proceed. It is required to choose a drying velocity consistent with the thickness of the material. The drying velocity imposes the vapour pressure gradient between the air and the surface of the moist product and hence the relation between the product humidity and the equilibrium moisture content (H product/H equilibrium).

8.1 In Table III, a numerical example is given for drying (from 40 down to 9%) of 25 mm thick oak at a temperature of 40°C. The heat pump is supposed to operate with an isothermal condenser (mounted in series with the circulation fans, according to the mounting in fig. 9) and the air circulated on the evaporator is submitted to Δh = 10 kcal/kg (maintained constant).

To the energy consumption required for air treatment one should add the energy necessary to heat up the load to the dryer temperature (20 kWh/m^3) and also the transmission losses.

TABLE III: Energy consumption for 25 mm thick oak planks dried from 40 down to 9% at 40°C with heat pump.

Timber moisture content (%)	Water to be evacuated (kg)	R. H. (%)	Specific consumption of heat pump (kcal/kg)	Condensing capacity per kg air (g/kg)	Drying time (at 120 m³/h air) (h)	Energy consumption per stage (kcal/m³ timber)
40 - 28	78.3	82	107.4	12.5	52.2	8,408
28 - 25	19.56	82	107.4	12.2	13.0	2,101
25 - 23	13.04	77	126.3	11.2	9.7	1,647
23 - 21	13.04	72	138.44	11	9.88	1,805
21 - 19	13.04	67	155.5	10.5	10.35	2,028
19 - 17	13.04	62	174	10	10.87	2,269
17 - 15	13.04	52	237	8.3	13.1	3,091
15 - 13	13.04	44	290	7.6	14.3	3,782
13 - 11	13.04	32	488	5.4	20.1	6,364
11 - 9	13.04	20	1,042	3	36.2	13,588
TOTAL	202.18				189.77	45,083 i. e. 52.42 kWh/m³

These are approximately compensated by the power absorbed by the compressor driving motor, provided the dryer is well insulated.

Note: In table III, calculation of drying time (column Nr. 6) is not indispensable. It is only referred to in order to verify whether the flow rate treated by the heat pump (in the practical case 120 m^3/h) has been chosen adequately and is consistent with the normal drying time for this timber thickness.

8.2 Calculation of energy consumption based on efficiency grades of the best dehumidifier tested (Nr. 1 - Fig. 12). This apparatus is operated at the same steady flow rate passing successively on the evaporator and on the condenser. The results are summarized in Table IV. This energy consumption is 38% higher than for a nearly optimized system (cf. Table III).

8.3 A comparison of dryer consumptions for oak sawed into 25 mm thick planks to be dried from 40 to 9% accordingly to different processes is indicated in Table V, taking into account the energy consumption for timber heating and the transmission losses.

Conclusions in Table V were acquired through drying of hard wood that has to be dried within a relatively long time with a small moisture gradient between the timber and its equilibrium moisture content (H timber/H equilibrium = 1.8 in Table III). The same conclusions concerning the relative energy consumption with a heat pump dryer and a ventilation dryer are valid for drying products that may be dried at higher moisture gradients, f. i. Sipo or Afrormosia for which H timber/H equilibrium = 2.6 is admissible. These conslusions are no longer valid for soft wood, which could better be dried in a continuous tunnel dryer instead of in a batch type system. This is not due to the fact that heat pumps do not adapt to tunnel dryers, but to the following reasons:

(a)   the higher allowable operation temperature for soft wood (up to 100°C in these tunnels). At the present time there are no heat pumps available on the market being able to operate at those temperatures. In the future, this will change since higher temperature equipments exist in the laboratories and in pilot plants and will become industrially available in the near future;

TABLE IV: Energy consumption for drying of 25 mm oak from 40 down to 9%, drying at 40°C by marketed heat pump.

H timber (%)	Water to be evacuated (kg)	R. H. air 40°C (%)	Specific energy consumption H. P. (kcal/kg)	Energy consumption (kcal/m³ timber)
40/28	78. 3	82	146. 2	11,447
28/25	19. 56	82	146	2,856
25/23	13. 04	77	154. 8	2,019
23/21	13. 04	72	180. 6	2,355
21/19	13. 04	67	197. 8	2,579
19/17	13. 04	62	223. 6	2,916
17/15	13. 04	52	288. 1	3,757
15/13	13. 04	44	369. 8	4,822
13/11	13. 04	32	765. 4	9,981
11/9	13. 04	20	1,500	19,560
				62,292 i. e. 72.43 kWh/m³

TABLE V: Energy consumption for 25 mm thick oak dried from 40 down to 9% according different processes.

Nr	Process	Energy consumption (kWh/m3 timber)				
		Heating	Transm.	Ventil. or dehumid.	Total (kWh/m³)	Relative consumption
1	Drying at 40°C by optimized H. P. (nearly isothermal condenser; K_{ex} = 0. 5*; 5 K temp. drop ** on exchangers)	20	(20)***	52. 4	72. 4	1
2	Drying at 40°C by marketed H. P.	20	(20)***	72. 4	92. 4	1.28
3	Ventilation drying at 40°C	20	20	203	243	3. 36
4	Ventilation drying at 40/60°C	28. 2	20. 6	172. 3	220. 5	3.05

*K_{ex} exergetic efficiency of the heat pump.
**   Temperature drop between the air and the evaporator on the one side and the air and the condenser on the other side.
***   Transmission losses are covered by the power absorbed by the driving motor of the compressor.

(b) power rates to be realised in tunnel dryers are very high. Soft wood may be dried with a much higher moisture gradient between the timber surface and the drying air. These high power values cannot be achieved by the hermetically sealed motor compressors and the available modular groups.

9. GENERAL CONCLUSIONS CONCERNING POSSIBLE ENERGY SAVINGS

9.1 The best "dehumidifier" mounted heat pumps on the market for the moment profitable replace the conventional hard wood ventilation dryers. They allow a primary energy saving as compared to conventional ventilation dryers even if the ventilation dryers are equipped with a static heat exchanger.

9.2 There is also an opportunity to improve considerably the efficiency of "dehumidification" dryers by better mounting schemes of the components and by an automatic control of the Δh the air cooled on the evaporator is submitted to as a function of the climatic conditions in the dryer. Research has to be carried out on these subjects.

10. INVESTMENT COSTS

The investment costs of heat pumps ($i_{H.P.}$) are normally given per kW motor power - they range from 10,000 to 25,000 F/kW (F = Belgian francs). The investment costs per unit of heat delivered at the condenser are expressed by:

$$\frac{i_{H.P.}}{C.O.P. \times 860} \tag{2}$$

They are a function of the coefficient of performance (C.O.P.) of the heat pump. An approximate value of the COP:

$$C.O.P. \simeq K_{ex} \cdot \frac{\theta_{condenser} + 273 + 5}{\Delta\theta + 10} \tag{3}$$

K_{ex}: exergetic heat pump efficiency (in the following computations = 0.5)
θ condenser: condenser temperature ($^{\circ}C$)
$\Delta\theta$: temperature difference evaporator-condenser (K)

For the same installed heating power, the difference in investment costs between a heat pump and a fossil fuel heated system (i_f) is:

$$\left(\frac{i_{H.P.}}{C.O.P. \times 860} - i_f \right) . \ P_i \tag{4}$$

P_i: installed power (kcal/h)
i_f: investment costs of fuel fired system $[(F/(kcal/h)]$

The difference in energy costs of the two systems may be written as:

$$\left(\frac{P_f}{\eta \times 10,000} - \frac{P_e}{C.O.P. \times 860} \right) Q \tag{5}$$

η : efficiency of a fuel fired system
Q : useful energy consumption (kcal)
$Q = P_i . t$
t : equivalent time that the full installed power is used (h)
pf : price of fuel (F/kg)
P_e: price of electric energy (F/kWh)

The supplementary investment costs of the heat pump system (4) must be compensated by a saving of energy costs expressed by (5). The working time (t_a) at full power (P_i) for reaching a compensation is given by:

$$t_a = \frac{Q}{P_i} = \frac{\dfrac{i_{H.P.}}{C.O.P. \times 860} - i_f}{\dfrac{P_f}{\eta \times 10,000} - \dfrac{P_e}{C.O.P. \times 860}} \tag{6}$$

If $t_a = 0$ a heat pump system as marketed is no more expensive as a fuel fired system. This can be realised if the C.O.P. $=$ (C.O.P.)$_a$.

$$(C.O.P.)_a = \frac{i_{H.P.}}{i_f \times 860} \tag{7}$$

244

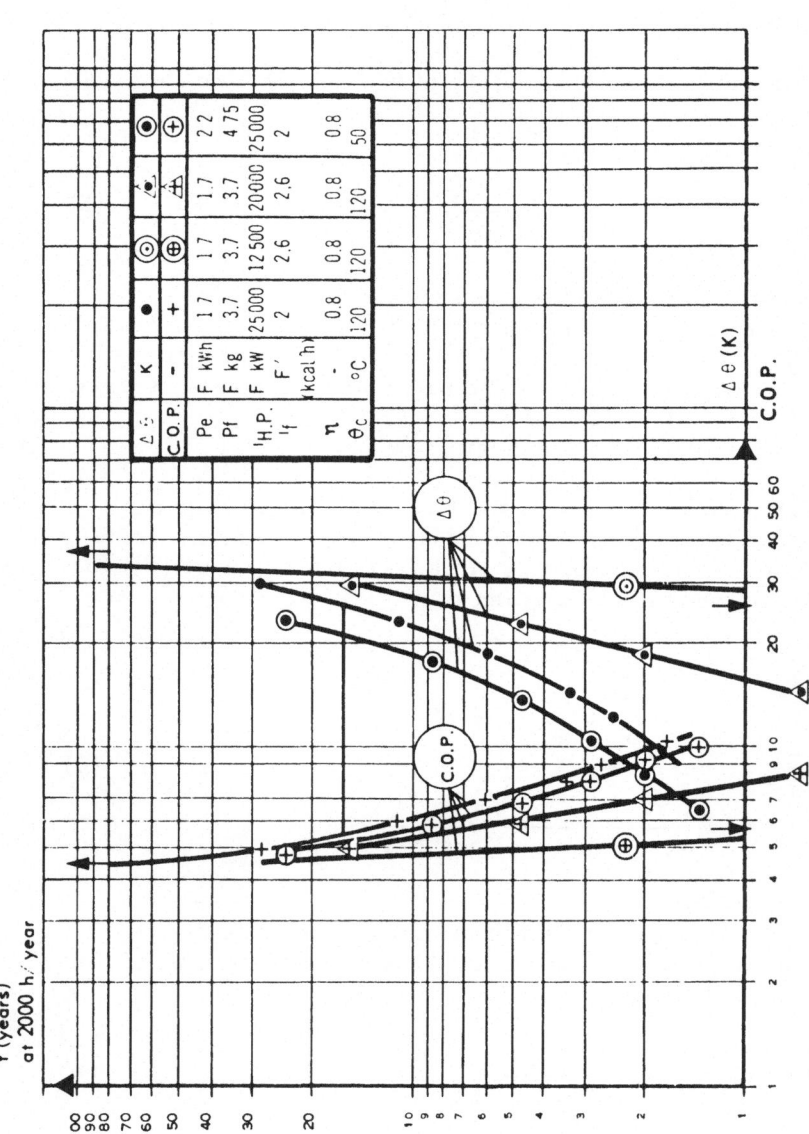

Fig. 13. Amortizement time of the heat pump equipment as a function of the COP and of the temperature difference ($\Delta\theta$) that must be covered.

The temperature difference that can be covered in this case can be deduced from the expression (3).

$$(\Delta\Theta)_a = \frac{K_{ex} \cdot (\Theta_{cond} + 278) \times i_f \times 860}{i_{H.P.}} - 10(K) \qquad (8)$$

For instance:

$i_f = 2$ F/(kcal/h); $i_{H.P.} = 25,000$ F/kW; $K_{ex} = 0.5$; Θ cond = 100°C substituted in (7) leads to $(C.O.P.)_a = 14.5$; and in (8) to $(\Delta\Theta)_a = 3$ K. This temperature difference is too low to be useful. At the present day equipment prices, heat pump systems are more expensive than fuel fired systems. Heat pump equipments must be justified by a saving on energy costs. These savings are a function of the energy consumption and of the working time (t) [cf. (6)].

The working time for paying off the supplementary investment costs by a saving on energy costs as expressed by (6) has been calculated for several hypotheses concerning the investment and fuel prices. The results are summarized on figure 13.

Fuel prices of 3.7 F/kg correspond to heavy industrial fuel and 4.75 F/kg to gas oil for domestic use. The two adapted electricity prices are again related to domestic (last column on figure 13) and industrial uses. The investment costs for fuel fired systems of 2 and 2.6 F/(kcal/h) correspond to systems without and with cooling tower with natural draft (this last system has not to be considered for drying applications).

Regardless of the hypotheses concerning the numerical values of p_e, p_f, $i_{H.P.}$, i_f, η and Θ_c it appears clearly that the temperature difference $(\Delta\Theta)$ that economically can be covered by heat pumps does not exceed a few tens of degrees (... 30 K ...).

The left series of curves indicating the C.O.P. shows that the C.O.P. should be higher than 4 to allow an economic operation of the heat pumps at the present-day investment and energy prices.

The two preceding conditions concerning C.O.P. and $\Delta\Theta$ are fulfilled for optimised heat pump air dehumidifiers in a certain number of applications like f.i. hard timber drying.

REFERENCES

1. " La déshumidification de l'air et le sechage par pompe à
   chaleur", Rapport Laborelec 8. 1127. f
2. " Le sechage de bois resineux", Rapport Laborelec 8. 3148

CONSIDERATIONS CONCERNING THE LARGE-SCALE USE OF HEAT PUMPS

K. F. Ebersbach

Forschungsstelle für Energiewirtschaft in München.

The heat pump is a special kind of heat generator, for it enables us to use the heat content of environmental air, water or soil, i. e. the practically unlimited energy that is available in many places to provide the energy required for space heating. Moreover, by means of the heat pump the low-temperature waste heat produced by technical processes can be made usable. So it seems quite natural that such an "energy-saving" technique should have been welcomed, sometimes over-optimistically, by many of those who, prior to October 1973, saw no need for the rational use of energy.

The fact is that, on the one hand, heat pumps have been in use for years in swimming pools, offices and other buildings and that there will be new fields for the application of heat pumps in the future. On the other hand, there are also limits to their wider use. These limits, as well as the further possibilities of heat pump application, will be pointed out in this lecture.

The most important of those limits at present is obviously the temperature attainable by a heat pump process. Domestic space heating requires a technique that guarantees maintenance-free and safe operation. Under these circumstances, however, temperatures of more than 60°C cannot at present be exceeded, while on the other hand, conventional hot water heating systems require (in the FRG) a maximum water temperature of 90°C. To replace an oil-fired boiler, for instance, by a heat pump it is therefore necessary to enlarge the radiators to correspond to the lower maximum temperature. This is not a problem of heat pump technique, but a problem that limits heat pump application.

This example shows that only low temperature heat demand can be covered by heat pumps, but it means also that the lower the temperature needed, the more economical the heat pump can be, especially when there is at the same time a small difference between the temperature needed and the temperature of the heat source. For this difference is decisive for the coefficient of performance.

We saw that, in general, one cannot simply replace any other heat generator by a heat pump without at the same time modifying the energy consumption. This means, for instance, the installation of ceiling or floor heating or of convectors with an additional fan, because these heating systems need a lower temperature than radiators.

Moreover, during recent years in the FRG, combined systems of heat pumps and boilers have been developed and partially operated. Space heating is done completely by the heat pump with outdoor temperatures of down to +5°C or even 0°C, where the water temperature attainable by the heat pump is still sufficient to meet the total heat demand. At lower outdoor temperatures, the heat demand is covered by the boiler, which means of course that the boiler capacity has to correspond to the maximum heat demand at the lowest temperature. With such systems it is possible to cover up to 90% of the annual space heating consumption by the heat pump, if the heating water temperature is governed by the outdoor temperature.

In a study sponsored by the Bundesminister für Forschung und Technologie, Bonn, the Forschungsstelle für Energiewirtschaft, Munich, has dealt with the principal techniques of such systems. One of the most interesting combinations was as follows:

One of two similar houses is equipped with conventional radiators of 90/70°C, the other with a floor heating system. At the lowest outdoor temperature, the heat demand of the first house is fully supplied by an oil-fired boiler, that of the second one by a water/water heat pump. At rising outdoor temperatures - and, consequently, falling temperatures of the heating water - the heat pump, in addition to meeting the decreasing heat demand of the second house, gradually replaces the boiler for the first house. Thus, the heat pump can be operated at about its maximum capacity during long periods of the winter.

Another important point for heat pump application is the

temperature of the heat source, especially if the heat source is outdoor air.

The outdoor air temperature varies during the heating period to a very large extent, and while of course the heat demand reaches its maximum at the lowest outdoor temperature, the co-efficient of performance of the heat pump decreases with the falling temperature of the heat source. The resultant problems of control techniques and design are not yet solved, at least as regards small-scale installations.

Another problem is that air humidity can cause freezing-up of the evaporator of the heat pump at low air temperatures. A solution to this problem would greatly favour the extended application of the heat pump.

In the future, waste heat recovery will offer a new field for heat pump application. Such waste heat recovery can be done in many industrial establishments as well as in private households, as the following examples show.

(a) Industry

Table 1 shows the daily waste heat output of a brewery with a beer production of about 800,000 hl per year. The waste heat is derived from the brew kettle vapour, from the wort cooling and from the bottle-rinsing machines. Moreover, the heat output of the refrigeration machines is available. It can be seen that, comparing the heat consumption of the brewery (105 MWh per day) with the waste heat output (64.5 MWh per day), about 60% of the heat demand could be covered by using waste heat, partially by direct heat exchange, partially by means of heat pumps.

Fig. 1 shows the pattern of the daily demand for the brew kettles, the bottle-rinsing machines and for water heating. The same figure shows the waste heat output which should according to an investigation by the Forschungsstelle fur Energiewirtschaft, be available for recovery.

The main heat recovery system shown in fig. 2, consists of two parts: firstly a direct heat exchanger using part of the waste heat from the wort cooling system for generating hot water at about 70°C, secondly a heat pump system generating hot water at about 50°C by using the waste heat from the bottle-rinsing machine and part of the heat from the refrigerating machines.

250

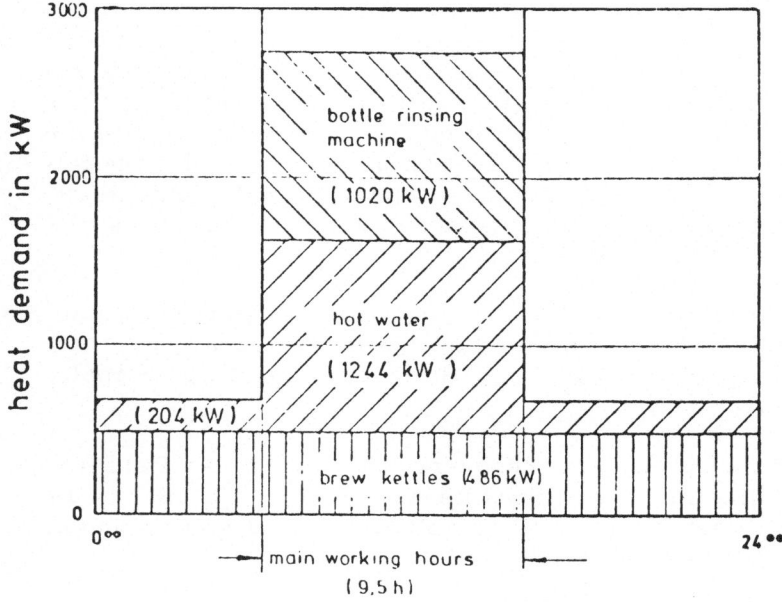

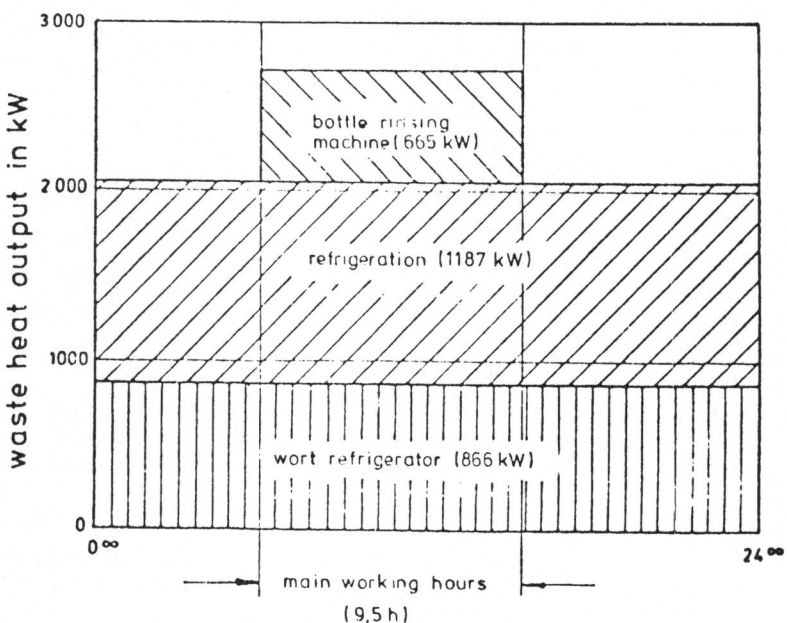

Fig. 1.   Average pattern of heat demand and waste heat output in a brewery.

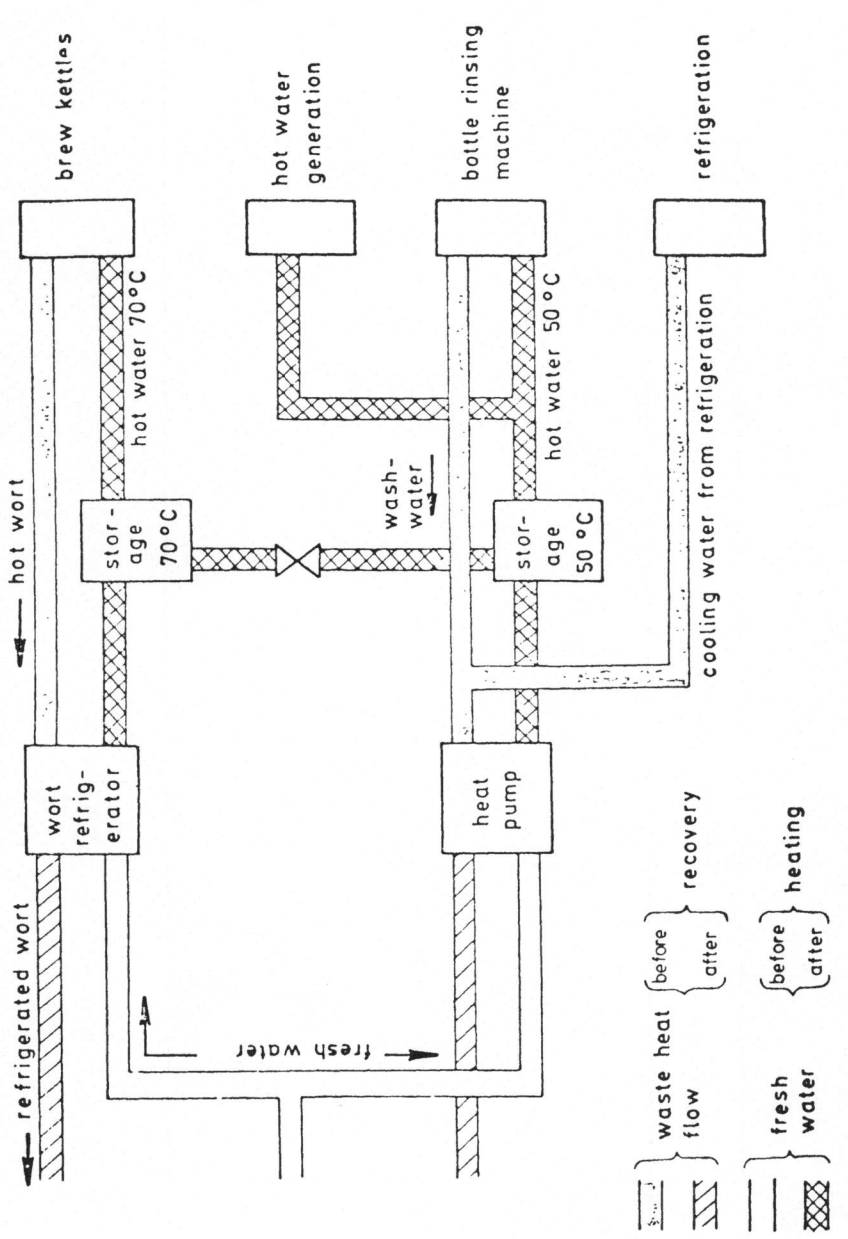

Fig. 2.   Basic system of heat recovery in a brewery.

TABLE 1

RECOVERY OF WASTE HEAT IN A BREWERY (ANNUAL PRODUCTION 800,000 hl)

Origin of waste heat	Heat consumption MWh/day	Utilizable waste heat MWh/day	Temperature range °C
Brew kettles	54.7	20.7	50-70
			80
Bottle rinsing machine	11.6	7.0	40
Refrigeration	-	29.0	40
Total for brewery	105.0	64.5	-

While the water at $70^{\circ}C$ is led to the brew kettle, the water at $50^{\circ}C$ is used for the rinsing machine and for other requirements. Both systems are equipped with a hot water storage device to compensate for differences in the pattern of heat demand and waste heat output.

Thus, the fuel consumption of the brewery can be reduced by about 35%.

(b) Household

In a family of four, the average annual electricity consumption of the water heater is about 2,000 kWh, that of the washing machine about 400 kWh, and that of the dishwasher about 600 kWh, i.e. a total of 3,000 kWh. The waste heat contained in the waste water from these three sources is about 2,000 kWh, for an average waste water temperature of $23^{\circ}C$. Utilization of this waste heat by heat pumps could reduce considerably the energy consumption.

Fig. 3 shows the principle energy flow of a heat recovery plant containing a heat pump with an average coefficient of performance of 2.5.

The main problems are:

(1) The very considerable variations in the heat demand pattern as well as that of the waste heat output.

(2) The non-synchronism of the two patterns.

(3) The often very high peaks of the waste heat output.

These facts are shown in fig. 4 (simplified waste heat output pattern) and fig. 5 (demand and output patterns of a washing machine). Consequently storage devices are necessary both on the heat demand and the waste heat sides.

Another problem is that a maintenance-free heat recovery technique must be developed.

Again, the first three points are not, and the fourth only partially heat pump technique problems.

For example, the extraction of heat from the waste water, which contains various kinds of dirt, might need techniques which considerably increase the cost of heat recovery.

254

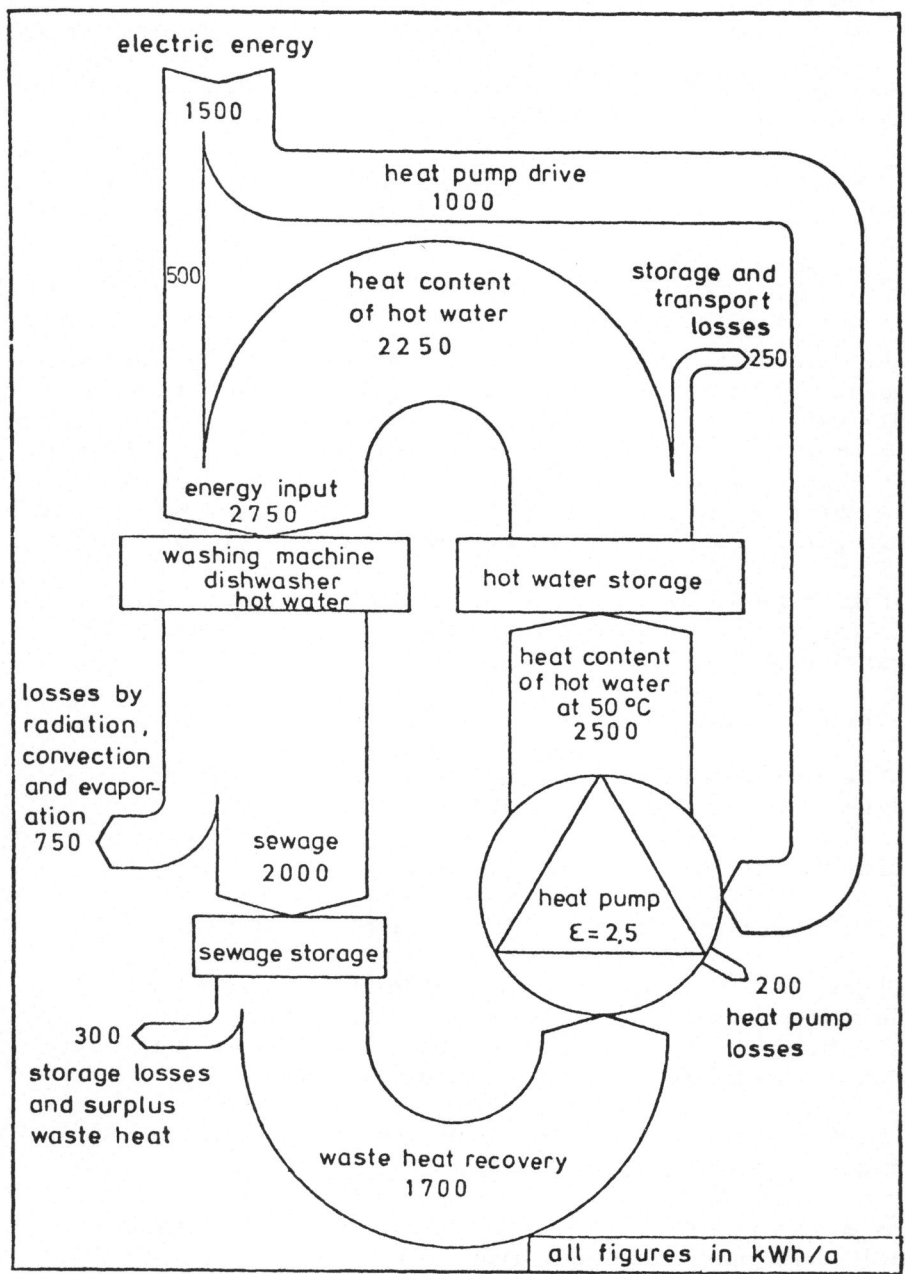

Fig. 3.   Heat recovery from household sewage.

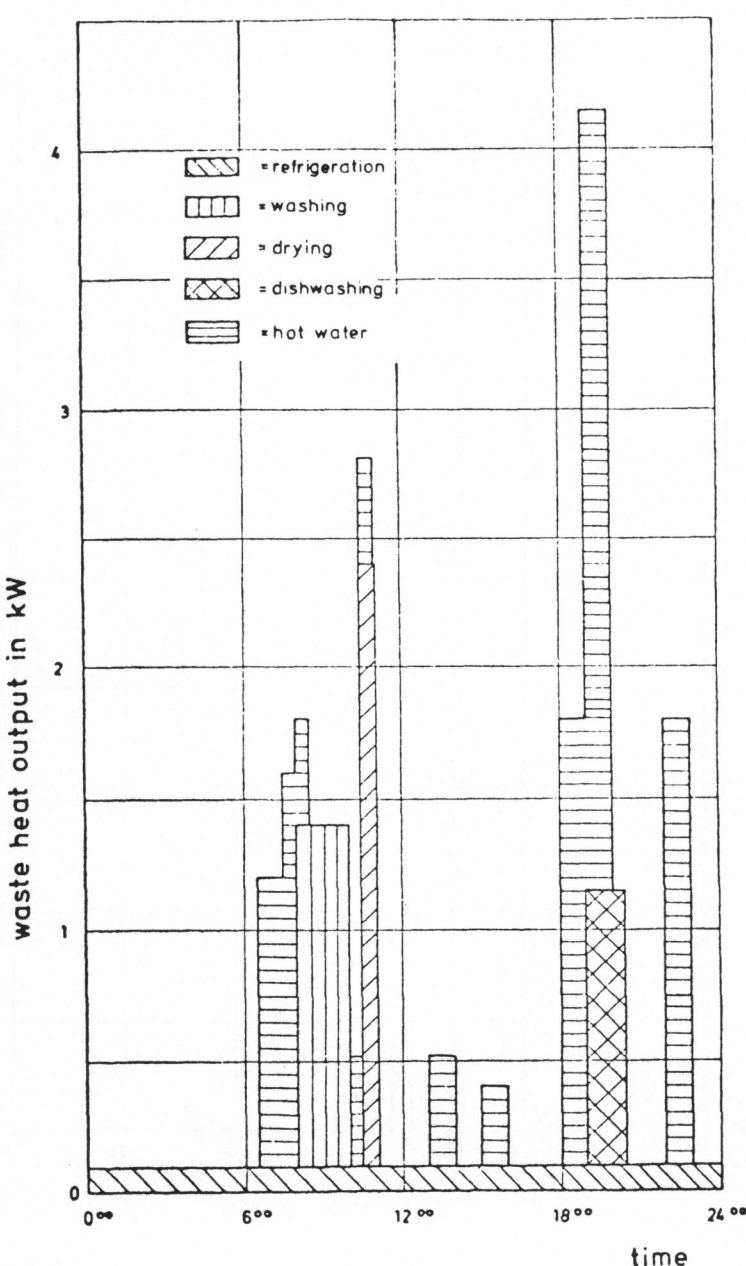

Fig. 4. Average pattern of waste heat output in a household.

256

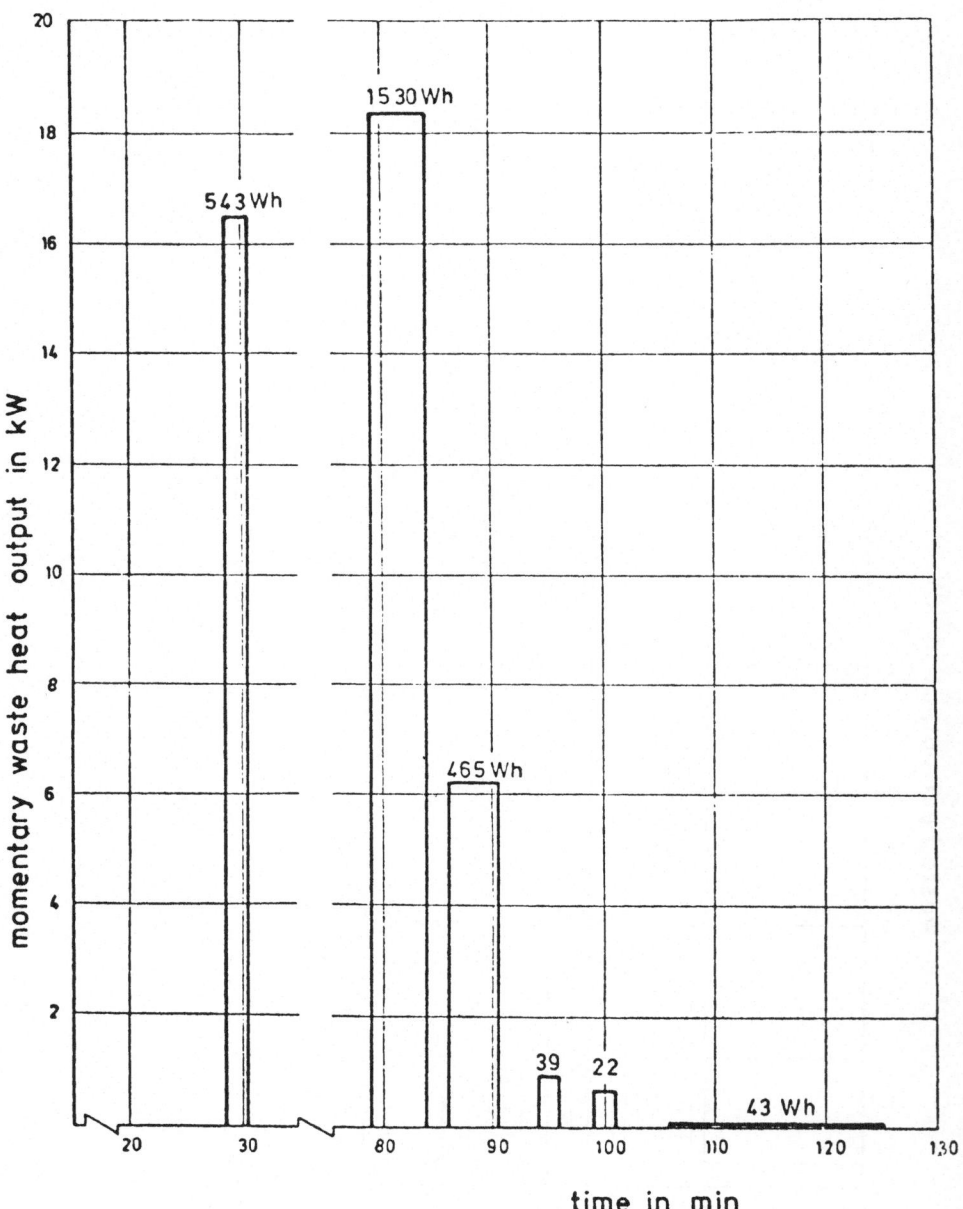

Fig. 5.   Waste heat pattern of a washing machine.

A heat recovery plant for a seven-family apartment house is now being designed and will be built in 1976 in Essen (FRG).

To conclude, it must be pointed out that economical heat pump application needs, above all, studies and analyses of the energy demand, for if we cannot say exactly where, when and for which purpose energy is consumed, the heat pump cannot contribute to the energy supply.

POTENTIAL OF AIR-TO-AIR HEAT PUMPS FOR ENERGY CONSERVATION IN RESIDENTIAL BUILDINGS

G. E. Kelly

Mechanical Systems Section, Center for Building Technology, National Bureau of Standards, Washington, D. C. 20234

ABSTRACT

Information is presented on the dynamic performance of a 5 ton air-to-air heat pump, which was installed in a residence in the Washington D. C. area. The effect of part load operation on the heat pump's cooling and heating coefficients of performance (COP) was determined. When the pump operated in the heating mode at outdoor temperatures below $40^\circ F$ ($4.4^\circ C$), a considerable discrepancy was found to exist between the measured performance and the performance data supplied by the manufacturer. This discrepancy is apparently due to the adverse effects of ice build up and defrosting of the outdoor coil. The seasonal heating coefficient of performance of the heat pump was estimated and then traced back to the power plant to obtain an "effective seasonal heating COP" which is then compared with the performance which might be expected from fossil fuel heating equipment.

INTRODUCTION

With approximately 11% of the total energy consumed in the United States[1] going into the space heating of residences, there is an urgent need for information on the energy effectiveness of the various systems used for residential heating. These systems include the traditional fossil fuel heating equipment, such as gas or oil-fired furnaces and boilers, electrical resistance heating systems, such as electric furnaces and baseboard units, and air-to-air heat pumps. Unfortunately, at the present time there exists very little reliable data on the part load and seasonal performance of either the fossil fuel or air-to-air heat pump systems.

In an attempt to obtain some quantitative information on the dynamic performance of air-to-air heat pumps, a 5 ton heat pump was installed in a 20 year old house in the Washington D. C. area. This heat pump experiment was part of a larger programme to experimentally measure the energy savings which could be achieved through retrofitting. The approach used was to measure the heating and cooling requirement of an old home and then to make various energy-saving modifications[2]. By measuring the energy requirement of the house after each modification an estimate of the energy savings resulting from each change could be made. The house studied was a single story frame rancher which was built in the early 1950s and had a floor area of approximately 2,500 sq. ft. (232.25 m^2). The house was referred to as the "Bowman" house, since Bowman was the name of the people who lived in it prior to it being acquired by the National Bureau of Standards.

The Bowman house was originally heated by an oil-fired furnace and this was used to determine the pre-retrofit heating requirement of the house during the 1973-74 winter. The furnace was then removed in the Spring of 1974 and a heat pump installed in its place, with little modification being made to the existing duct system. The heat pump was instrumented and used to measure the pre-retrofit cooling requirement of the house during a test period in the Summer of 1974. During the 1974-75 Winter, various stages of the post-retrofit heating test was carried out. First, weather stripping and caulking were applied around the doors and windows and the heating requirement of the house was then determined. Next storm windows were installed and the heating requirement of the house was remeasured. The third and last stage of the post-retrofit heating test consisted of adding additional insulation to the 4 inches already present in the attic and the blowing of insulation into the walls; the new heating requirement of the house was then measured. The post-retrofit cooling requirement of the Bowman house, which has yet to be determined, will be measured some time during the current 1975 summer.

The Bowman house data presented in this paper pertains to measurements made during the pre-retrofit cooling test, the first stage of the post-retrofit heating test after caulking and weather stripping and after the second stage of the post-retrofit heating test (after the installation of storm windows). The

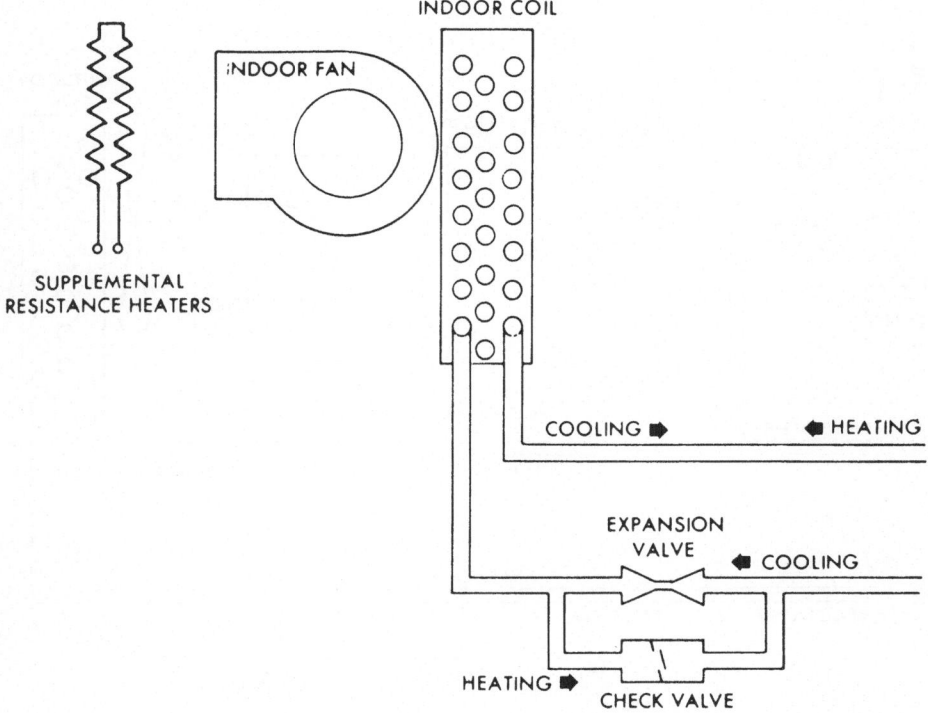

Fig. 1.   Schematic of heat pump's indoor unit

performance of the air-to-air heat pump after the placement of
additional insulation in the attic and the blowing of insulation
into the walls will not be discussed, since this modification
resulted in a greatly oversized heat pump with different performance characteristics.   The pre-retrofit cooling test lasted
for a period of four days and 373 hours of heat pump data was
obtained for the combined first and second stages of the post-
retrofit heating test.   Additional data was not obtained during
the cooling test because of the difficulty of making these measurements and lack of desirable weather conditions.   The amount of
data obtained during the post-retrofit heating test was limited by
the fact that the heating requirement of the house was also
measured using straight resistance heat, instrumentation failures,
and the need to perform three stages of retrofitting and testing
during one heating season.

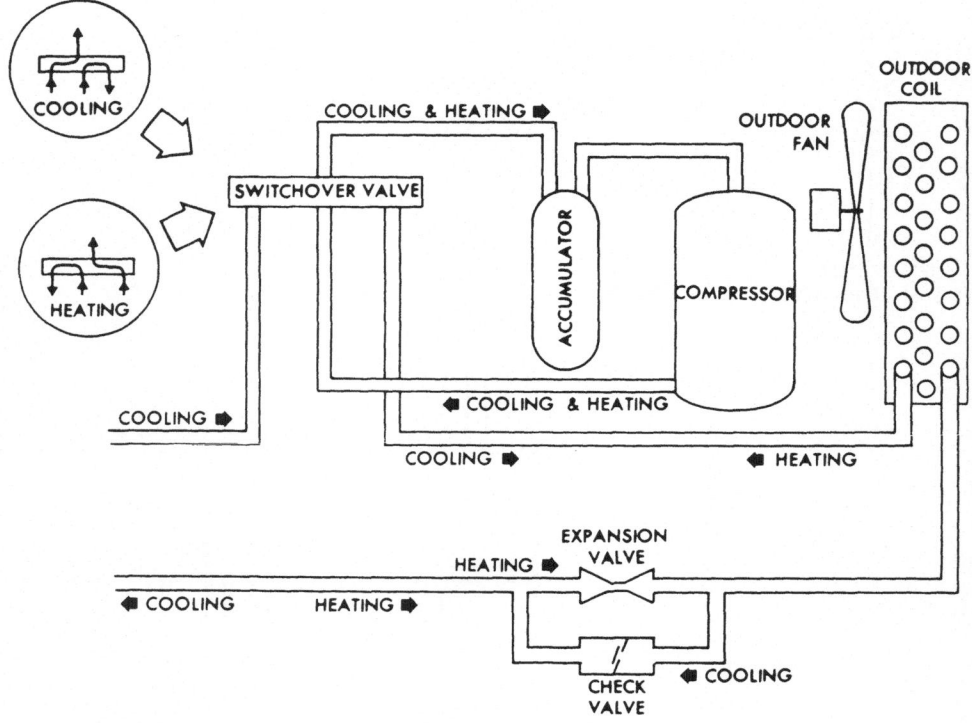

Fig. 2.   Schematic of heat pump's outdoor unit

EXPERIMENTAL EQUIPMENT

The 5 ton heat pump installed in the Bowman house was a split system employing thermostatic expansion valves. It was purchased from a heat pump manufacturer and was sized to meet the calculated pre-retrofit cooling requirement of the house using a procedure recommended by ASHRAE[3]. Figure 2 is a schematic of the outdoor section, which contained the outdoor coil, outdoor fan, compressor, accumulator, switch-over valve and the expansion valve used during the heating process. The indoor section, which is shown in figure 1, contained the indoor coil, indoor blower, supplemental resistance heaters, and the expansion valve employed when the heat pump was used to cool the house.

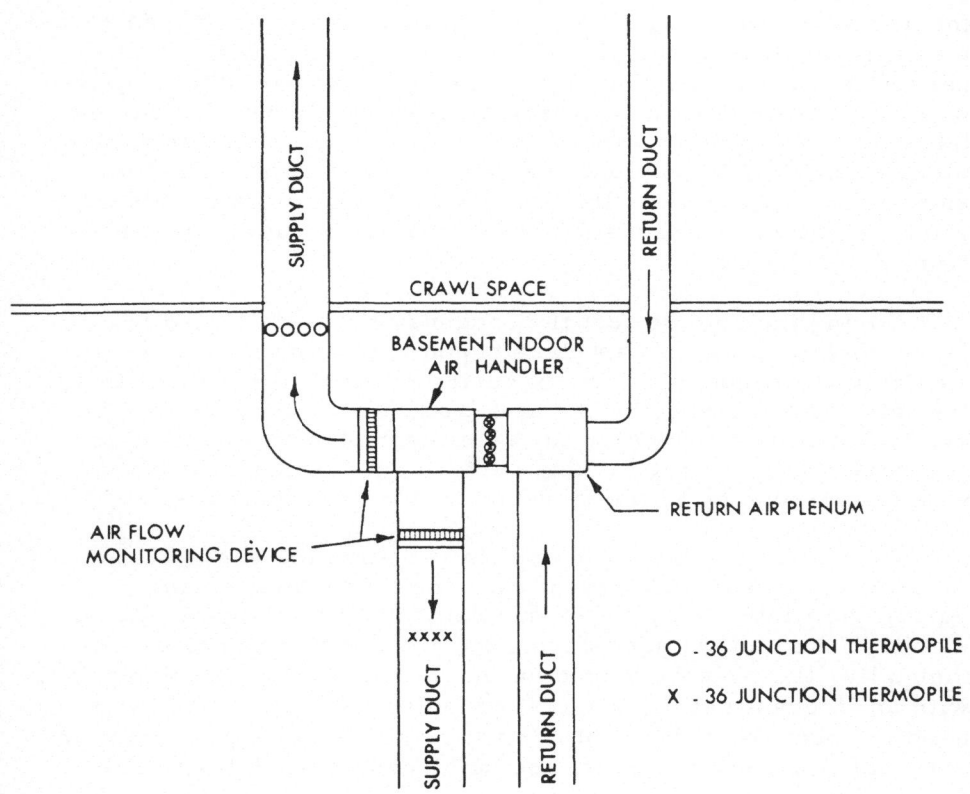

Fig. 3.   Schematic of duct system in the Bowman house

The heating duct system is shown schematically in figure 3. Since half the house was over a basement and the other half over a crawl space, there were separate supply and return ducts for each half of the house. The returning air was fed into a return air plenum and then into the indoor air handler, where it was either heated or cooled. Upon leaving the indoor air handler, the air entered the two supply ducts where air flow monitoring devices were used to measure the mass flow rate of air delivered to each half of the house. Each of these air flow monitoring devices were purchased commercially and consisted of a honeycomb straightener and a rake of pitot tubes located at centres of equal areas. By measuring the difference between the average total and the average static pressure in each supply duct and knowing the barometric pressure and the temperature of air in each duct, the mass flow rate of air delivered to each half of the house could be calculated. Two 36 junction copper-constantan thermopiles, which were located as shown in figure 3, were used to measure the temperature rise or drop in the air delivered to each supply duct as it passed through the air handler. The output of each thermopile was fed into an electronic integrator, which allowed for calculation of the average temperature difference between the return plenum and each supply duct over a period of time. The main supply ducts, as well as the branches leading to the individual room registers, were insulated in order to cut down the losses from these ducts.

During the cooling test the condensate was collected in a large container which was placed upon a mechanical scale. By weighing the condensate hourly during the summer cooling test, the latent cooling done by the heat pump was determined. The relative humidity sensor in the return air plenum was used to compare the measured cooling performance of the heat pump with that preducted in the manufacturer's specifications.

Watt-hour meters were used to measure the energy input to the compressor and outdoor fan, the indoor blower and, during the winter test, to the supplemental resistance heaters. Although the meters used during the cooling test had to be read manually, the ones employed on the heating test were equipped with an internal set of contacts. A pulse counter-printer was used to count the number of contact closures. A signal from an external source would cause the pulse counter-printer to print out and, to reset to zero.

Strip chart recorders and copper-constantan thermocouples were used to continuously record the air temperature in the return air plenum and in both supply ducts. These recordings allowed for constant monitoring of the heat pump and provided information on the length of on-time and off-time. In addition, the return air temperature measurement was necessary in order to accurately compare the measured heating and cooling done by the heat pump with the manufacturer's performance data.

The outdoor temperature and relative humidity were also recorded hourly. The copper-constantan thermocouple, used to measure the outdoor temperature, and the humidity sensor were housed in a small outdoor enclosure in order to protect them from rain, snow and direct sunlight.

EXPERIMENTAL PROCEDURE

In both the cooling and heating mode of operation, the performance of the heat pump is described by its coefficient of performance or COP. When the heat pump is used to cool the house, the coefficient of performance, which in this case may be referred to as the Cooling COP, is the ratio of the total heat extracted from the return air to the total energy supplied to the heat pump. When the heat pump is used for heating, its coefficient of performance or Heating COP is the ratio of the total heat delivered by the heat pump system to the total energy input to the system. The heat pump system is defined so as to include the first step of the supplemental resistance heaters when this step operates during a defrost period. During periods of operation not involving defrost, the input and output to the supplemental heaters are not included in the calculations to determine the Heating COP. This definition of the heat pump system was adopted because it was felt that the heat pump should be penalized for the energy required to temper the supply air during the periods that heat is being extracted from the return air in order to defrost the outside coil. During defrost, the first stage of supplemental resistance heaters in the Bowman house installation was capable of just cancelling the cooling effect of the indoor coil, which resulted in a supply air temperature approximately equal to the return air temperature and neither heating nor cooling being done to the interior space.

When the heat pump cools the interior space, the total cooling, $Q^C(\tau)$, done over a time period τ is given by the equation (1).

$$Q^C(\tau) = Q_s^C(\tau) + Q_L^C(\tau), \qquad (1)$$

where $Q_s^C(\tau)$ is the total sensible cooling and $Q_L^C(\tau)$ is the total latent cooling done in the time τ. The quantities $Q^C(\tau)$, $Q_s^C(\tau)$ and $Q_L^C(\tau)$ in addition to depending on the time period τ, depend upon the outdoor dry bulb temperature, the indoor wet and dry bulb temperatures, the rate of air flow through the indoor unit, the performance of the heat pump's various components, and the cooling load of the house.

The quantity $Q_s^C(\tau)$ is determined using:

$$Q_s^C(\tau) = \overline{C_p} \sum_{i=1}^{2} \overline{\dot{m}_i} \ \overline{\Delta T_i} \ (\tau) \qquad (2)$$

where $\overline{C_p}$ is the average specific heat of the air-water mixture leaving the indoor unit for typical on-cycle conditions,

$i = 1, 2$ pertains to the supply ducts passing through the basement and crawl space;

$\overline{\dot{m}_i}$ is the average mass flow rate of the air-water mixture in supply duct i during on-periods

$$\overline{\Delta T_i}(\tau) \equiv \int' \Delta T_i(\tau) \, dt,$$

time
period τ

$\int'$ in the equation defining $\Delta T_i(\tau)$ indicates that integration is to be carried out during periods when the heat pump is operating;

$\Delta T_i(\tau)$ is the absolute value of the change in dry bulb temperature experienced by the air entering supply duct i at time τ as it passed through the indoor unit.

It has been assumed that there is no cooling of the interior living space during the off-cycles, since the indoor blower and heat pump ceased operating at the same time and the location of the indoor unit in the basement prevented any natural convective cooling.

During each of the four days in the pre-retrofit cooling test, the mass flow rate in each supply duct was periodically checked while the heat pump was on. These values were then averaged to find an average air flow rate, $\overline{\dot{m}}$, for each supply duct which was used in equation (2) to determine the sensible cooling provided by the heat pump. The standard deviations of these measured mass flow rates were approximately 2% and 3% of the average mass flow rates for the two supply ducts. The quantities ΔT_i were measured with thermopiles and then integrated over time using electronic integrators. During the off-periods, the signals to the integrators were shorted which effectively set $(\Delta T)_i$ equal to zero and thus was equivalent to integrating only over the on-periods.

The total latent cooling provided by the heat pump over a time period τ was determined using the equation:

$$Q_L^c(\tau) = \Delta W(\tau) \ (h_g(T) - h_f(T)) \tag{3}$$

where $\Delta W(\tau)$ is the weight of condensate collected in this time period, and $h_g(T)$ and $h_f(T)$ are the specific enthalpies of saturated water vapour and saturated liquid, respectively, at the average evaporator coil temperature T.

The Cooling COP was calculated using:

$$\text{Cooling COP} = \frac{Q^c(\tau)}{I(\tau)} \qquad\qquad (4)$$

where $I(\tau)$ is the total measured energy input to the heat pump during the period of time τ, used to determine $Q^c(\tau)$. For evaluating the Cooling COP, data was collected hourly on $\overline{\Delta T_i}$, ΔW and I and these hourly values were then added together to obtain values for time periods τ which either contained an integral number of complete on-off cycles or consisted of all the consecutive hours in a 24 hour period for which a cooling load existed.

Two cooling tests were also run with the heat pump operating in a steady state manner. One test was for a three hour period and the other was for a one hour period. In order to achieve approximately steady indoor conditions, the doors and some of the windows in the house were partially opened and the heat pump allowed to operate until conditions stabilized. Data was then collected on $\overline{\Delta T_i}$, ΔW and I and the steady Cooling COP was calculated for each of the four one-hour periods.

When the heat pump operated in the heating mode, the total heat $Q^H(\tau)$, delivered to the interior living space during the time period τ, is given by

$$Q^H(\tau) = \sum_{i=1}^{2} \int_{\substack{\text{time} \\ \text{period } \tau}} C_{pi}(t)\dot{m}_i(t)\,\Delta T_i(t)\,dt = \qquad (5a)$$

$$= \overline{C_p}\sum_{i=1}^{2}\overline{\dot{m}_i}\,\overline{\Delta T_i(\tau)} + \langle C_p\rangle\sum_{i=1}^{2}\int_{\substack{\text{off} \\ \text{period} \\ \text{in } \tau}}\dot{m}_i(t)\,\Delta T_i(t)\,dt \qquad (5b)$$

where $C_{pi}(t)$ and $m_i(t)$ are the specific heat and mass flow rate at time t of the air-water mixture entering supply duct i, $\overline{C_p}$, $\overline{\dot{m}_i}$, $\overline{\Delta T_i(t)}$ and $\Delta T_i(t)$ were defined for equation 2, and $\langle \dot{C}_p \rangle$ is the specific heat of the air-water mixture leaving the indoor unit for typical off-cycle conditions. The average mass flow rate during on-periods, $\overline{\dot{m}_i}$, was determined by periodically checking the mass flow rate in supply duct i throughout the heating test. The standard deviations of the measured mass flow rates were approximately 1.6% and 1.1% of the average mass flow rates for the two supply ducts. The quantities $\overline{\Delta T_1(t)}$ and $\overline{\Delta T_2(t)}$ were determined in the same manner as for the cooling test, with the signals to the integrators being shorted during off-periods. The convective heating, which corresponds to the second term on the right hand side of equation 5b, was estimated by running a series of tests to measure the change with time in $\dot{m}_i(t)$ and $\Delta T_i(t)$ after the heat pump shut off. The mass flow rate was measured at various cross sectional positions in each supply duct using a hot wire anemometer and $\Delta T_i(t)$ was obtained using two 36-junction thermopiles and strip chart recorders. It was found that the convective heating term could be approximated in this series of tests by an equation whose independent variable was the length of the off-period, provided that the off-period was less than 30 minutes. When the off-period exceeded 30 minutes a constant value was used. This approximation was then used to estimate the convective heating provided during off-periods throughout the entire post-retrofit heating tests, since it was impractical to measure this quantity for each individual cycle. Although the use of one approximation for all off-periods could introduce a small error, the effect of this error on the heating COP was small since the convective heating term was only a small percentage of the total heat term, $Q^H(\tau)$.

In the post-retrofit heating test, the quantities $\overline{\Delta T_1(\tau)}$ and $\overline{\Delta T_2(\tau)}$; and the energy inputs to the indoor blower, I_B; the outdoor fan and the compressor, I_o; and supplemental resistance heaters, I_R; were obtained for each on-off cycle. This was achieved by having the pulse counter-printer, attached to the watt-hour meters, and the two integrators, used to integrate the signals from the thermopiles, print out and reset to zero each time the heat pump came on. The total heat delivered $Q^H(\tau)$ during each on-off cycle was then computed and used to calculate the Heating COP for each cycle using the equation:

$$\text{Heating COP} = \frac{Q^H(\tau_1)}{I'(\tau_1)} \tag{6}$$

where τ_1 is length of the on-off cycle and

$$I' \equiv \begin{cases} I_B + I_o & \text{when no defrosting takes place during the on-off cycle} \\ \\ I_B + I_o + I_{R1} & \text{when a defrost process occurs during the on-cycle} \end{cases}$$

The quantity I_{R1} is the energy input to the first step of supplemental resistance heaters and is equal to I_R, $I_R/2$ or $I_R/3$, depending on whether one, two or three steps of resistance heaters were operating during the defrosting process.

Four steady state tests were also performed to measure the COP of the Bowman house heat pump when it was operating continuously in the heating mode. Two of the tests were done when the outdoor temperature was above 50°F (10°C) and they consisted of operating the heat pump for approximately one half hour until steady state conditions were reached and then measuring the energy input and heat output over a half hour test period. The same procedure was followed in the other two steady state tests, except that since they were performed at lower outdoor temperatures the heat pump was first made to go through a defrost cycle. This removed any ice which may have accumulated on the outdoor coil and resulted in a measured COP which represented the maximum performance achievable by the heat pump at the given indoor and outdoor test conditions. As in the steady state cooling tests, the doors and windows in the Bowman house were partially opened in order to achieve the approximately steady indoor conditions necessary for the performance of these four tests.

During the pre-retrofit cooling test, the post-retrofit heating test and the steady state heating and cooling tests, a considerable number of additional measurements were made. These measurements included hourly readings of the outdoor dry bulb temperature and continuous strip chart recordings of the dry bulb temperature and relative humidity of the return air as it entered the indoor unit. These latter measurements were used to obtain the average outdoor dry bulb and the average indoor dry bulb and wet bulb temperatures existing during periods for which Cooling and Heating COPs were calculated. These average indoor and outdoor conditions were then used in conjunction with the manufacturer's performance data to obtain the Cooling and Heating COPs which the manufacturer claimed the heat pump should achieve under these conditions, if it had been operated in a steady state manner. By dividing the measured Cooling or Heating COP by the manufacturer's respective steady state Cooling or Heating COP, an indication of the effect of part-load operation on the heat pump's performance was obtained.

The air flow rates encountered in the Bowman house during the cooling and heating tests were slightly less than the 2,000 cfm (56.65 m^3/min) for which the manufacturer's performance data was available. The average air flow rate during the cooling test was approximately 1,750 cfm (49.56 m^3/min) and correction factors supplied by the manufacturer were used to calculate the steady state capacity and compressor power consumption. The average air flow rate during the post-retrofit heating tests was around 1,680 cfm (47.58 m^3/min) and the correction factors employed in obtaining the steady state COP from the manufacturer's performance data were obtained by a small extrapolation of the correction factors supplied by the manufacturer. The difference in flow rates between the summer and winter was due in part to the fact that during the heating test some of the registers in the half of the house over the basement had to be partially closed in order to achieve a uniform temperature throughout the house. The reduction in air flow rate from 2,000 cfm (49.56 m^3/min) to 1,680 cfm (47.58 m^3/min) during the post-retrofit heating test resulted in an approximate 6% decrease in the Heating COP of the heat pump. The values presented in the section on Experimental Results on seasonal Heating COP, effective Heating COP and effective seasonal Heating COP may be approximately corrected to an air flow rate of 2,000 cfm (49.56 m^3/min) by multiplying them by 1.06.

Since there is not a perfect correlation between the hourly cooling or heating requirement of a house and the difference between the indoor and outdoor dry bulb temperatures, it was necessary to determine the actual cooling or heating load on the Bowman house heat pump in order to properly evaluate its dynamic performance. The equations used to define these loads were:

Cooling load =

$$\frac{\text{Cooling done by the heat pump in time } \tau}{(\text{Manufacturer's steady state cooling capacity}) \times \tau} \quad (7a)$$

Heating load =

$$\frac{\text{Heating done by the heat pump system in time } \tau}{(\text{Manufacturer's steady state heating capacity}) \times \tau} \quad (7b)$$

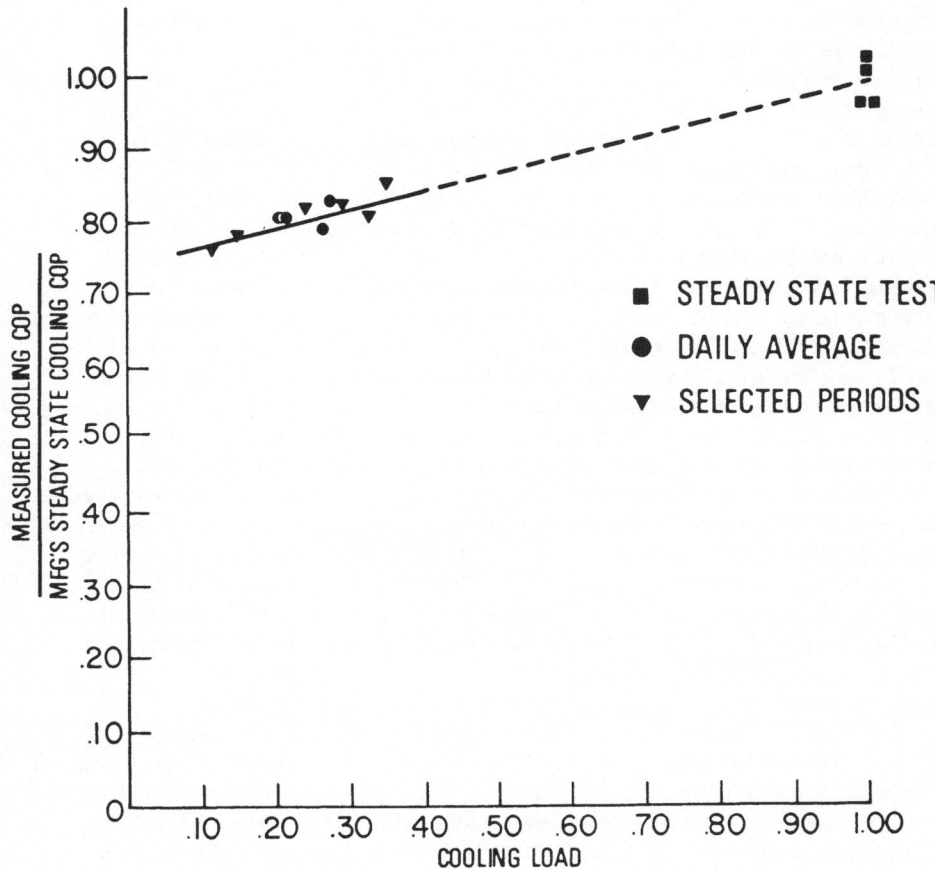

Fig. 4.　Variation in performance with load cooling operation

and τ is a time period or cycle over which the cooling or
heating COP was measured and it is to be understood that the
manufacturer's steady state cooling or heating capacities
correspond to the average indoor and outdoor conditions which
existed in the time period τ.　The numerator in equation 7b
includes the output from the first step of supplemental resistance heaters when this step operated during a defrost period.

EXPERIMENTAL RESULTS

　　The effect of part load operation on the cooling performance
of the Bowman house heat pump is shown in figure 4.　The

ordinate in this figure is the ratio of the measured cooling COP
to the cooling COP calculated from the manufacturer's steady
state performance data for the average indoor and outdoor
conditions existing during the measurement period. The
abscissar is the cooling load which was defined in equation 7a.
The four steady state, hour-long cooling test, corresponding to
a load of 1.0, had an average ordinate of 0.99, which indicates
good agreement between the measured steady state performance
of the heat pump and the manufacturer data. The data points
represented by circles correspond to time periods consisting of
all the consecutive hours in each 24 hour period for which a
cooling load existed. The triangular data points represent
time periods which were selected for analysis because the heat
pump's COP could be determined for a complete number of
on-off cycles and the cooling load was approximately constant.
As can be seen from figure 4, data were only obtained for loads
less than 0.35. This was due to unseasonably mild weather
which existed during and after the pre-retrofit cooling test.
A straight line is shown passing through the data in figure 4 with
the region where no data exists being indicated by a broken line.
Although the lack of data over the entire load range precludes
certainty as to the exact shape of the curve in figure 4, this
straight line approximation is believed to be correct because
similar results were obtained during the post-retrofit heating
test.

The heating COP of the Bowman house heat pump, as calculated using the manufacturer's performance data, is presented
in figure 5 as a function of outdoor temperature for three different return air temperatures. The data presented is for an
indoor air flow rate of 1,680 cfm (47.58 m^3/min), since this is
the average air flow rate existing during the heating test.
Correction factors for determining the heating capacity and compressor power consumption at this air flow rate were extrapolated
from the manufacturer's data as discussed in the section entitled
Experimental Procedures.

Figure 6 is a plot giving the heat pump's capacity (as
obtained from the manufacturer's performance data) at various
outdoor temperatures and three different curves showing heating
load vs outdoor temperature. The heating capacity is for a
return air temperature of 70°F (21°C) and an indoor air flow
rate of 1,680 cfm (47.58 m^3/min). Two of the heating load
curves were determined for infiltration rates of 1.0 and 0.5
indoor air changes per hour using procedures recommended by
ASHRAE[5] for calculating the heating requirement at the outdoor
design temperature. The third heating load curve was obtained

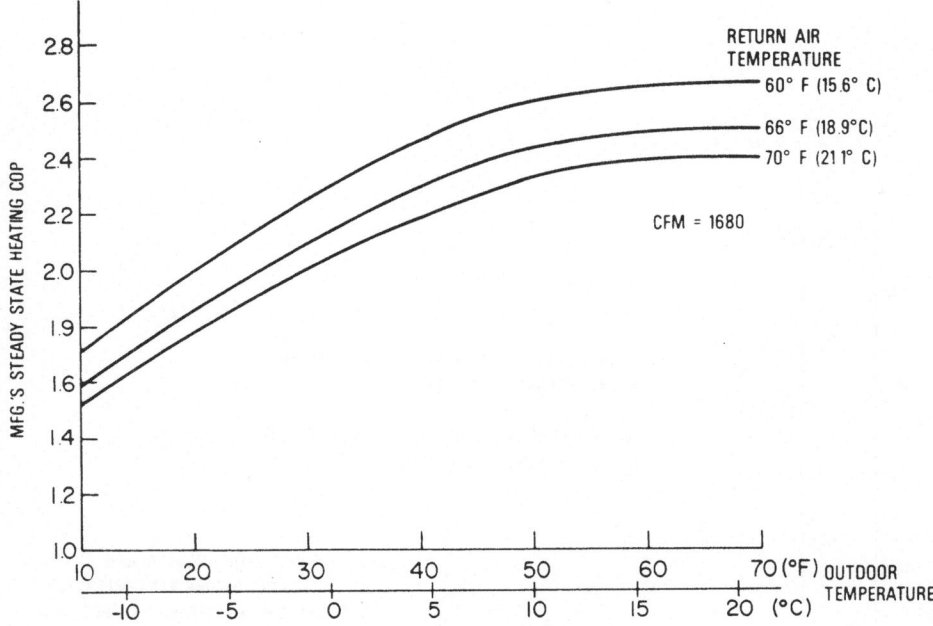

Fig. 5.   COP calculated from manufacturer's data
heating operation

by passing a least square fit straight line through experimental
data and an assumed point.   The experimental data corresponded
to 24 hour averages of the heat delivered to the Bowman house per
hour plotted against the average outdoor temperature existing in
each 24 hour period.   The data was corrected to an indoor temperature of 70°F (21°C), but was not normalized to any standard
wind or solar radiation conditions.   This lack of normalizing
wind and solar effects could be the reason why the heating
requirement of the Bowman house with and without storm windows,
appears to be the same for the data plotted in figure 6.   The
assumed point was a zero heating requirement at an outdoor
temperature of 67.5°F (19.7°C) and was arrived at by calculating
the outdoor temperature at which the heat generated by the
operating interior lights would just cancel the heat loss from the
house.   Although it was originally felt that the infiltration rate
would be at least one air change per hour, actual measurement
using a tracer gas[2] showed it to be around 0.5 air changes
per hour.   Even at the correct infiltration rate however, the
ASHRAE procedure for determining the heating load at the outdoor design temperature considerably over-estimated and the

274

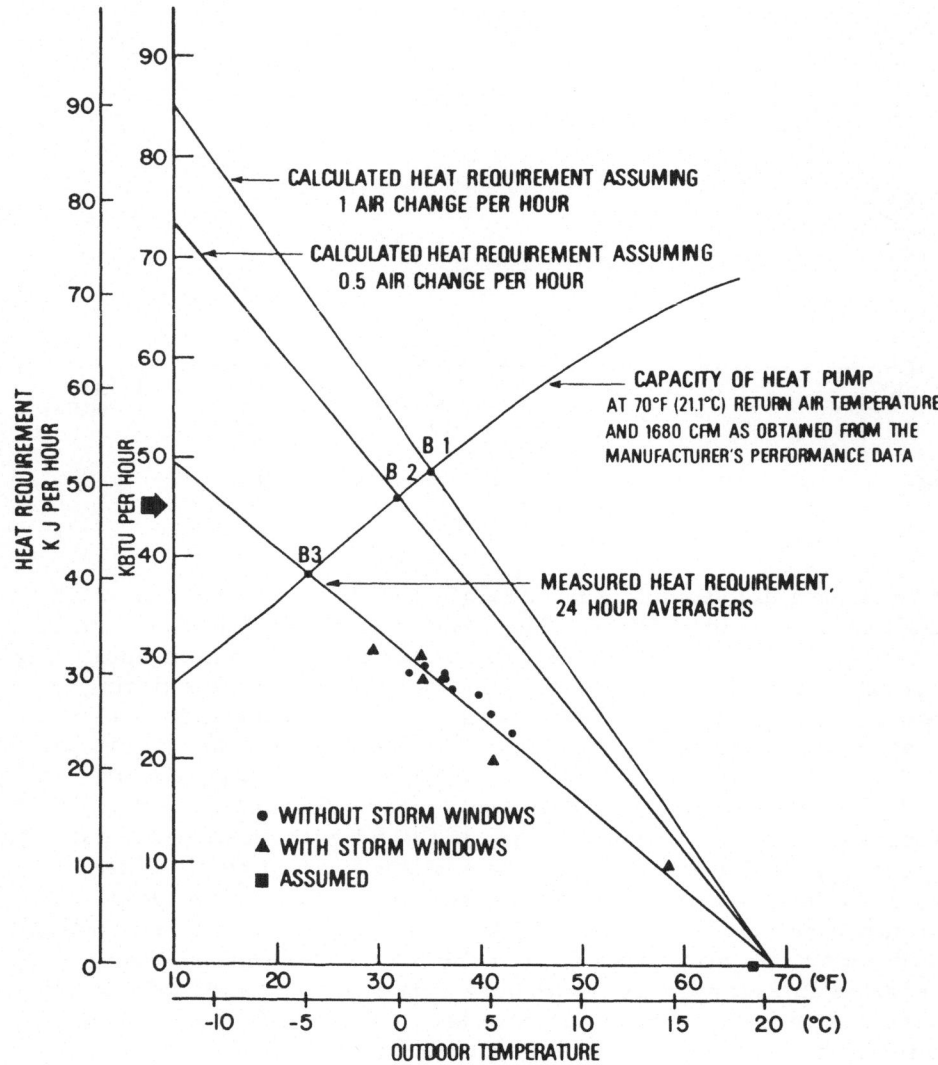

Fig. 6. Heat requirement of house vs outdoor temperature, indoor temperature = 70°F(21.1°C)

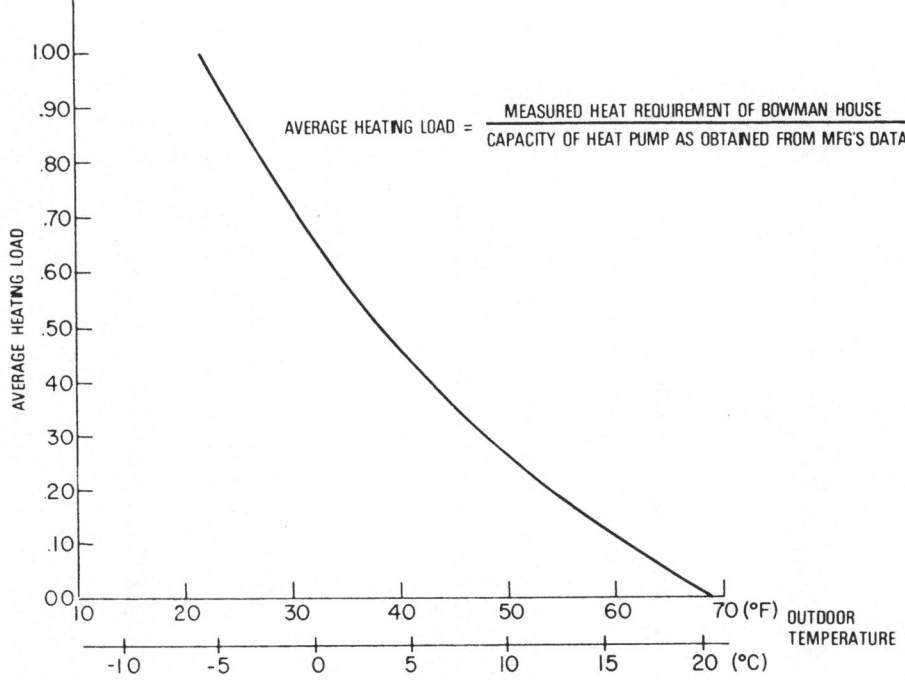

$$\text{AVERAGE HEATING LOAD} = \frac{\text{MEASURED HEAT REQUIREMENT OF BOWMAN HOUSE}}{\text{CAPACITY OF HEAT PUMP AS OBTAINED FROM MFG'S DATA}}$$

Fig. 7.   Average heating load vs outdoor temperature

heating requirement of the Bowman house. The points B_1, B_2 and B_3, which are referred to as "balance points", give the respective outdoor temperature at which the heat pump's output (without supplemental resistance heaters) would just equal the heating requirement if the heating load curve passing through each point represented the heating requirements of the Bowman house. Although it appears from the experimental data that B_3 should be the correct balance point, the actual balance point turned out to be in the neighbourhood of $30^\circ F$ ($-1.1^\circ C$) which is closer to B_2. This was due to the fact that the heating capacity of the heat pump in this temperature region was considerably reduced from the value given by the manufacturer due to ice build up on the outdoor coil and the need for repeated defrost.

The heating load, as defined by equation 7b, may be estimated from figure 6 by dividing the measured heating requirement at a given temperature by the manufacturer's steady state capacity at the same temperature. This has been done in figure 7, where the resulting curve should be considered as providing only an approximate relationship between heating load

276

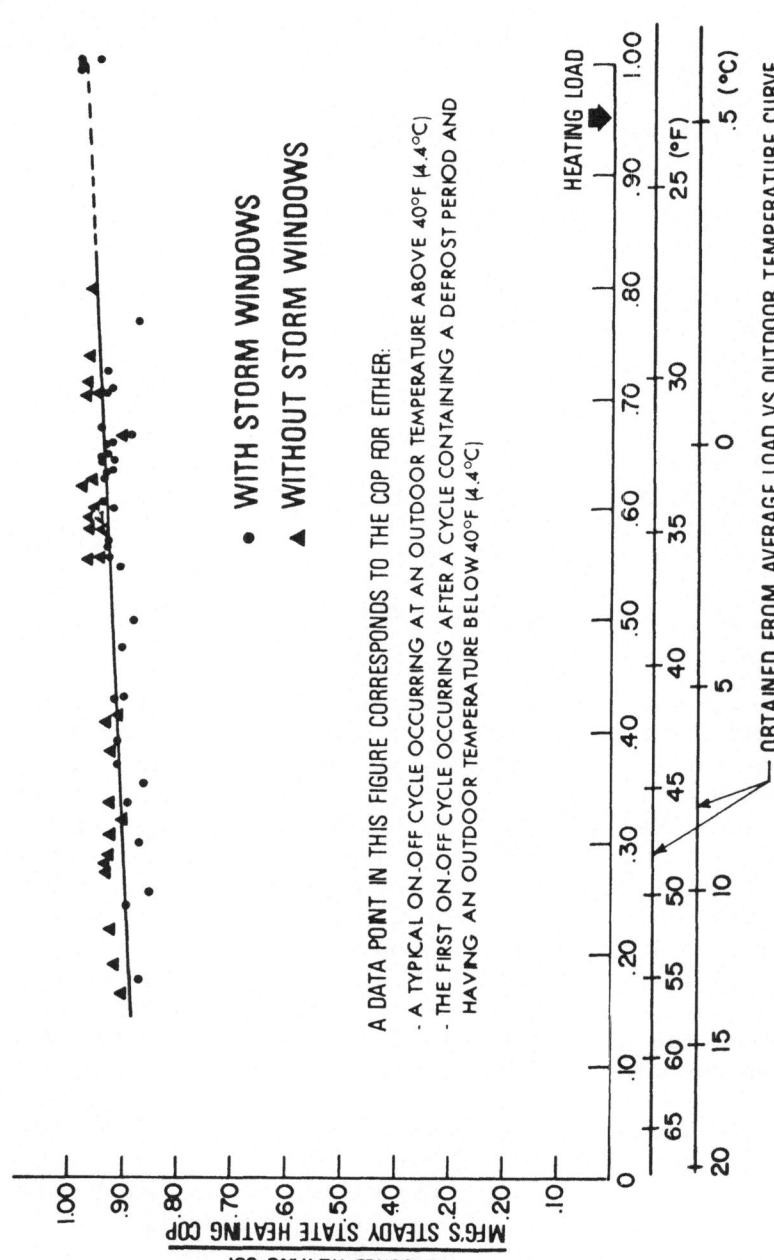

Fig. 8.   Effect of cycling on heat pump performance, heating operation

and outdoor temperature, since the heating requirement of a house is also a function of the amount of solar radiation received, wind speed and the thermal storage capacity of the structure. These variables will cause the house's daily average heating requirement to deviate slightly from the straight line passing through B_3 in figure 6 and could cause considerable deviation from this line for the average hourly heating requirement existing during periods of time less than 24 hours. In addition, it should be pointed out that since the definition of heating load involves the manufacturer's steady state capacity, which might be considered an ideal capacity, the heat pump's balance point was actually located somewhere between a load of 0.70 and 0.80. Thus the heat pump operated continuously at a heating load considerably less than 1.00.

Figure 8 shows the effect of cycling on the heating performance of the Bowman house heat pump. The data plotted in this figure are of two kinds. For outdoor temperatures above 40°F (4.4°C), cycles were selected which appeared to be representative of the heat pump's performance at various temperatures. The heating COP, heating load, and manufacturer's steady state COP (as calculated from the manufacturer's performance data) was determined for each cycle and plotted as shown. For outdoor temperatures below 40°F (4.4°C), the first cycle occuring after a cycle involving a defrost period was selected, analysed and plotted. Thus the data shown in figure 8, does not show the effect of ice build up and defrosting on the performance of the heat pump. Since ice build up does occur and defrosting is necessary, figure 8 cannot be used for predicting the performance of the Bowman house heat pump in the defrost region. It does however indicate that the linear approximation used in analysing the cooling data in figure 4 is probably a reasonable one.

Figure 9 shows the heating performance of the Bowman house heat pump as a function of heating load. For outdoor temperatures above 40°F (4.4°C) the data shown in figure 8 is replotted here. For temperatures below 40°F (4.4°C), the heating COP, heating load, and manufacturer's steady state COP were determined for periods of time which started at the end of an on-off cycle containing a defrosting process and terminated at the end of another on-off cycle containing a defrosting process. The minimum period of time analysed was 3 hours, while the maximum was almost 24 hours. These time periods always contained many on-off cycles and a number of periods contained multiple defrosting processes. Since considerable scatter existed in the data obtained for temperatures below 40°F (4.4°C),

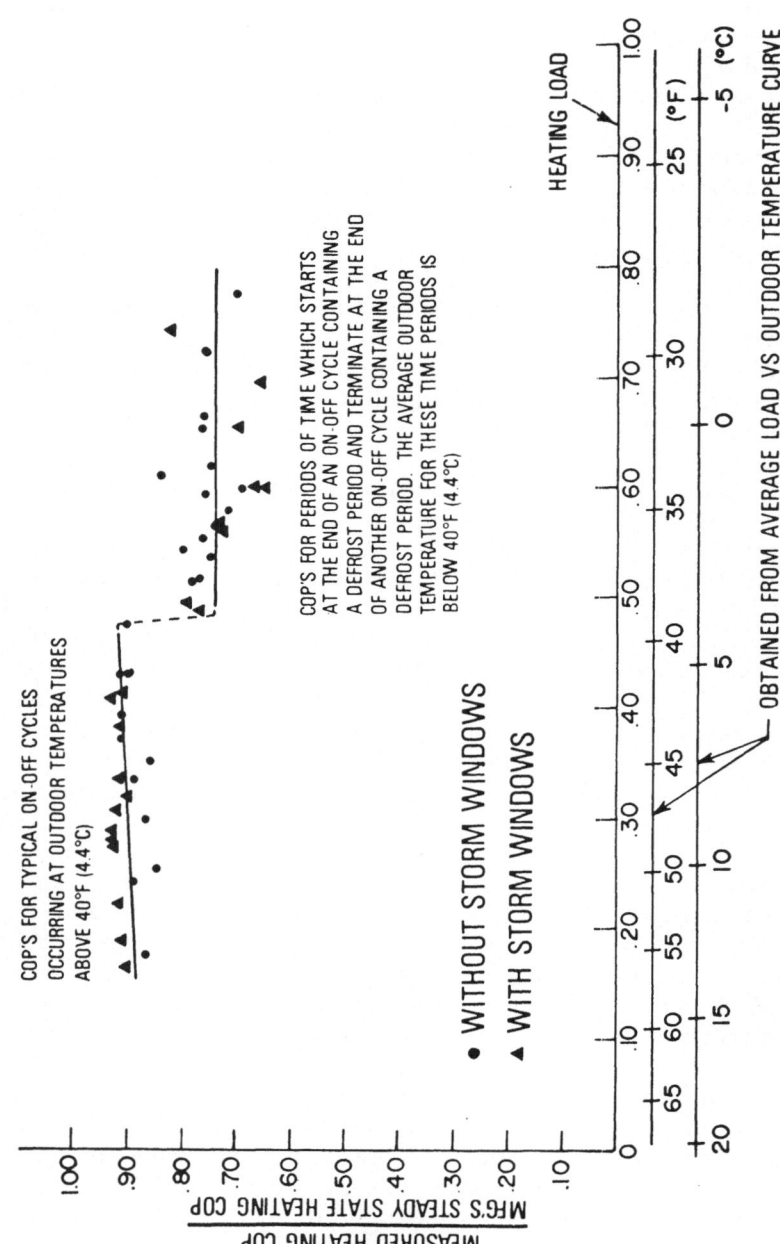

COP'S FOR TYPICAL ON-OFF CYCLES
OCCURRING AT OUTDOOR TEMPERATURES
ABOVE 40°F (4.4°C)

COP'S FOR PERIODS OF TIME WHICH STARTS
AT THE END OF AN ON-OFF CYCLE CONTAINING
A DEFROST PERIOD AND TERMINATE AT THE END
OF ANOTHER ON-OFF CYCLE CONTAINING A
DEFROST PERIOD. THE AVERAGE OUTDOOR
TEMPERATURE FOR THESE TIME PERIODS IS
BELOW 40°F (4.4°C)

HEATING LOAD

OBTAINED FROM AVERAGE LOAD VS OUTDOOR TEMPERATURE CURVE

● WITHOUT STORM WINDOWS
▲ WITH STORM WINDOWS

MEASURED HEATING COP / MFG'S STEADY STATE HEATING COP

Fig. 9.   Variation in heat pump performance with load heating operation the ordinate of all the data points were averaged and found to have a value of 0.74. A straight line parallel to the X axes was then passed through this data in figure 9 at this average height.

In order to determine if the solid lines drawn through the data in figure 9 were truly representative of the Bowman house heat pump's performance, all the heat pump data collected after the first and second stages of retrofit, were used to determine an average hourly heating COP of the post-retrofit heating test. The result was:

$$\text{(Average hourly heating COP)}_{\text{measured}} = \frac{1}{N} \sum_{i=1}^{N} (COP)_i = 1.71 \qquad (8)$$

where $(COP)_i$ = COP measured for hour i

and N = the total number of hours of data = 373

The average hourly COP was then predicted for the same period by determining the manufacturer's steady state COP for each hour, multiplying by a correction factor obtained from figure 9 and then averaging. It was found that:

$$\text{(Average hourly heating COP)}_{\text{predicted}}$$

$$= \frac{1}{N} \sum_{i=1}^{N} \left(\begin{array}{c} \text{Mfg's steady state} \\ \text{heating COP} \end{array} \right)_i \left(\begin{array}{c} \text{Correction} \\ \text{factor } (T_i) \end{array} \right)_i = 1.76 \qquad (9)$$

where the mfg's steady state heating COP is the heating COP calculated from the manufacturer performance data for hour i and the correction factor is the ordinate of the curve in figure 9 corresponding to average outdoor temperature, T_i, for hour i. The measured and predicted average hourly heating COPs were found to agree within 3%.

The seasonal heating performance of the Bowman house heat pump was estimated using the bin method and weather data for Andrews Air Force base which is near Washington, D.C. The equation used was:

$$\frac{\text{Seasonal}}{\text{heating COP}} = \frac{\text{Total energy delivered over the heating season}}{\text{Total energy input over the heating season}} =$$

$$= \frac{\displaystyle\sum_{j=1}^{M} 0_j^{HP} + \sum_{j=1}^{M} 0_j^{s}}{\displaystyle\sum_{j=1}^{M} I_j^{HP} + \sum_{j=1}^{M} I_j^{s}} = \frac{\displaystyle\sum_{j=1}^{M} 0_j^{HP} + \sum_{j=1}^{M} 0_j^{s}}{\displaystyle\sum_{j=1}^{M} \frac{0_j^{HP}}{(\overline{COP})_j} + \sum_{j=1}^{M} 0_j^{s}} \qquad (10)$$

where $\quad$ j $\quad$ indicates the jth temperature bin

$\qquad$ M $\quad$ is the total number of temperature bins used

$_0O_j^{HP}$ $\quad$ is the heat supplied in temperature bin j by the heat pump over the entire heating season

$_0O_j^{s}$ $\quad$ is the heat supplied in temperature bin j by the supplemental resistance heaters over the entire heating season

I_j^{HP} $\quad$ is the energy used by the heat pump in delivering heat $_0O_j^{HP}$ and is equal to $_0O_j^{HP} / (COP)_j$

$(COP)_j$ $\quad$ is the average seasonal heating COP of the heat pump for temperature bin j

I_j^{s} $\quad$ is the energy required by the supplemental resistance heaters to produce heat 0_j^{s} and is equal to 0_j^{s} , assuming jacket losses may be neglected

Two calculations were performed to determine the expected seasonal heating COP of the Bowman house heat pump. The first one approximated $(COP)_j$ by the steady state heating COP which was calculated from the manufacturer's performance data for temperature bin j. The second calculation took

$(COP)_j$ = (manufacturer's steady state heating COP for temperature bin j) (correction factor obtained from figure 9 for temperature bin j)

These calculations are shown in Table 1. Using the manufacturer's steady state data, the seasonal heating COP was predicted to be 2.16. Employing the results in figure 9 to correct for the effect of cycling, ice build up and defrost, the seasonal heating COP of the Bowman house heat pump turned out to be 1.74 or some 19% lower.

The procedure used in Table 1 to estimate the seasonal heating COP of the Bowman house heat pump yields the same result as the method recommended by the manufacturer of the heat pump for the case where $(COP)_j$ is calculated from the manufacturer's performance data. It is interesting to note, however, that this method indicates that the Bowman house heat pump supplied 98% of the total energy required over the heating

Table 1   Estimated seasonal COP of Bowman house heat pump

OUTDOOR TEMPERATURE (°F)	OUTDOOR TEMP. (°C)	HEAT REQUIRED PER HOUR (KBTU PER HOUR)	MFG'S CAPACITY BELOW BALANCE POINT, CFM=1680 RETURN AIR = 66°F (18.9°C)	SUPPLEMENTAL HEAT REQUIRED PER HOUR (KBTU PER HOUR)	SEASONAL HEATING ANDREWS AFB (HOURS)	TOTAL HEAT SUPPLIED BY HEAT PUMP* (KBTU)	TOTAL SUPPLEMENTAL HEAT SUPPLIED KBTU)	MFG'S COP RETURN AIR = 66°F (18.9°C) CFM = 1680	CORRECTION FACTOR (MEASURED COP/MFG.'S COP)	COP PREDICTED FROM HEAT PUMP DATA	INTERMEDIATE STEP	INTERMEDIATE STEP
	A	B	C	$D = B - C$	E	F	G	H	I	$J = H \times I$	$K = F \div H$	$L = F \div I$
67	19.4	1.27			848	1080		2.50	0.87	2.18	432	495
62	16.7	5.50			761	4190		2.50	0.87	2.18	1680	1920
57	13.9	9.73			701	6820		2.48	0.88	2.18	2750	3130
52	11.1	14.0			670	9380		2.44	0.89	2.17	3840	4320
47	8.3	18.2			649	11800		2.41	0.90	2.17	4900	5440
42	5.6	22.4			744	16700		2.33	0.91	2.12	7170	7880
37	2.8	26.7			720	19200		2.25	0.74	1.67	8530	11500
32	0.0	30.9			639	19700		2.16	0.74	1.60	9120	12300
27	-2.8	35.1			.353	12400		2.04	0.74	1.51	6080	8210
22	-5.6	39.4	38.6	0.8	205	7910	164	1.92	0.74	1.42	4120	5570
17	-8.3	43.6	34.3	9.3	111	3810	1030	1.83	0.74	1.35	2080	2820
12	-11.1	47.8	30.1	17.7	47	1410	832	1.65	0.74	1.22	855	1156
7	-13.9	52.0	26.9	25.1	11	296	276	1.54	0.74	1.14	192	260
2	-16.7	56.3	23.4	32.9	1	23	33	1.42	0.74	1.05	16	22
SUM OF COLUMNS SHOWN.						114,719	2,335				51,765	65,025

* $F = \begin{cases} B \times E \text{ above balance pt.} \\ C \times E \text{ below balance pt.} \end{cases}$

SEASONAL HEATING COP
Predicted using MFG's data** $= \dfrac{\Sigma F + \Sigma G}{\Sigma K + \Sigma G} = \dfrac{114,719 + 2,335}{51,765 + 2,335} = 2.16$

** Σ_x = Sum of Column X

SEASONAL HEATING COP
Predicted using results in figure 9 ** $= \dfrac{\Sigma F + \Sigma G}{\Sigma L + \Sigma G} = \dfrac{114,719 + 2,335}{65,025 + 2,335} = 1.74$

season and the supplemental heaters supplied only 2%. This is really an under-estimation of the supplemental heat required because the method assumes that the capacity of the heat pump in the defrost region is the same as that given in the steady state performance data. This assumption results in a balance point at B_3 as shown in figure 6. As pointed out earlier this is not the case since the capacity of the heat pump in the defrost region is affected adversely by ice build up and defrosting and this results in the actual balance point being closer to B_2. Thus more supplemental resistance heat is required than accounted for in Table 1 and the estimated seasonal heating COP should actually be slightly less than 1.74. The resulting change would, however, be small and for simplicity, this effect is not included in the calculations presented in Table 1.

Table 2   Effective heating COP for Bowman house heat pump

OUTDOOR TEMPERATURE	2°F	12°F	22°F	32°F	42°F	52°F	62°F
	−16.7°C	−11.1°C	−5.6°C	0°C	5.6°C	11.1°C	16.7°C
COP CALCULATED FROM MFG.'S DATA	1.42	1.65	1.92	2.16	2.33	2.44	2.50
COP PREDICTED USING RESULTS PRESENTED IN FIGURE 9	1.05	1.22	1.42	1.60	2.12	2.17	2.18
EFFECTIVE COP = (0.29) (COP Predicted USING RESULTS PRESENTED IN FIGURE 9)	0.30	0.35	0.41	0.46	0.61	0.63	0.63

Results presented are for a CFM of 1680 and a return air temperature of 66 °F (18.9 °C)

COMPARISON WITH PRESENT DAY FOSSIL FUEL HEATING EQUIPMENT

In order to compare the energy effectiveness of the Bowman house heat pump with fossil fuel heating equipment, it is necessary to trace the energy consumption of the heat pump back to the primary source energy used by the power plant. This may be done by defining an effective heating COP which is given by:

$$\text{Effective heating COP} = (0.29) (\text{heat pump's heating COP}) \qquad (11)$$

where 0.29 takes into account the fact that only about 29% of the heat energy of the fuel burned at the power plant in making electricity will reach the heat pump[6]. Table 2 lists the effective heating COP of the Bowman house heat pump at various outdoor temperatures, assuming that the heat pump's heating COP may be obtained by correcting the manufacturer's steady state COP using the results presented in figure 9. Equation 11 may also be used to determine an effective seasonal heating COP, where:

$$\text{Effective seasonal heating COP} = (0.29) (1.74) = 0.50 \qquad (12)$$

Thus for every unit of energy consumed at the power plant, only about 0.5 units of heat were delivered on the average to the interior living space of the Bowman house. Even if this value of the effective seasonal heating COP, were to be corrected to the manufacturer's recommended indoor air flow, the effect would only be to increase it by approximately 6% to a value of 0.53.

If an attempt is made to compare this estimated seasonal performance of the Bowman house heat pump with that of fossil fuel equipment, it is found that considerable controversy exists concerning the seasonal efficiency of residential gas and oil-fired heating systems. Numerous studies exist which report seasonal efficiencies from 35% to 65% for this type of equipment[6]. Many of these studies, however, involved the comparison of houses equipped with fossil fuel heating equipment with those using electric resistance heating and often considerable differences existed between the two types of houses in the amount of installed insulation, tightness with respect to air infiltration and occupant usage. Many other studies calculated the heating requirements of the house at various outdoor temperatures, neglect the heat contributed by lighting, appliances, occupants and solar, and used monthly fuel bills to determine seasonal efficiencies. These procedures either under-estimated or over-estimated the seasonal efficiency of residential fossil fuel heating equipment and have greatly contributed to the controversy.

Recently, another approach to determining the seasonal performance of fossil fuel heating equipment has been undertaken by Honeywell with their development of a computer model for such equipment. The model is capable of calculating the seasonal losses, including those due to the presence of H_2 in the fuel, the heat contained in the products of combustion and excess air going up the flue, infiltration resulting from the combustion process and draught control, and off-cycle draughts passing through the heat exchanger. Figure 10 summarizes the results obtained when this model was used to calculate the seasonal efficiency of a typical gas-fired furnace operating in Minnesota[7]. A seasonal efficiency of 57.1% was obtained with off-cycle losses constituting the largest single loss of 12.7%. Honeywell is presently in the process of verifying this model by comparing its output with results obtained from experiments conducted in the laboratory and field. Once this process is completed, such a model could be a powerful tool for determining the average seasonal efficiency of different types of fossil fuel heating equipment.

Recognizing the problems involved with all the studies which have been conducted on the seasonal efficiency of fossil fuel residential heating equipment, it seems safest to conclude that:

CALCULATED SEASONAL LOSSES

10.3 % — PRESENCE OF HYDROGEN IN THE FUEL

11.8 % — DRY FLUE GAS LOSS

1.1 % — LOSS DUE TO HEATING COMBUSTION AIR TO ROOM TEMPERATURE

4.0 % — LOSS DUE TO HEATING DRAFT DIVERTER TO ROOM TEMPERATURE

3.0 % — PILOT LIGHT LOSS

<u>12.7 %</u> — OFF-CYCLE LOSSES

42.9 % — TOTAL SEASONAL LOSS

CALCULATED SEASONAL EFFICIENCY = 57.1 %

Fig. 10.   Typical seasonal performance of a gas-fired furnace

(a)   there is likely to be a wide spread in seasonal efficiencies even among the same type of heating equipment, and

(b)   there probably exists a large percentage of gas and oil-fired residential furnaces and boilers which have seasonal efficiencies in the neighbourhood of 50%.

Assuming this to be true, the estimated seasonal performance of the Bowman house heat pump was equivalent to the performance of many residential gas- and oil-fired heating systems.

DISCUSSION

The effect of part load operation and performance of the Bowman house heat pump was found to be greatest when the heat pump operated in the cooling mode. At 30% cooling load, the ratio of the heat pump's cooling COP to the manufacturer's steady state cooling COP was reduced to approximately 82% of the measured steady state value (see figure 4); at 30% heating load the ratio of the heating COP to the manufacturer's steady state heating COP was roughly 94% of the measured steady state value (see figure 9). A large part of the difference is cooling and heating performance at part load was due to the fact that the heat pump was located in the basement and, in the heating mode, the heat contained in the indoor unit would tend to be transferred to the interior living space by convection during the off cycles. In the cooling mode, however, the coolness was trapped in the indoor unit during off cycles and most of it was eventually transferred to the basement. The situation might have been reversed, with the heating COP being more affected by part load operation, if the indoor unit had been located in the attic. A possible modification which should improve both the cooling and heating COP in basement and attic installations, respectively, would be to have a time delay on the indoor blower which would allow it to operate for several minutes after the compressor shuts off.

In the heating mode of operation, the largest adverse effect on the performance of the Bowman house heat pump was due to ice build up on the outdoor coil and the resulting requirement for defrosting this coil. For outdoor temperatures below 40°F (4.4°C), the average ratio of heating COP to manufacturer's steady state heating COP in figure 9 was 77% of the measured steady state value in figure 8, which does not contain the effect of ice build up or defrosting.

The seasonal heating COP of the Bowman house heat pump was estimated to be 1.74 using the results presented in figure 9 and weather data which was representative of the Washington D. C. area. This turned out to be approximately 19% lower than the seasonal heating COP calculated using the manufacturer's performance data. This difference in seasonal performance was due to the effect of part load operation, ice build up and defrosting.

When the estimated seasonal heating performance of the Bowman house heat pump was traced back to the power plant, it was found that only about a half a unit of heat energy was actually delivered to the interior living space for every unit of energy consumed at the power plant. This is comparable to the probable performance of many gas and oil-fired residential heating units. Assuming that this estimated seasonal performance of the Bowman house heat pump is representative of residential heat pumps in areas having heating seasons similar to Washington D. C., there does not appear to be any clear-cut advantage to choosing either a heat pump or a fossil fuel heating system in these areas for the purpose of saving primary source energy. For areas which have a considerably colder heating season than Washington D. C., it is likely that gas or oil-fired residential heating systems would be more energy effective than heat pumps, while heat pumps would probably be better than gas- or oil-fired units in areas having winters which were milder than Washington D. C.

ACKNOWLEDGEMENTS

The author would like to express his appreciation to Mr. John Bean for the tremendous job which he did in reducing the raw data. The help of Mr. Douglas Burch in carrying out this investigation is also gratefully acknowledged.

References

1. Newman, Dorothy, K., and Day, Dawn, The American Energy Consumer, Ballinger Publishing Co., Cambridge, 1975.
2. Burch, D. et. al., Experimental Study of Retrofitting Existing Residences, to be published.
3. ASHRAE Handbook of Fundamentals, 1973, ASHRAE, N. Y. 1972, Page 440.
4. Method of Testing for Rating Unitary Air Conditioning and Heat Pump Equipment, ASHRAE Standard 37-69, American Society of Heating, Refrigeration and Air Conditioning Engineers, New York, 1969.
5. ASHRAE Handbook of Fundamentals, 1973, American Society of Heating, Refrigeration and Air Conditioning Engineers, New York, 1972, Chapter 21.
6. Hise, E. C., Seasonal Fuel Utilization Efficiencies of Residential Heating Systems, Oak Ridge National Laboratory Report, ORNL-NSF-EP-82.
7. Bonne, U., and Johnson, A. E., Thermal Efficiency in Non-Modulating Combustion Systems, Proceedings and Addendum to the Proceedings of the Conference on Improved Efficiency in HVAC Equipment and Components for Residential and Small Commercial Buildings, Purdue University, 7th-8th October, 1974.

TEMPERATURE LEVELS OF HEAT GENERATION IN
CONNECTION WITH HEAT TRANSPORTATION. TECHNICAL
AND ECONOMIC ASPECTS OF LONG DISTANCE HEAT
CONVEYANCE AND DISTRIBUTION

Allan Haag,

Industriverken Malmö, Sweden.

1. SCOPE OF DISTRICT HEATING TECHNOLOGY, fig. 1.

1.1 Generation

Irrespective of where a symposium on district heating is
held, almost all papers presented deal only with production
matters. It seems to be thought that the difficult questions,
those dealing with distribution and transforming of heat from
one temperature or pressure level to another, do not need to be
covered. Such questions are, however, of the absolute greatest
importance, and I shall here first present what is really involved
in the conception of district heating. The picture summarizes all
the main points which must be considered if a district heating
system is properly to perform its task.

As far as the actual production plants are concerned it is
found that the hot water temperature levels are dependent upon
the design of the plants. In a straight hot water boiler plant
there is, for example, freedom to choose a very high outgoing
temperature, and the upper temperature limit is really only
governed by material strength limits. It can be shown mathema-
tically that it is most economical to operate at the highest possible temperature throughout the year as this will result in the
lowest pipe sizes and pumping powers. Further, a hot water
boiler needs to operate at a high temperature in order to avoid
corrosion.

If the heat is produced in a back pressure turbine it must be borne in mind that the lower the temperature in the heating condenser the greater the electricity production. It is thus very relevant to question the suitability of the straight back pressure turbine as the best machine for the combined production of heat and electricity. As I shall show further on in this paper an extraction turbine with, for example, three extraction stages is much better than a straight back pressure turbine with only one exhaust. As the personnel requirements are approximately the same for both large and small back pressure plants the important conclusion can be drawn that there is a natural lower limit below which no adequately economic plant can be justified. The extraction turbine eliminates these disadvantages to some extent and should therefore, in my opinion, be the only type of turbine which should be considered in the future for combined production. In fact, such machines should not merely be considered, but must be used for this purpose, for if care is also given to dimensioning the pipe distribution system then no better combination can be found which gives greater community benefits.

These points have been exploited in the survey on transmitting the heat from the nuclear power station at Barsebäck, where - if the system is allowed to be realized - the turbine will be a three-stage extraction machine having an outgoing hot water temperature of 165°C. I shall return to this large system later.

If it is desired to utilize the heat from a gas turbine an exhaust boiler or exhaust heat exchanger must be used. All outgoing hot water temperatures considered, both past, present and future, can be obtained with a gas turbine plant, making it a functionally ideal arrangement in that any required rate of electricity production between minimum and maximum can be obtained irrespective of whether heat requirements are at a maximum or not. This is not possible with steam turbine. A gas turbine is, however, shackled by the high prices of fuels which it can use.

Something should also really be said here on diesels, but due to their high maintenance costs and very restricted outputs they are, to my mind, almost without interest in district heating applications. See figs. 2-6.

1.2 Regional distribution.

Only exceptionally can it be said that regional distribution is involved in today's district heating systems. If the various

problems, both political and technical, now besetting nuclear
power can be overcome and further nuclear power stations can
be built in Sweden, then it will be relevant to talk about regional
distribution, for such power stations must be sited quite a long
way from larger communities. Regional heat distribution is not
really anything new at all, for in, for instance, Rumania and
Bulgaria such distribution has existed to some degree for about
15 years. This has, however, up to now been a question of heat
from fossil-fueled power stations. Barsebäck is a nuclear power
station, and quite a big one at that, and so the combination of
electricity and heat from it may be the more interesting.

Due to the expense it will be necessary to run large mains
for regional heat distribution above ground as far as possible,
just as has been done in the eastern European countries. (See
figs 7-12.) Many problems, however, still remain to be properly
solved: for example, the method of supporting the pipes, the
distance between isolating valves, the type of insulation to be
used etc.

It can definitely be said that future efforts must be directed
in a considerable degree to solving regional distribution problems,
while few problems remain to be solved in connection with production plants.

1.3 Local distribution, fig. 13.

Our town of Malmö has about 260,000 inhabitants and the
local distribution system is shown in the figure. It is so dimensioned that, with the necessary extensions, a nominal connected
heat load of 2,000 MW_{th} in 1990 can be met. Large pipes are
run in concrete culverts, while smaller ones are of the prefabricated Swiss PAN-ISOVIT type. Considerable care must be
given to dimensioning, for installation costs are high in Sweden.

1.4 Subscriber stations, figs. 14-16.

The temperatures and pressures used in the local distribution systems are usually too high for direct connection to consumer equipment: e.g. radiators. Heat exchangers are therefore
used and one or more are installed in each building. The design
is different depending on whether it is a house or a block of flats
which is concerned. Fig. 14 shows the design for a private house
and figs. 15-16 for a block of flats. Normal tap water draw-off
requirements can be met by connecting the tap water heat

exchanger to the outlet from the radiator circuit heat exchanger but during high draw-off rate periods primary hot water must be led directly to the tap-water heat exchanger as shown in the figure.

1.5 Additional uses

If new houses are built for heating by means of air at about 30°C the return circuit primary water at a temperature of about 60-70°C can be used, thus dropping its temperature further. A requirement, of course, is that the new building development is so sited that such a connection can be arranged. Swimming pools and pavements can also be heated by the return water, but the shortage of fuels may make such applications rare in future.

I have tried to let my introduction cover a wide range in order to show that all the parts must be mastered in order to be able to guarantee that the whole will be economically advantageous and technically efficient.

2.   THE SWEDISH DISTRICT HEATING SITUATION, figs. 17-18.

Since the introduction of district heating to Sweden at the end of the 1940's the total connected load has risen as shown in fig. 17. Two forecasts have been made, one a minimum forecast and one a maximum. Which of these two which will be attained depends on whether legislation on energy use is introduced in Sweden or not. May we plan and operate power and energy matters in the best possible way, for then the greatest benefit to the community would be achieved.

During the last two decades, we have primarily built multi-family houses, but the situation changed in 1971 as shown in fig. 18. More dwelling units are nowadays built in the form of single-family houses than as multi-family houses, and this has partly led to new requirements for distribution system design. A distribution system for single-family houses is namely much more expensive per connected kWh_{th} than a distribution system for multi-family houses.

3.   HEAT DEMAND IN MALMÖ. Development and
      forecast, fig. 19.

Ever since the start of district heating in Malmö in 1951 the demand for new connections has been considerable.

The picture shows the development and the forecast for the future. The steep rise of the curve prompts the question, "Could this development have been foreseen at the beginning of the 1950's? ". The answer is No! The situation has been similar in other Swedish towns and communities. As we in Sweden have designed all our systems for NT16 pressure rating and 120°C maximum outgoing temperature, it can be seen that the trunk mains rapidly became too small. My major point of criticsm against the misleading philosophy which is now being spread that we should install systems for lower temperatures than 120°C is that it will be very expensive in the long run, for we have seen what has happened previously in Sweden. When the demand has increased we have been forced to build new trunk mains from the production plants simply in order to be able to pump out the hot water. If we instead had built mains designed for NT25 and 170°C outgoing temperature we would have been able to send out twice the heat power that we can today and for almost the same investment in the distribution system!

I would like to express my hope that the result of this conference will not be a statement that you recommend lower temperatures for the transfer of hot water over medium and long distances, for such a conclusion is basically wrong!

4.   STRATEGIC PLANNING OF LOCATION OF PRODUCTION
      PLANTS,   fig. 20.

Many district heating engineers seem to think that all production plants should be grouped in a single station in order to obtain the maximum benefits. If it cannot be shown at the planning stage that the heat sink can justify a combined heat and electricity power station in the long run then this idea is quite wrong.

It is probable that the idea is wrong in any case, even for a combined heat and electricity power station, if the distribution system is built without ring mains, for a mains failure can then lead to terrible difficulties. In Malmö we have two combined power stations, which run in parallel on one and the same hot water distribution system. We have also a number of fixed hot-water-only boiler plants for peak lopping and stand-by requirements. There is also a refuse incineration plant connected to the system.

A serious mains failure in 1969 showed that our strategy was correct, for when one of the trunk mains from one of the power stations burst due to corrosion we could shut down the power station and only that part of the system within which the failure had occurred was effected. Full heat demand on the rest of the system was met with the help of the hot water boiler plants. It is easy to realize what would have happened if the whole system had been built up on the radial principle.

5.   THE DILEMMA OF DISTRICT HEATING,   figs. 21-22.

Combined production of electricity and heat in a district heating power station is at the mercy of the seasonal temperature changes. If, namely, a straight back pressure turbine is installed and the winters are mild, it is not then possible to produce as much electricity as had been planned. Figs. 21-22 show this dilemma quite clearly in that the heat load factor does not amount to more than about 62% of the total nominal connected load.

The conclusion can thus be drawn again that combined production should take place in extraction turbines with, for example, three hot water condensers with by-pass connections and a cold condenser for maximum electrical power. From this follows naturally the conclusion that such a power station must be sited in suitable proximity to a large water recipient.

It should be clear by now that the planned transmission of 950 MW_{th} of hot water from the nuclear power station at Barsebäck to the towns of Malmö and Lund will meet all these technical requirements, for - if permission is granted to carry out the project - there will be three extractions and a cold condenser on the turbine. Transmission will also be based on varying outgoing hot water temperatures up to a maximum of 165°C. I will describe other finesses further on.

The introduction of heat pumps on a large scale in areas where district heating already is established is thus not economically justifiable. If the primary aim is to use all the heat energy proceeded in a thermal power station, then the potential heat load should not be reduced by installing heat pumps taking their heat from the outer air with load peaks from the electrical system. This way must surely be quite wrong.

It seems to me as if the use of heat pumps in large towns and for straight heating applications for houses is a technical aim in the wrong direction!

The only justifiable application seems to me to be the use of the cooling water discharged from a condensing power station as the heat source. It is likely, however, that the use of heat pumps is not justifiable even in such contexts other than in exceptional cases.

6. DURATION CURVES, figs. 23-26.

The figures show various load duration curves. It can be seen that the heat load curve is quite "peaky", from which the conclusion can be drawn that it is quite uneconomic to build a back pressure plant capable of meeting the entire heat load. If a straight back pressure plant is to attain a reasonably long utilization time then it is only economically justifiable to employ a machine having an electrical rating of about 25-30% of the total nominal connected heat load. On the other hand, the operating costs for small back pressure plants are high in relation to their capacity. This again indicates an extraction turbine plant. Fig. 23 shows the principles of extraction at various temperatures, and fig. 26 shows the practical arrangement with two hot water condensers.

7. LOAD VARIATION OVER 24 HOURS AND THE NEED FOR ACCUMULATION, figs. 27-29.

In Sweden, as in all other countries, the heating load is considerably less at night than during the daytime. Fig. 27 shows how the load in Malmö varied during the coldest day in 1973.

Reliable performance statistics from many years' operation give us a good basis for maintaining that accumulation should be employed in all combined systems in order to save both fuel and money. Furthermore, the systems can naturally be run with somewhat smaller production machinery, but this is only another way of saying that money can be saved. It is my definite opinion that the future will show that many accumulator installations are economically justifiable, for it seems to me to be impossible, with the present pattern of social development, to try and smooth out the load curves. People want, to a greater degree, to dispose their daily routines such that they sleep at night and work during the day: a tendency which makes shift working less and less attractive. Fig. 28 shows the principle for accumulation and discharge, and fig. 29 shows the practical arrangement as we have it in Malmö.

8. TEMPERATURE LEVELS AND CORRESPONDING CAPACITIES IN MAINS, figs. 30-33.

As I have previously mentioned, the maximum outgoing mains temperature in Sweden is 120°C. The mains have been designed for this temperature with respect to strength and expansion. The corresponding minimum outdoor temperature is -16°C for Malmö and -20°C for towns somewhat further north. In the northern part of the country the systems are designed on the basis of still lower outdoor temperatures.

Most house radiator systems are designed for 80-60°C and this results in a return temperature of the primary water of 70-80°C in winter. With outdoor temperatures between 0 and +20°C the primary water return temperature can be held to about 60°C which, with today's subscriber stations, is sufficient to meet the temperature requirements of tap water supply. This is illustrated in fig. 30.

The supply temperature of 120°C is really only used in Scandinavia, and depends on a number of factors. In other countries, West Germany, Rumania, Bulgaria, USSR etc., the temperatures are higher. Fig. 31 shows the outgoing temperatures in a West German system. The picture is taken from the magazine Fernwärme-International and shows that a supply mains temperature of 160°C is used. I think that it would be sensible if future new district heating installations were based on a supply temperature of 170°C, as this would enable the investment in the distribution system to be reduced.

Various sources often mention optimum water flow velocities. One can really wonder what is meant by this. Naturally, for any given pipe, an optimum once! The aim with dimensioning a pipe system must thus be to find those sizes which give the lowest cost during the whole write-down time. All future costs should therefore be discounted, based on future expansion, to the point in time when the decision to build a district heating system is taken. Only those dimensions worked out in this manner, and no other, are optimum! It can easily be seen that this work requires not only a computer but also considerable patience.

Fig. 32 shows the relationship between pipe size, transferred heat power, temperature differences and the constant water velocity. (Based on present costs and the way of reckoning as used up till now the optimum water velocity is usually about 3 metres per second, at 400 mm pipe dimension.)

Fig. 33 shows the relationship between pipe diameter, transferred heat power and temperature differences with a constant specific pressure drop per unit length of pipe. The velocity has been set at 3 metres per second with a pipe diameter of 400 millimetres. Figs. 32-33 show that in this latter case the transferred heat power is about 50-75% greater for a given pipe diameter than in the former case. The conditions applying in fig. 33 are more like the recommended means of calculation for new systems than those previously used.

9.   DESIGN CULVERTS, figs. 34-39.

As a result of a considerable amount of investigations, calculations and operating experience (including culvert failures) we have arrived at the design of culvert for large dimension pipes shown in fig. 34. As can be seen from the picture, the pipes are hung in flexible pendulum supports partly to deal with expansion and partly to prevent any ground water which leaks into the culverts from saturating the mineral wool insulation. With the old types of pipe supports, where the pipes were carried on steel cradles rising from the bottom of the culvert, ground water could easily rise up the supports and cause damage, such as corrosion etc.

All service connections should be made at the top of a trunk main, as otherwise dirt can easily collect in the bends and cause blockages.

Fig. 35 shows an 800 millimetre main built in 1972-73.

For smaller dimensions we use a pre-fabricated type of main, which I showed in my introduction and which is also shown in figs. 36-37. The design gives an absolutely tight and corrosion-free main, but unfortunately it is not possible with present day chemical techniques to produce a polyurethane foam for use at higher temperatures than about 120-130°C. We have reason to believe, however, that foam qualities will be available within the near future capable of withstanding higher temperatures. Pre-fabrication ensures that the installation costs are reasonable.

Careless design can lead to impressive but catastrophic failures. Fig. 38 shows what happens if pipes are not supported from above, but rather carried from underneath, in a culvert. Fig. 39 shows a bellows failure: stress corrosion due to poor design. It cannot be too strongly stressed that incompetent designers and contractors can result in considerable - very considerable - costs for the system owner.

10. COSTS FOR CULVERTS, figs. 40-41.

In connection with the Barsebäck investigation, we have worked out the costs for quite large culverts together with our Danish friends. Fig. 40 shows how these costs devolved for pipe installation in culverts, in ditches (on pedestals) and above ground (on pedestals). It can be seen that there is a certain difference between the Danish and the Swedish calculations.

Fig. 41 shows the cost structure for smaller culverts of prefabricated design. It can be seen that the peripheral costs, rather than that of the actual pipe itself, are those which dominate. There is thus little scope for reducing the total price by haggling over the price of the pipes, unless the peripherals can also be reduced at the same time.

11. HELENEHOLM POWER STATION, figs. 42-44.

The Heleneholm power station in Malmö is described in the brochure which I have distributed here to this conference. Figs. 42-44 show the power station in isolation and how it fits into its surroundings. The power station is particularly impressive at night with its floodlighting. It is built right in the middle of a housing area, and no-one reacts to it any longer. We have two STAL-LAVAL turbines, which can be connected in series, having an electrical output of 140 MW_e and a heat output of 285 MW_{th}. The steam data is 120 kp/cm^2 and 530°C. We have at present three boilers and two direct steam/water heat exchangers, the latter providing standby capacity in the event of turbine loss. The station has been built in two stages, and a third stage with a further boiler is to be built. We are proud of our power station and think that it is an architectural success. During the halcyon days of low oil prices we produced condensing power, the principle of which is shown in fig. 45, and the practical arrangement in figure 46. In these days of high oil prices it does not pay to produce condensing power, as electricity prices have not risen in step with oil prices due to the effect of electrical energy supplied from Swedish nuclear power stations.

12. SPECIFIC PLANT COSTS, fig. 47.

The cost development during the last few years has been such that small back pressure plants can hardly no longer be built if they are to be profitable. For smaller towns it would therefore seem as if only straight hot water boiler plants can be economically justified.

As said previously, the annual personnel costs for a power station are approximately the same whether the plant has an output of 100 MW_e or 25 MW_e. Further, the overhaul costs are much the same for both sizes, so in my view it should be possible to say that no power station less than about 100 MW_e with associated hot water output should be built in the light of today's costs and electricity prices unless the operational costs can be reduced. In this size range it should probably also be profitable to install extraction turbines and cold water condensers. A 100 MW_e installation needs a nominal connected heat load of about 400-500 MW_{th} for a utilization time of 3,000-4,000 hours per year at full capacity, and less than this should not be planned for. Fig. 47 shows clearly how the specific plant costs vary as a function of plant size.

13. DISTRICT HEATING FROM THE BARSEBÄCK NUCLEAR POWER PLANT, figs. 48-50.

The potential district heating load available in southern Sweden, i. e. in Malmö, Lund and Helsingborg, has grown to a considerable value during the last 25 years. The nominal connected heat load in Malmö is now more than 1,000 MW_{th}, with an expansion up to 2,000 MW_{th} by 1990 being expected, as mentioned previously. Althouth Lund has only 60,000 inhabitants, against Greater Malmö's 260,000 inhabitants, there is still the same essential need for district heating in the former town in the interests of good environmental conditions.

The distance between Barsebäck and Malmo is 26 kilometres and between Barsebäck and Lund 17 kilometres. The heat production capacity planned for Malmö is 700 MW_{th} and for Lund 250 MW_{th}. The hot water will be transported in a 1,200 millimetre diameter pipe from the power station to the junction where the Lund and Malmö branches separate. The pipe diameters beyond this point will be 600 millimetres to Lund and 1,000 millimetres to Malmö.

The complete system can be in operation by 1983/1984 at the earliest, which is when the third unit at Barsebäck nuclear power station is planned to come into full operation.

The distance in the other direction between Barsebäck and Helsingborg, with 85,000 inhabitants, is 40 kilometres. A fourth town, Landskrona, lies along this route with 30,000 inhabitants. The planned production of heat for Helsingborg in 1990 is 300 MW_{th} and for Landskrona 100 MW_{th}. If all four load towns are to be fed from the nuclear plant and require full load simultaneously, it will be necessary for the fourth unit at Barsebäck to be in operation. Figs. 48-50 show the complete system: notice how the service road can be used as a bicycle track.

14. THE HEATING AND ELECTRICAL SYSTEM, figs. 51-56.

It is hoped that unit number three in the Barsebäck nuclear plant will be a combined unit for the simultaneous production of heat and electric power. It is preliminarily planned to be put into operation in 1983 at the earliest.

If heat is not supplied from the Barsebäck plant, the economic evaluation has been based on the assumption that all future extensions of the heat production plants in the four towns will be in the form of hot water boiler plant. This assumption has been used in the study and the question of building back pressure plants, with combined electricity and heat production, in Helsingborg, Lund and Malmö will be studied further on.

Fig. 51 (Bl) shows the schematic arrangement of the turbine with three double extractions to three hot water condensers and a cold condenser.

Fig. 52 shows the arrangement in more detail, and fig. 53 shows the relationship between heat and electricity production.

Fig. 54 shows for how long a period each year the various extraction stages (hot water condensers) are connected. The figure also shows the variations in supply and return temperatures during the year.

Fig. 55 shows the mass flows to Lund and Malmö, and fig. 56 shows the electrical output throughout the year.

15. HEAT DEMAND, fig. 57.

Fig. 57, showing the expected rate of growth of the four towns, is used as a basis for planning the transmission system. When fully built out, a fourth unit will be required at Barsebäck.

The Malmö load is the dominant one, and exceeds the total consumption in the three other towns together.

16. PUMP STATIONS, figs. 58-60.

With the long distance involved here there will naturally be need of further pumping stations besides that situated at the power station. We reckon on needing to install one pump station in Malmö and one in Helsingborg. A further pump station is also required between Landskrona and Helsingborg due to the height differences.

Heat exchanger stations will be built at the supply points to all four towns for transforming the 165°C water temperature to 120°C. These heat exchangers will also separate the two pressure classifications, NT25 and NT16, on each side of them.

Fig. 58 shows the principle arrantement of the pump stations and heat exchanger stations. Fig. 59 shows the pressure gradients between Barsebäck and Lund and Malmö, while fig. 60 shows the pressure gradients between Barsebäck and Landskrona and Helsingborg. A total height difference of about 90 metres will be encountered in the area of Glumslöv and will require a pumping station, as previously mentioned, which is planned to be built at Örja.

Careful studies must be made of what may happen if any of the pumps stop due, for example, to power failure. Similarly, the availability and reliability of the valves must be studied.

17. PROFIT CRITERIA, figs. 61-62.

The total investment costs for the whole project are about 440 and 720-810 million Swedish crowns respectively. Profitability has been judged against the alternative of production in oil-fried plants.

Fig. 61 shows that the Barsebäck-Lund-Malmö project is profitable if oil prices are greater than 23 kronor/Gcal and 2 27 kronor/Gcal for above-ground and underground trunk mains respectively.

Fig. 62 shows how other parameters affect the profitability, and it can be seen that the two most significant are culvert pipe-laying along the whole route and steel price variations. A 10% increase in the nuclear power station plant price, however, is relatively insignificant.

CONCLUSION

If permission is granted by Parliament and Government to
build the third stage of Barsebäck, then not only will about
500,000 tons of oil per year be saved but also the provision of
district heating to Malmö and Lund will be solved for a considerable time into the future.

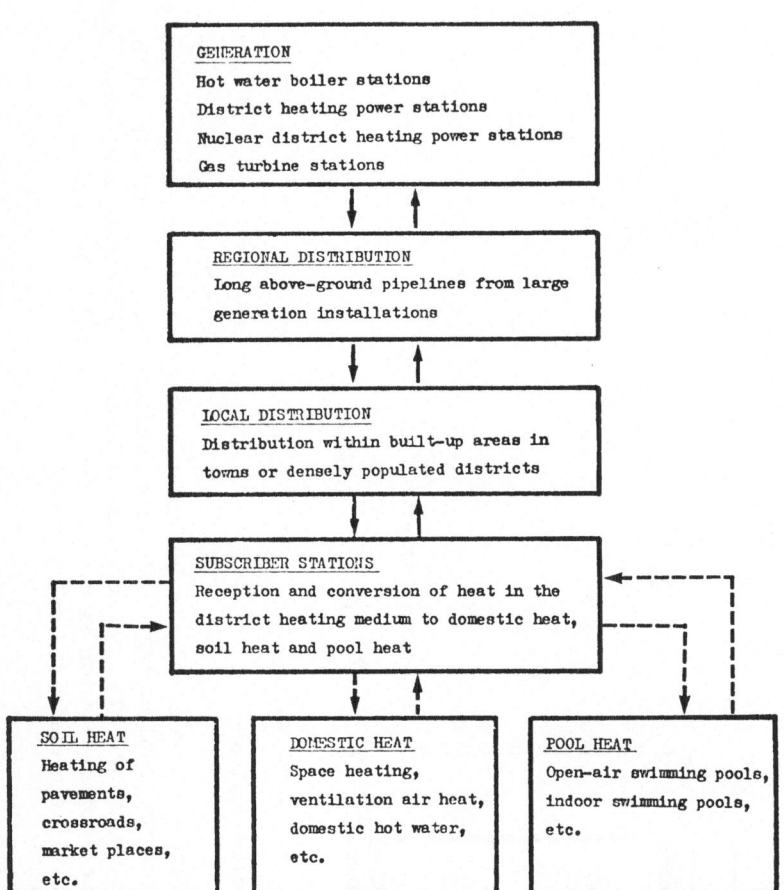

Fig. 1.   Scope of district heating technology.

302

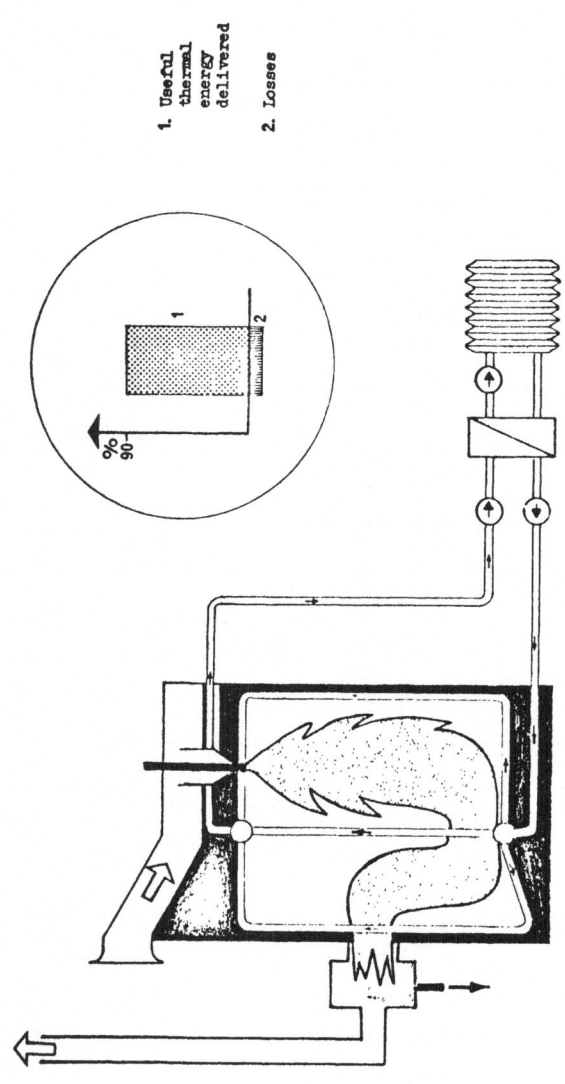

1. Useful thermal energy delivered

2. Losses

Fig. 2. Hot water boiler station.

In a hot water boiler station between 80 and 90% of the electric power and fuel energy supplied are utilized for producing the thermal energy. District heating water is circulated through the boiler and is transported to the subscribers, where heat is extracted by means of heat exchangers and the district heating water boiler station is thus merely to supply thermal energy and this type of station is therefore known as a district heating station.

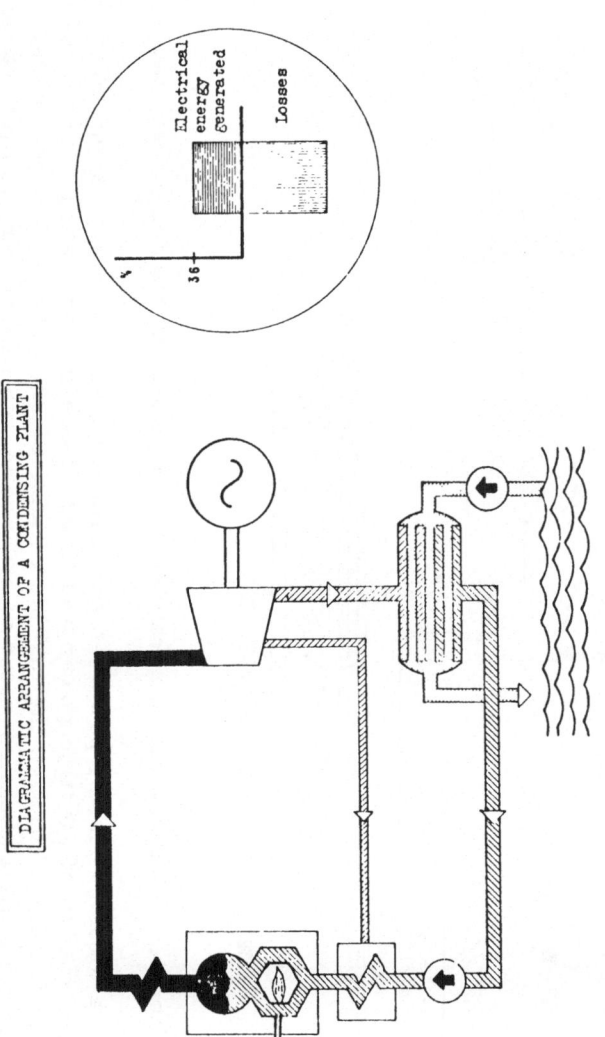

Fig. 3.   Diagrammatic arrangement of a condensing plant.

In a condensing power plant, only about one-third of the energy supplied by the fuel is converted to electrical energy. Most of the remaining two-thirds of the energy supplied is dissipated to the cooling water in the turbine condenser.

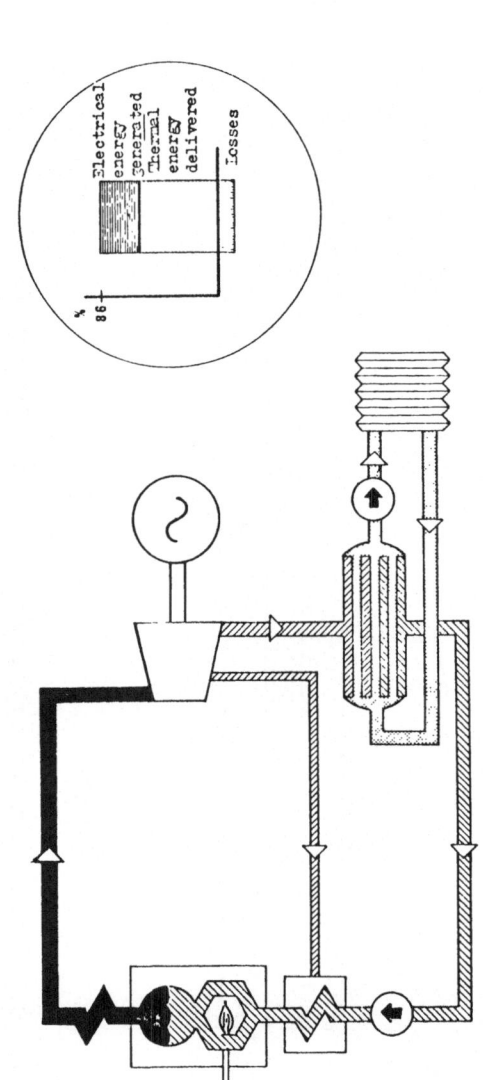

Fig. 4.   Diagrammatic arrangement of a district heating power station.

In a district heating power station, about 86% of the energy supplied in the fuel is utilized for the generation of electric power and thermal energy for heating. In this case, district heating water is circulated through the turbine condenser and acts as the cooling water. The heat thus absorbed by the district heating water is utilized by the subscribers to meet various heat demands. The heat required to meet these demands is extracted by means of heat exchangers, located between the radiator and condenser symbols in the figure.

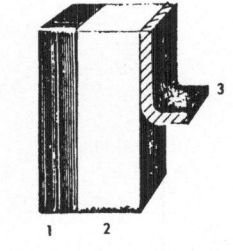

1. Extracted Electrical Energy
2. Extracted Heat
3. Losses

In cases where the heat power demands vary within wide limits it may be advantageous to install a combined district heating and condensing plant. One of two alternatives may be chosen: —

1. The plant may include a water cooler in which the returning water is cooled. This solution is suitable when the power peaks are of very short duration.

2. A district heating turbine of the pass-out condensing type may be chosen (see figure below). This type of machine is economically justifiable if during extended periods, the heat demand is insufficient to generate the whole of the power demand. In such cases the condensing portion of the turbine must be carefully optimised, taking into account the duration and magnitude of the power and heat demands.

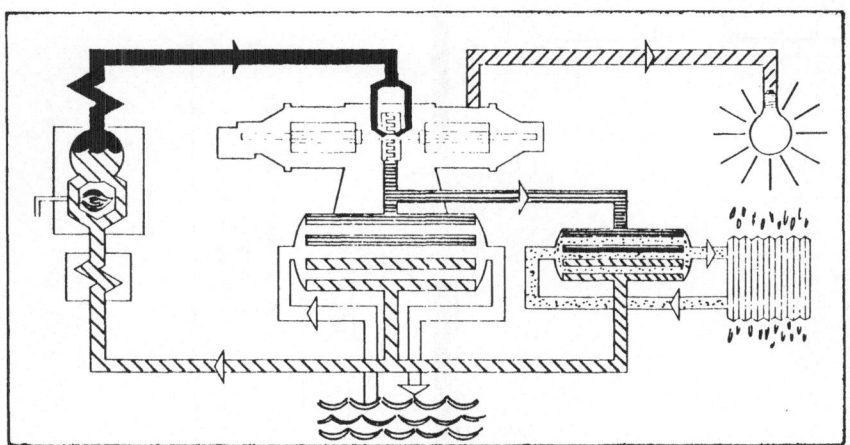

Fig. 5.   Combined plants.

306

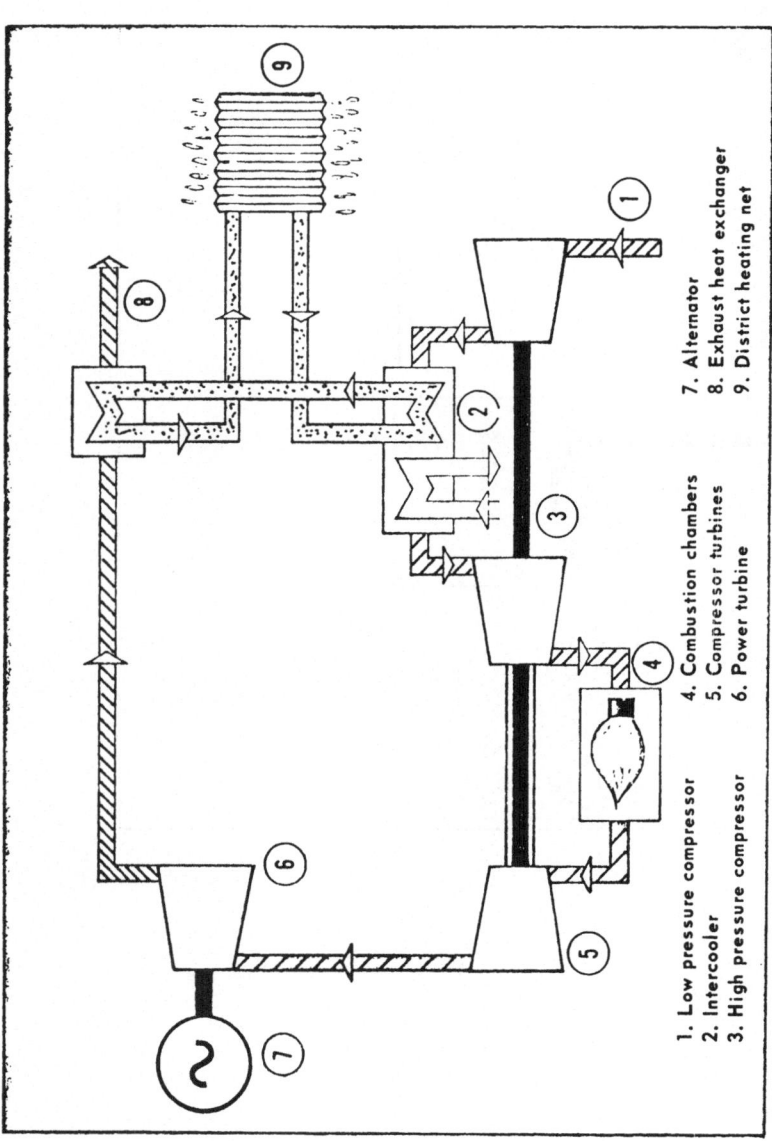

1. Low pressure compressor
2. Intercooler
3. High pressure compressor
4. Combustion chambers
5. Compressor turbines
6. Power turbine
7. Alternator
8. Exhaust heat exchanger
9. District heating net

Fig. 6.   District heating by gas turbines.

Fig. 7.   Regional heating pipes.

Fig. 8.   Regional heating pipes.

Fig. 9.   Regional heating pipes.

Fig. 10.   Regional heating pipes.

Fig. 11. Regional heating pipes.

Fig. 12. Regional heating pipes.

310

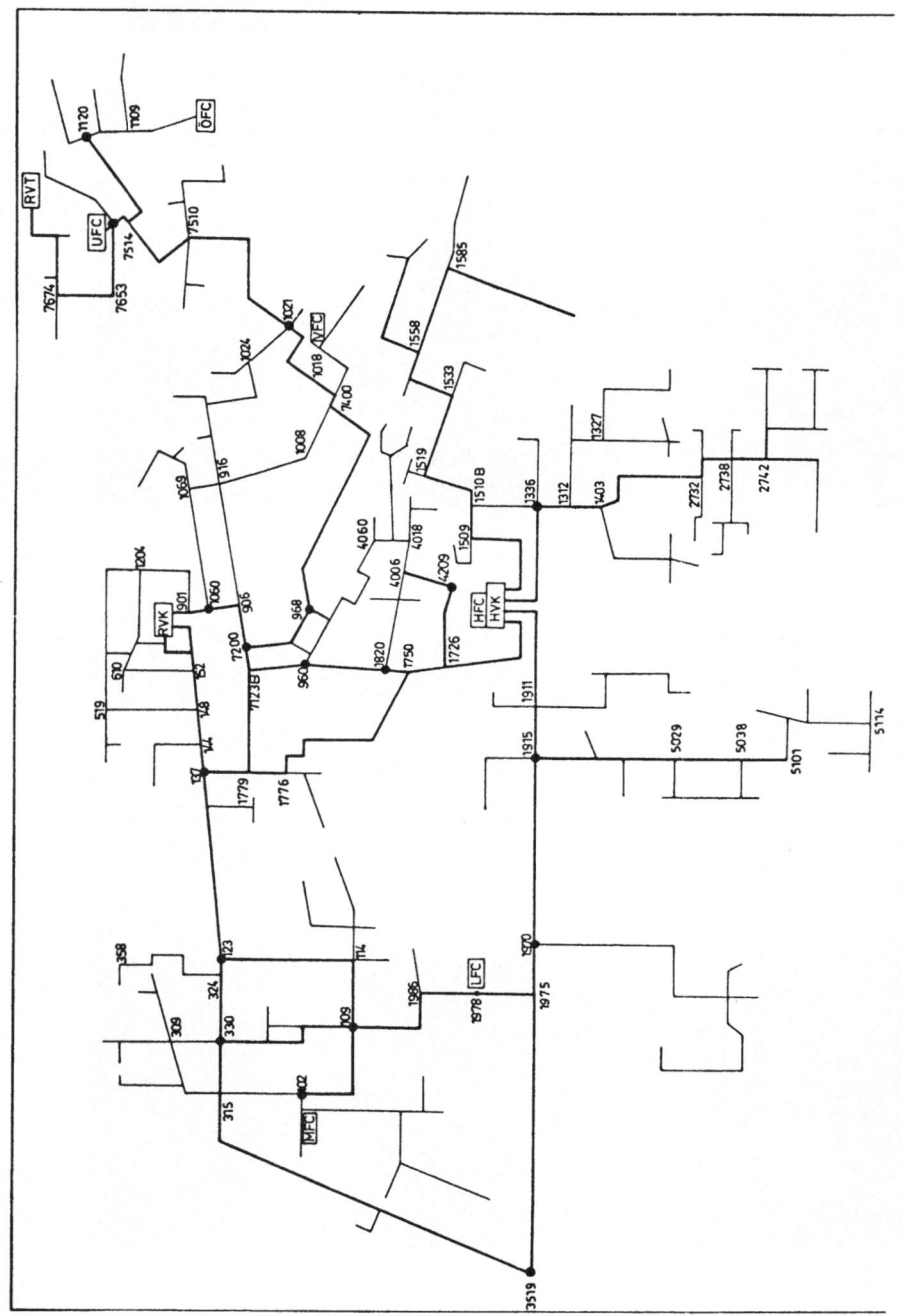

Fig. 13.   Schematic presentation of the district heating network in Malmö.

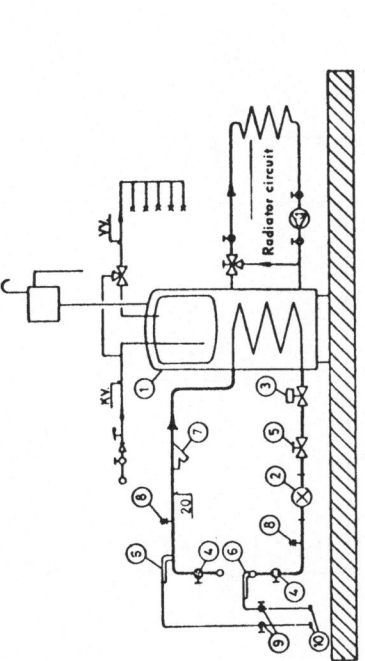

Item No.	Description
1	Heat exchanger
2	Hot water flow meter, 20 mm nom. bore with 1" B.S.P. thread. Supplied by the district heating power station. Lenght without couplings : 220 mm RSK 3225, 3227 couplings with flat seal also required.
3	Return temperature limiter. Samson type 1A.
4	Isolating valve, 20 mm nom. bore valve
5	Control valve (Not required in certain cases)
6	Air collector, 20 mm nom. bore, 250 mm long.
7	Filter, RSK 4830/10, 20 mm nom. bore.
8	Contact thermometer, RSK 4251, for 20 mm nom. bore pipe.
9	Air vent valve, 15 mm nom. bore ball valve.
10	Tundishes, RSK 1076, red, 15 mm nom. bore.

Parts of the above components may be included in certain prefabricated Units.

VV = Hot water
KV = Cold water

Fig. 14.   Fundamental diagrammatic arrangement of district heating installations in private houses with hot water and radiator circuits.

312

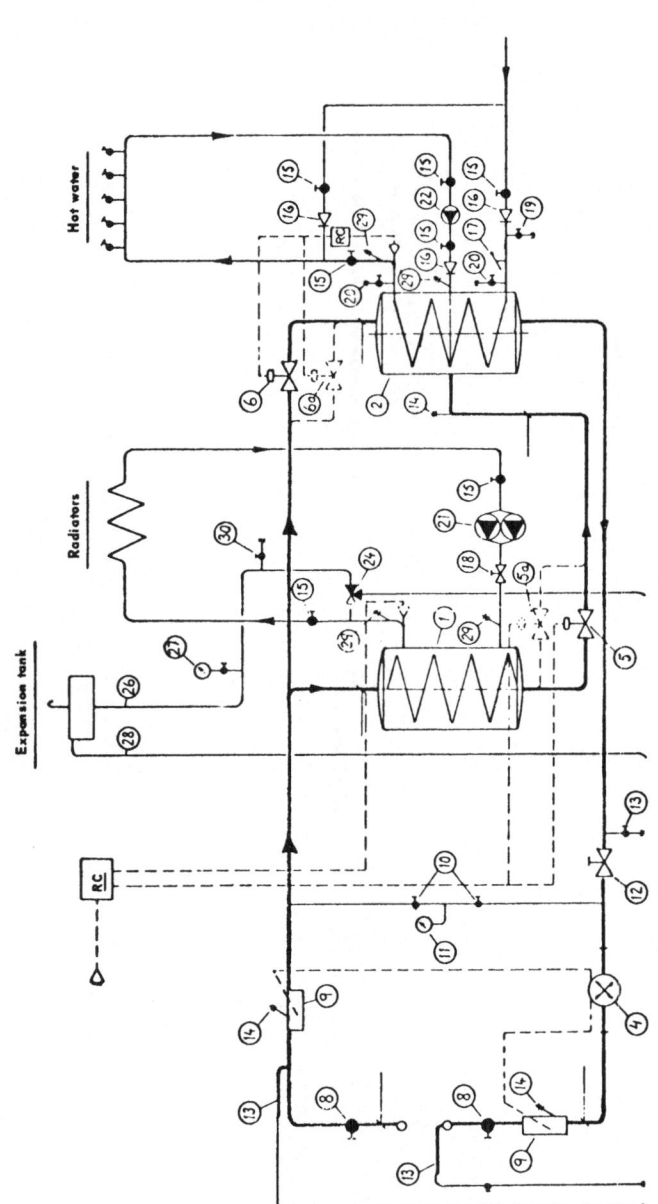

RC = Control unit

Fig. 15.   Fundamental diagrammatic arrangement of heat converter without air supply unit.

313

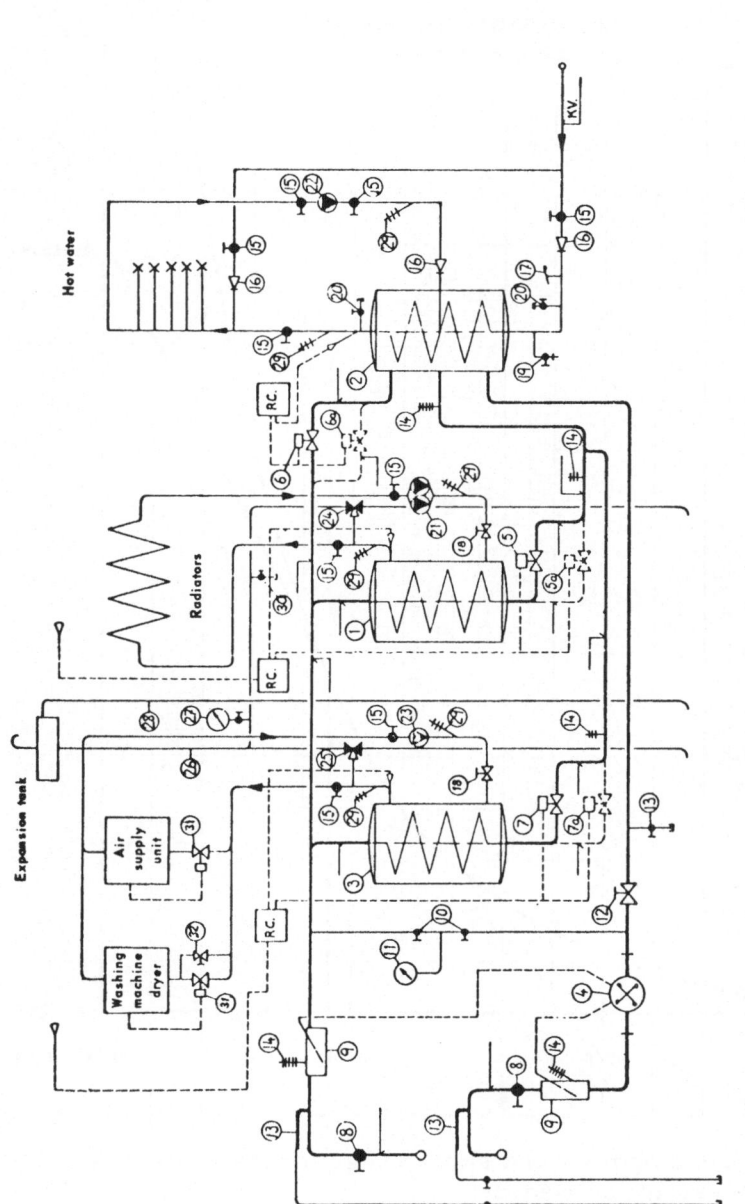

RC = Control unit

Fig. 16. Fundamental diagrammatic arrangement of an installation with heat converter and dryer or hot air supply unit.

314

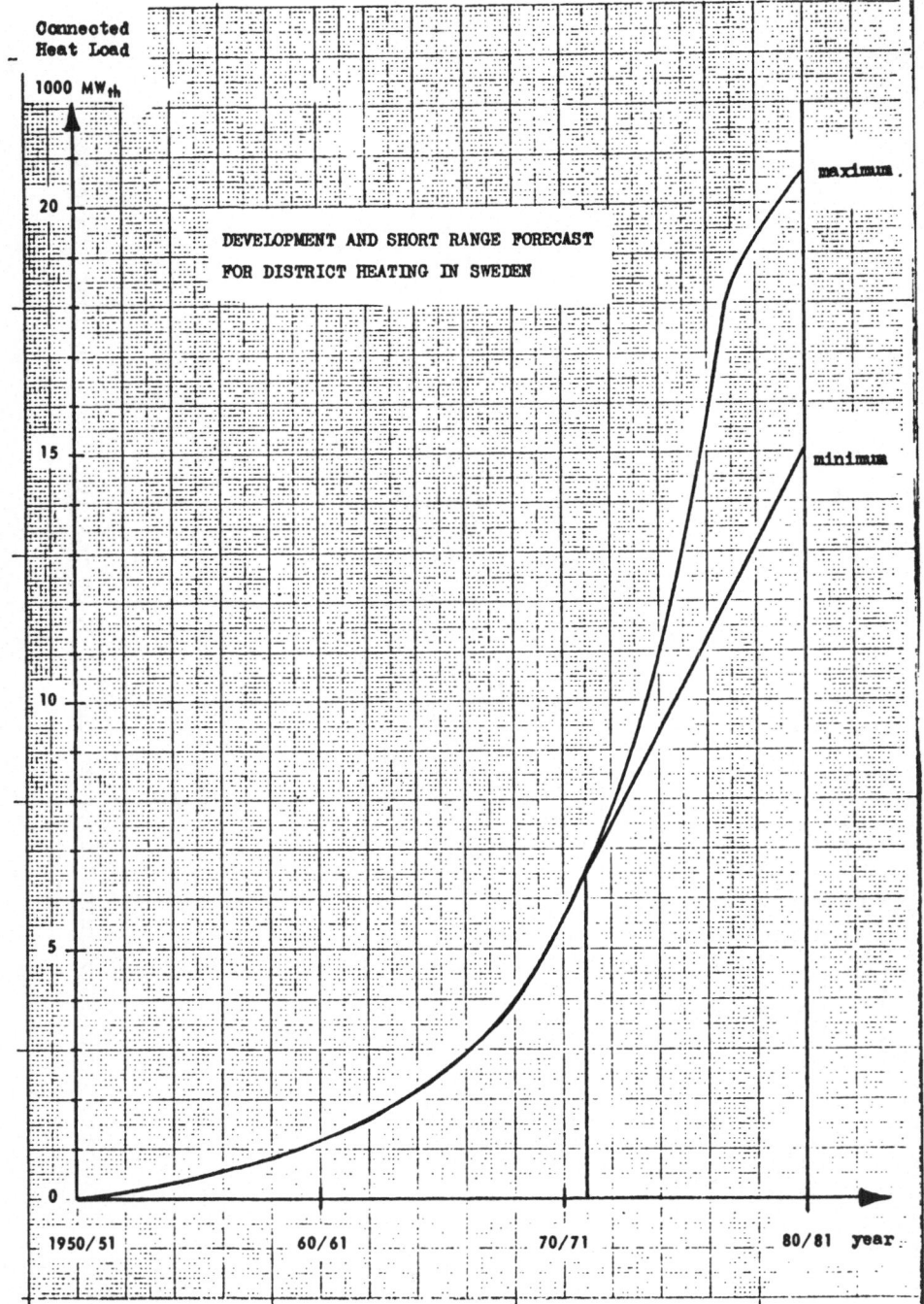

Fig. 17.   Development and short range forecast for district
heating in Sweden.

315

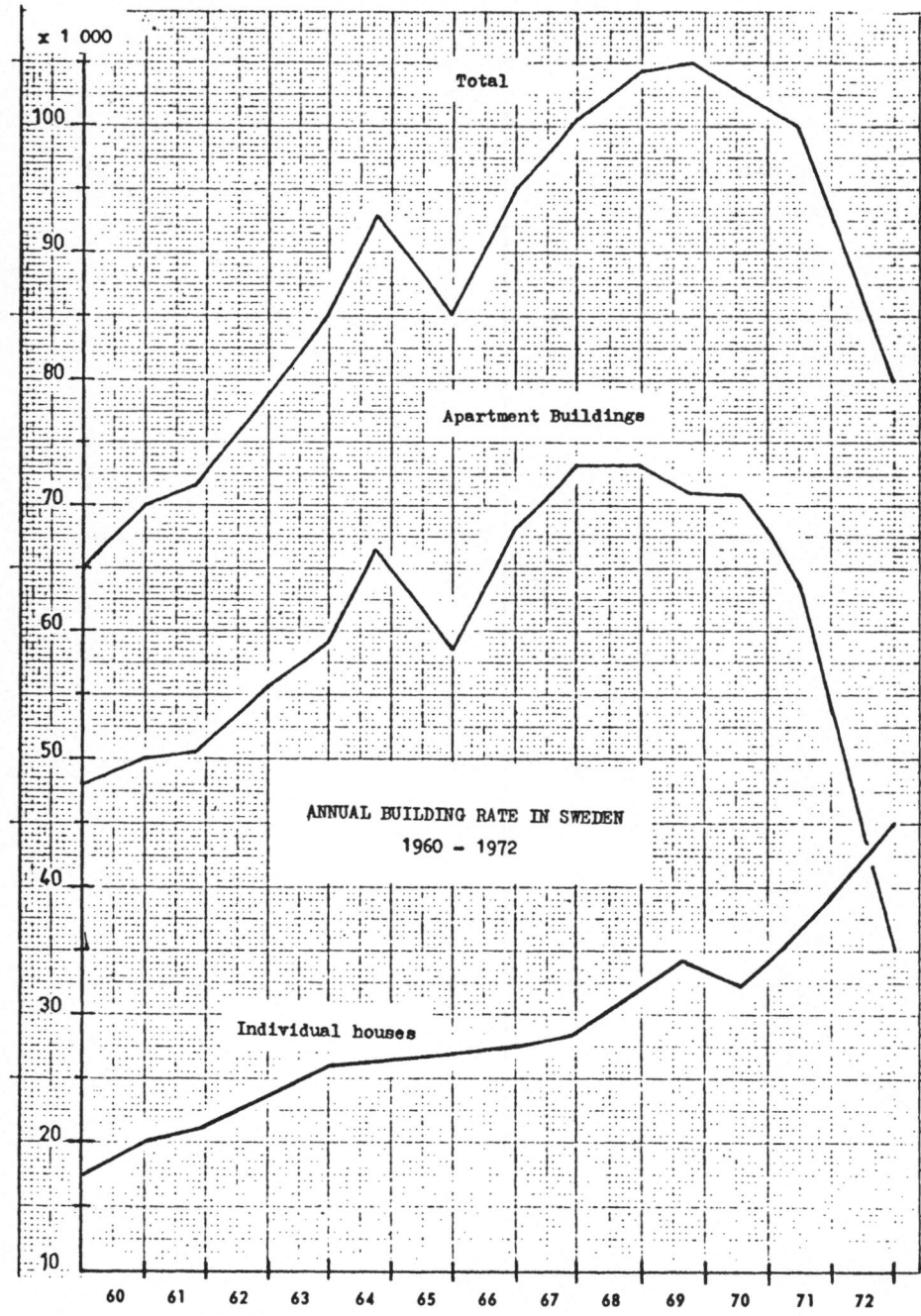

Fig. 18.   Annual building rate in Sweden 1960-1972.

316

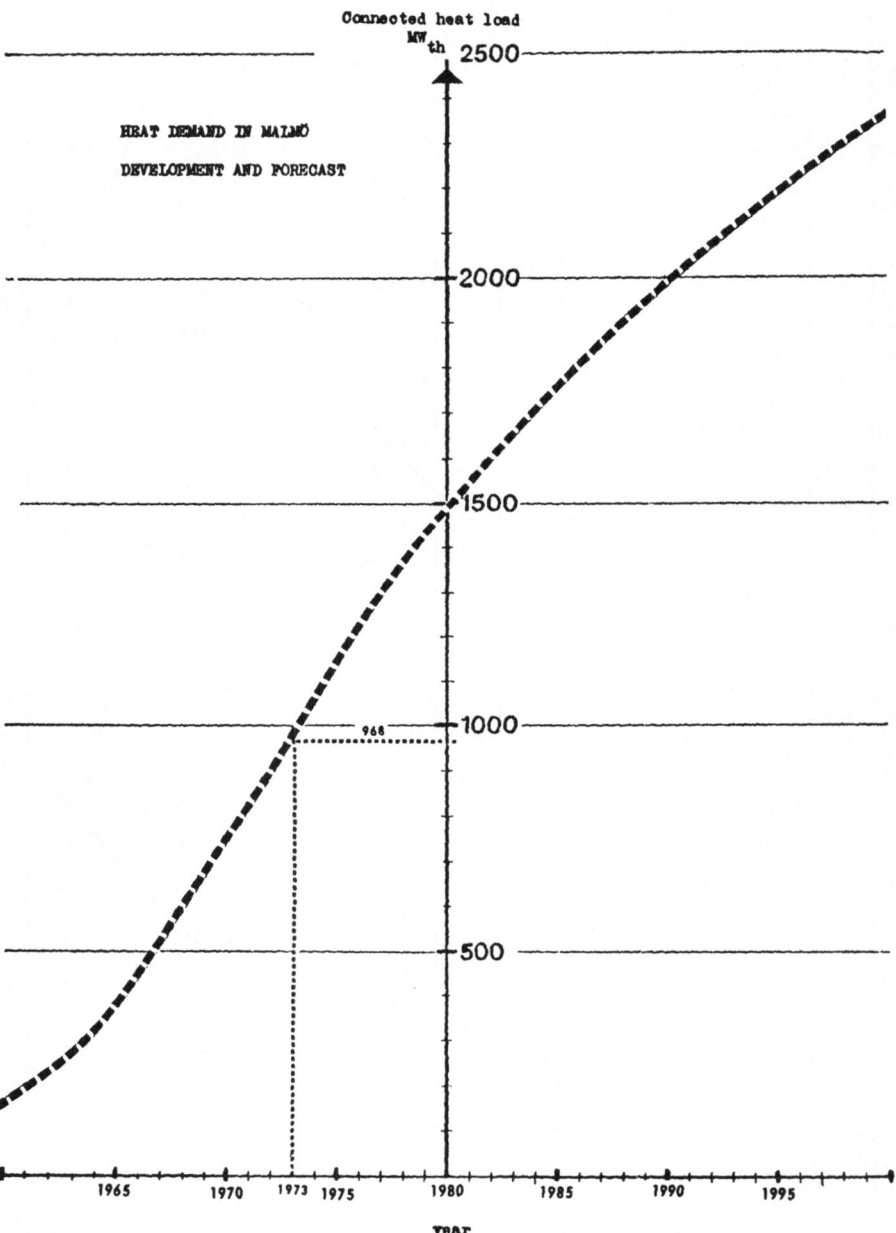

Fig. 19. Heat demand in Malmö - development and forecast.

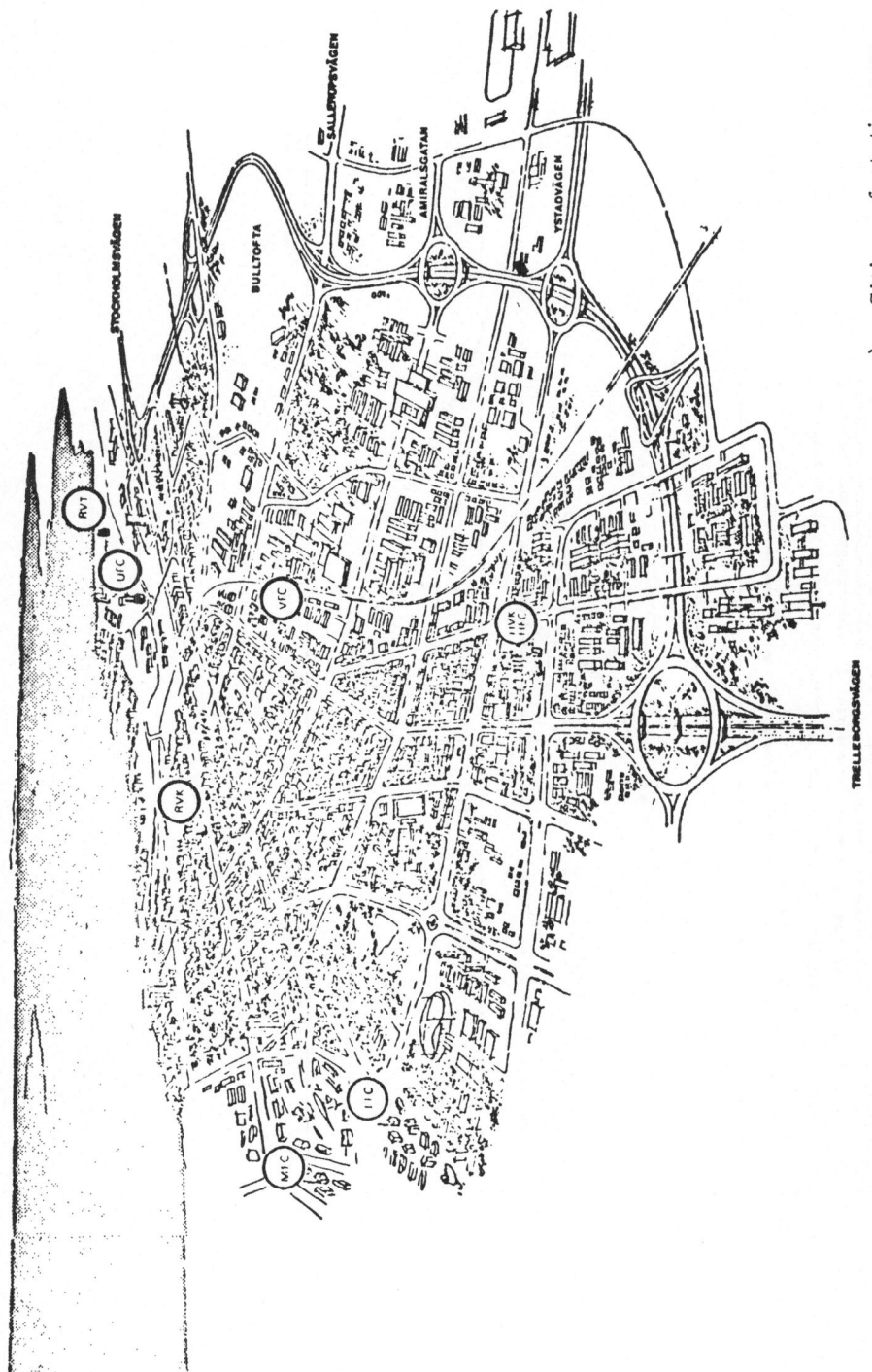

Fig. 20.   Malmö Kraftvarmeverk (Malmö district heating power system) - Siting of stationary
           productions sources

318

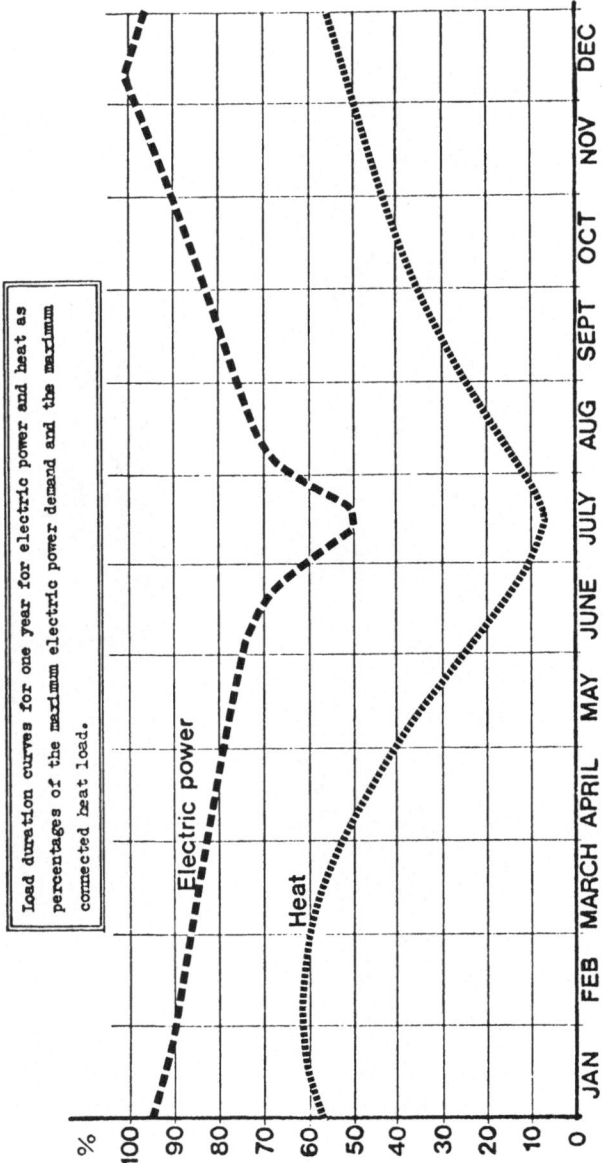

Fig. 21. Load duration curves for one hear for electric power and heat as percentages of the maximum electric power demand and the maximum connected heat load.

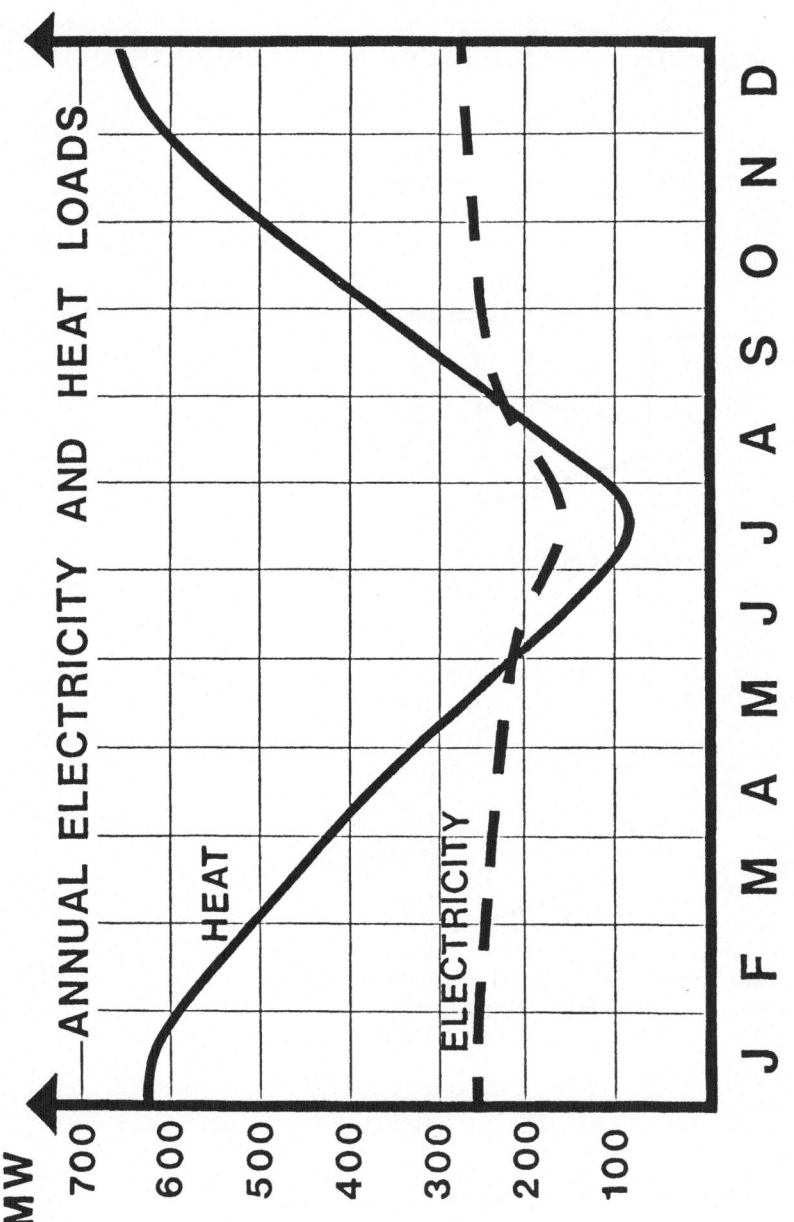

Fig. 22.   Annual electricity and heat loads.

320

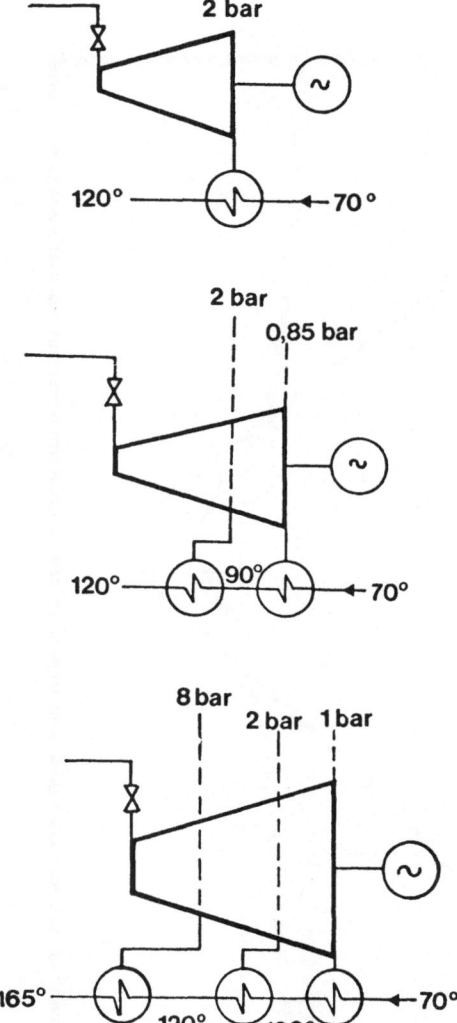

Fig. 23.   One, two and three-stage heating.

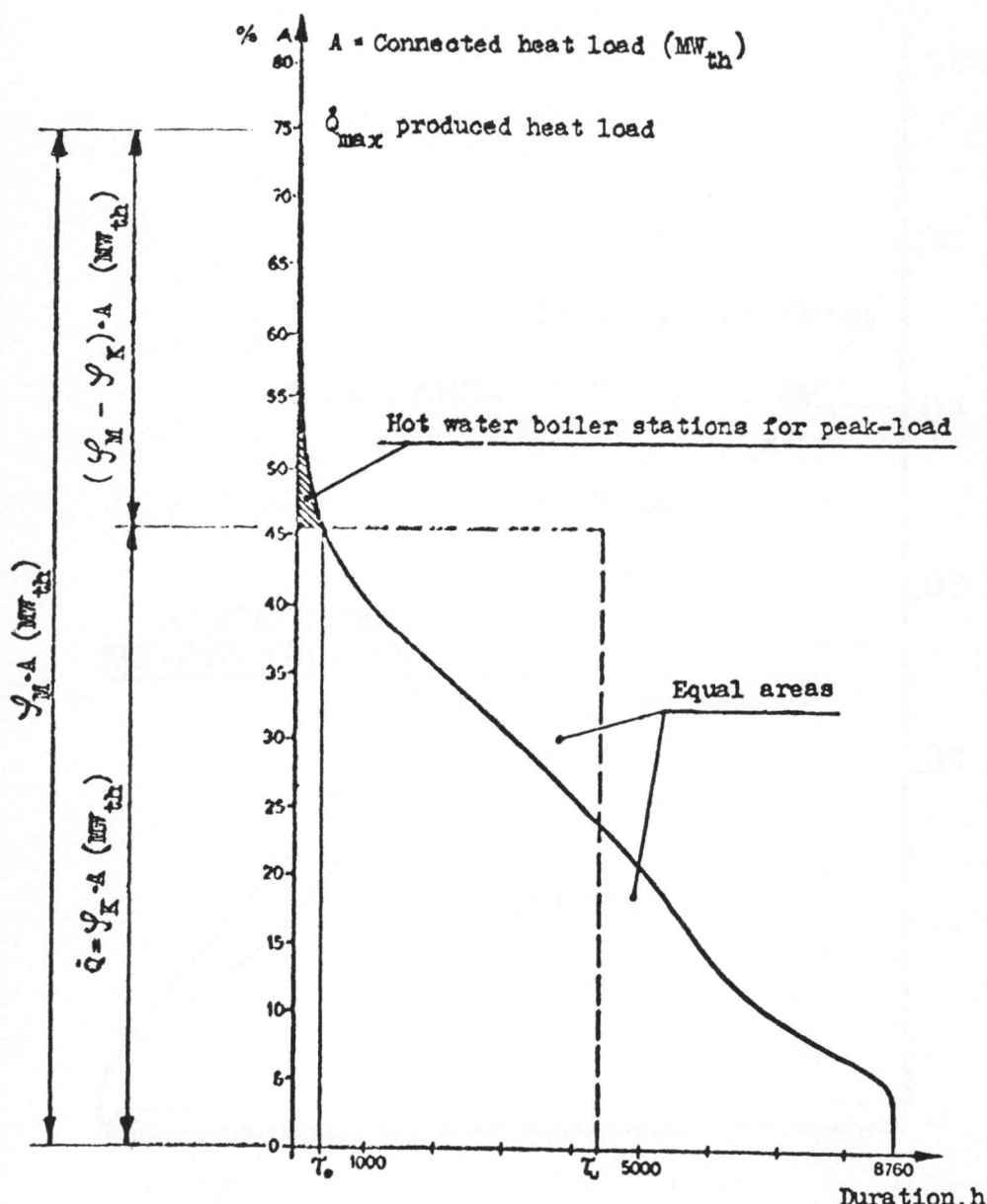

Fig. 24. Annual heat load duration curve.

322

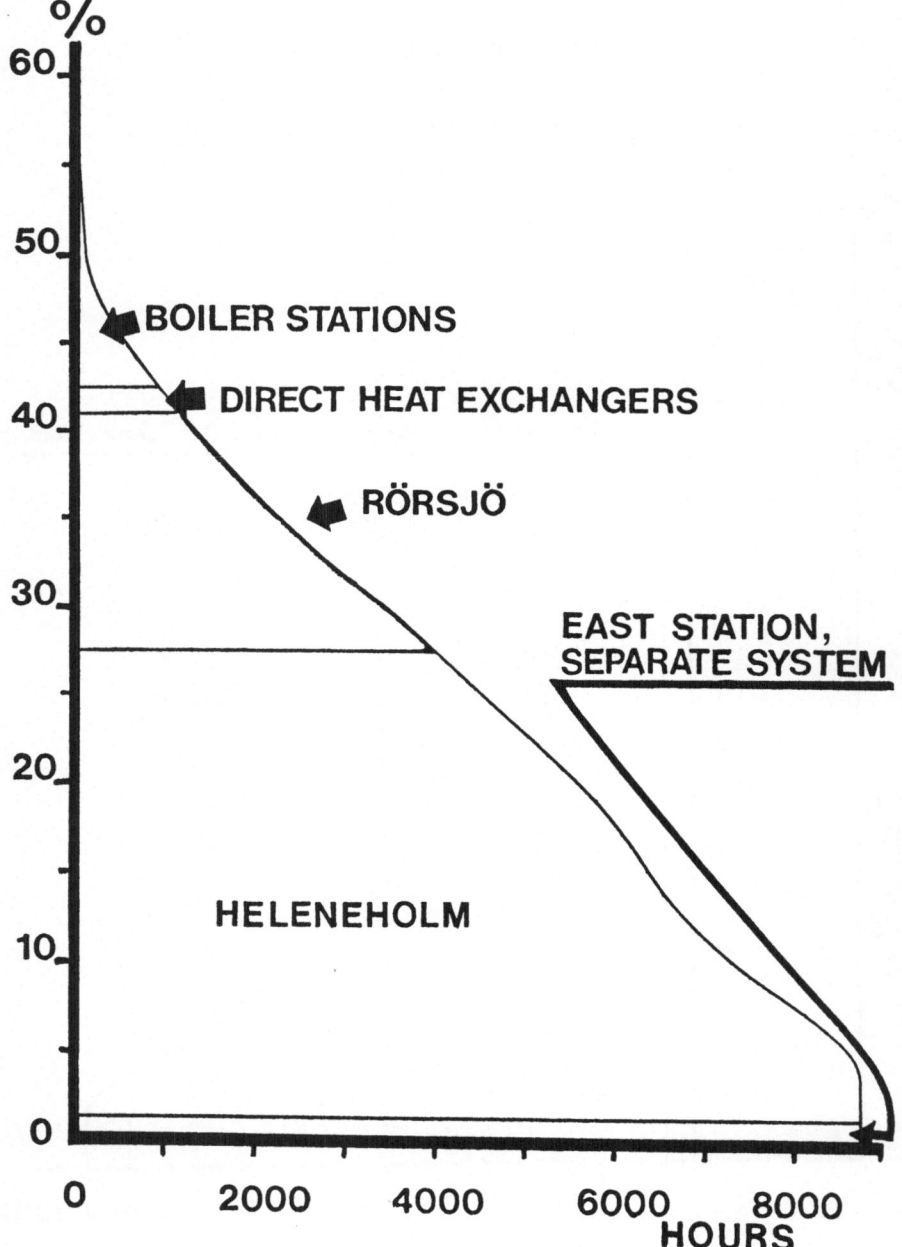

Fig. 25.   Annual load duration curve, heat.

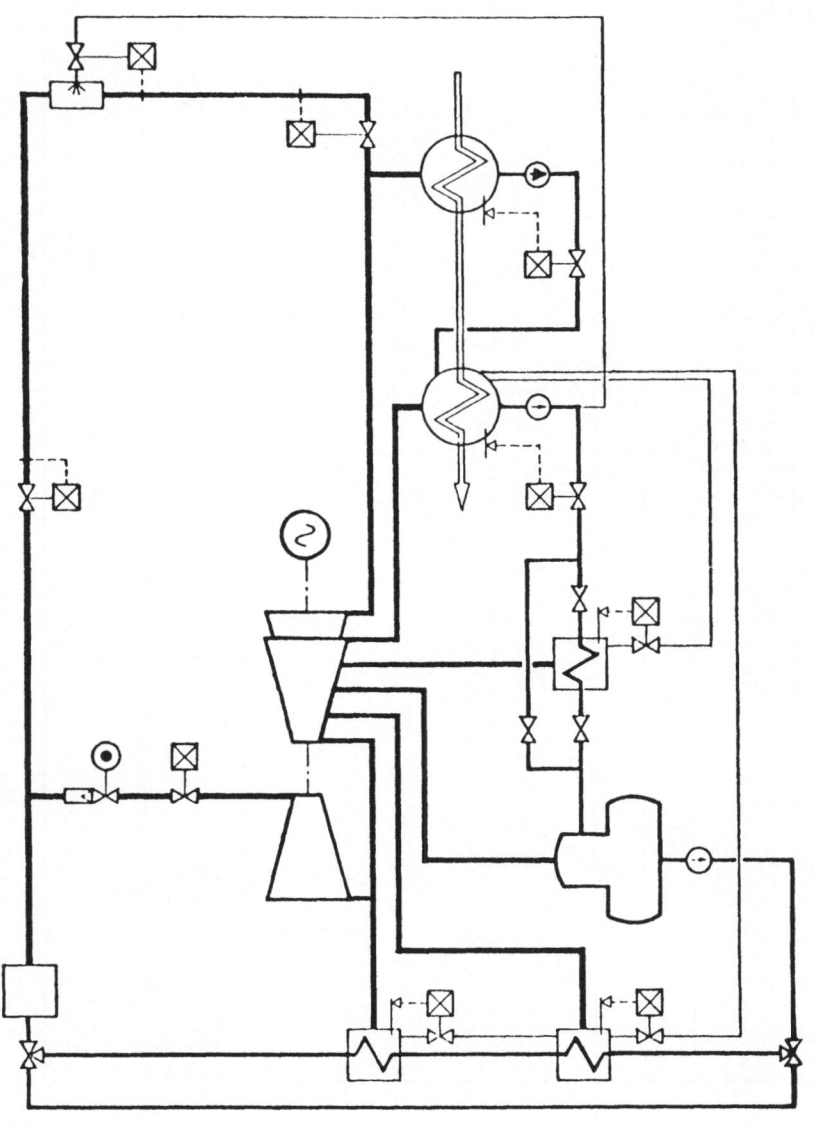

Fig. 26.   District heating power station with two condensers in series.

324

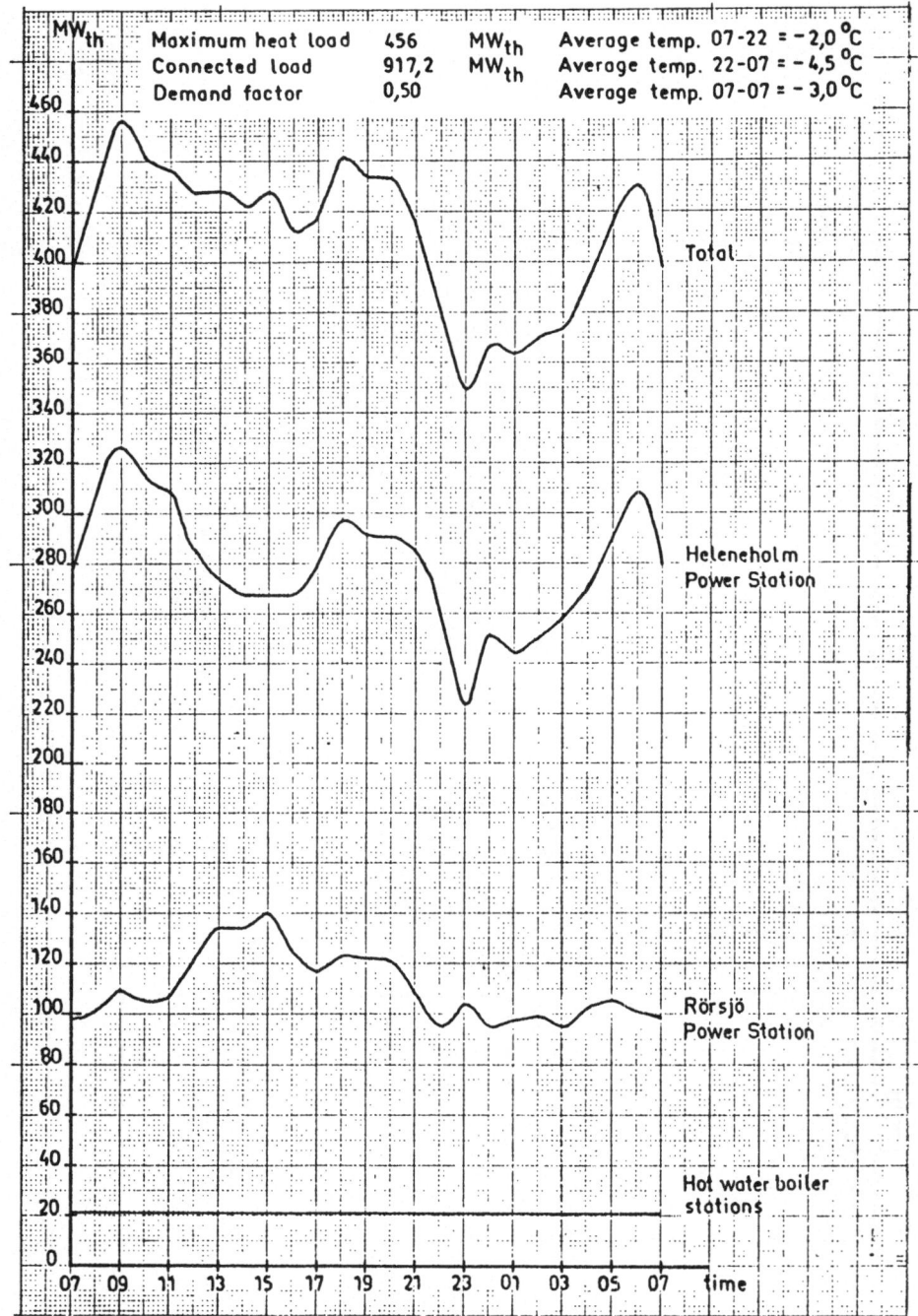

Fig. 27.   Maximum heat demand during February 1973.
24-hour load duration curves.
Maximum day: Monday 26th February.

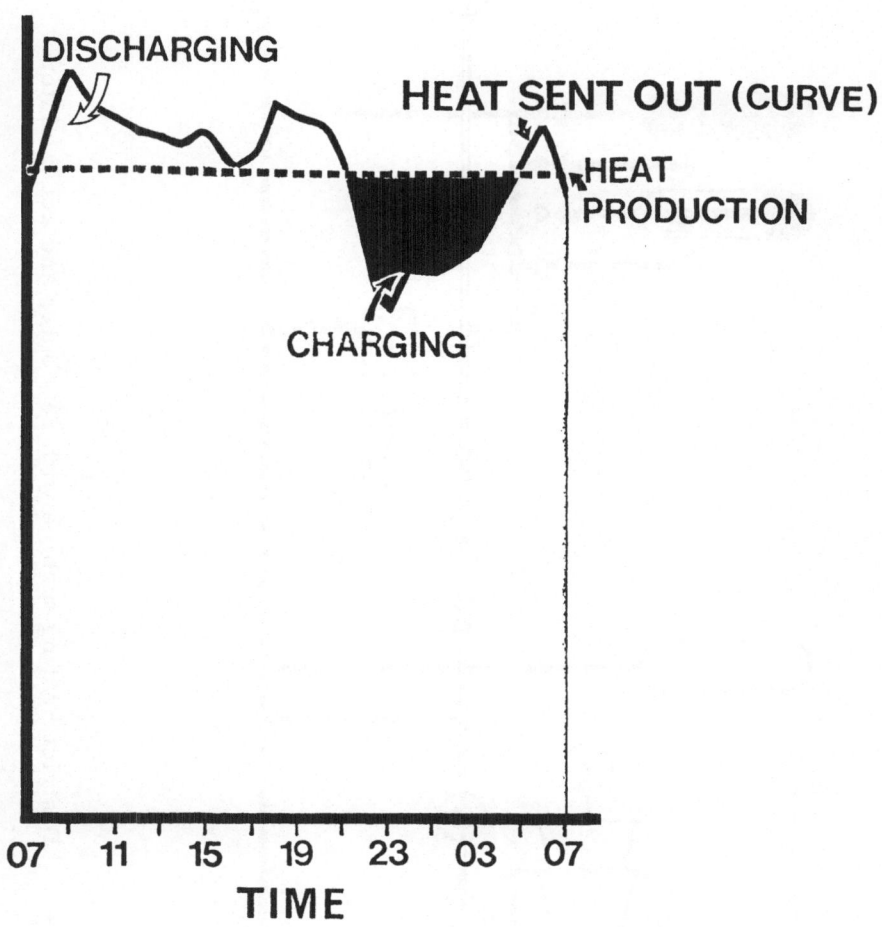

Fig. 28. Use of accumulators.

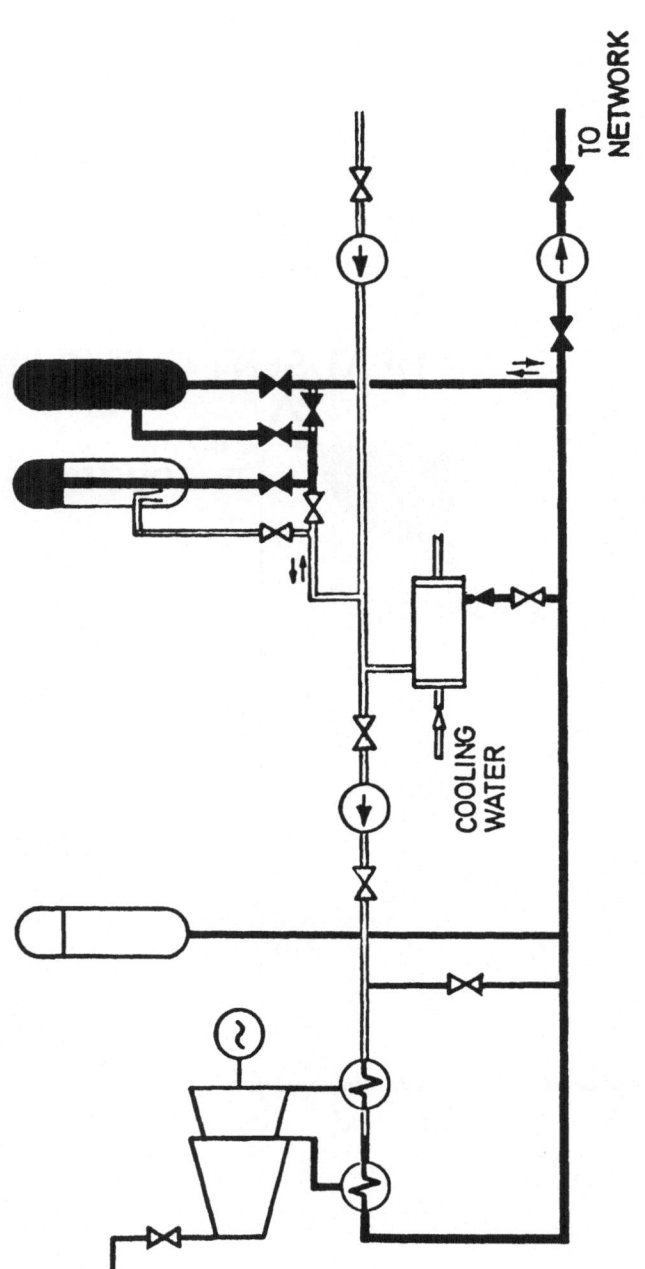

TO
NETWORK

COOLING
WATER

Fig. 29.   District heating turbine with re-cooler and accumulators.

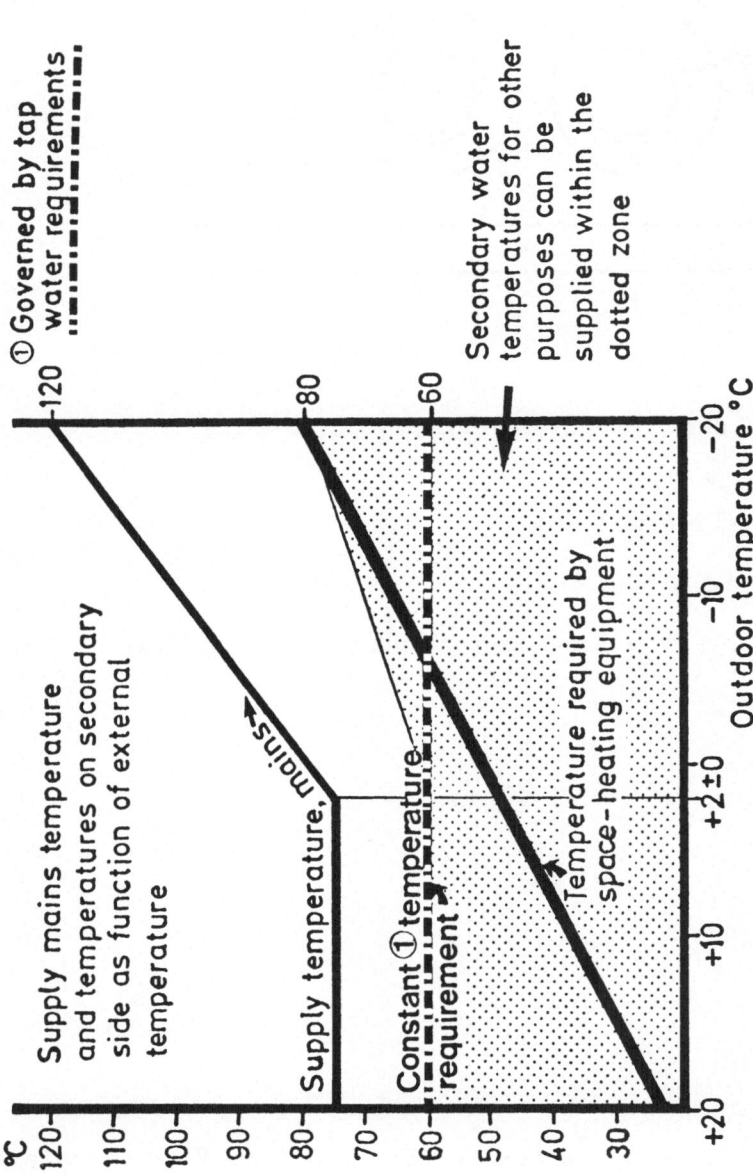

Fig. 30.   Supply mains temperature and temperatures on secondary side as function of external temperature.

328

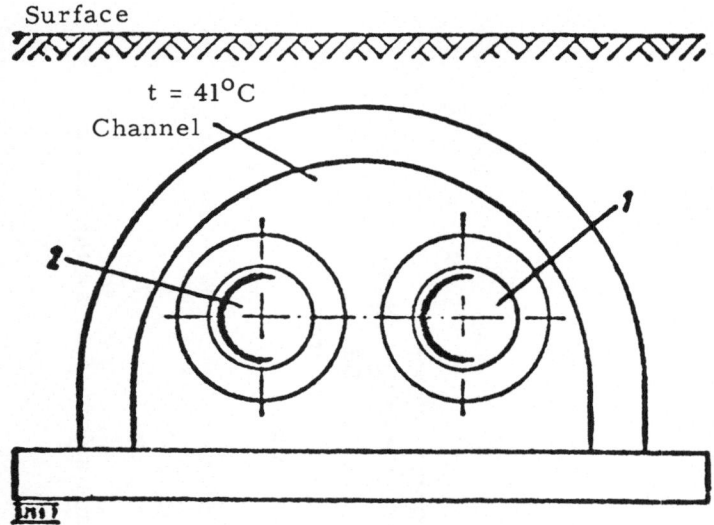

Surface

t = 41°C

Channel

2

1

Hot water duct

t_λ = 23°C

1 Arrival NW 300, 160°C

1 Return NW 300, 80°C

Reference: Fernwärme international - FWI, Jg. 3(1974), Heft 2

Fig. 31.   Main temperatures. (Example from West Germany)

Water velocity = *w* m/s
Thermal capacity = $\dot{Q}$ kW$_{th}$

$$\dot{Q} = (\varrho \cdot w \cdot \frac{\pi}{4} \cdot d_2) \cdot c_p \cdot \Delta t = 3124.6 \quad \cdot w \cdot d^2 \cdot \Delta t \quad (kW_{th})$$

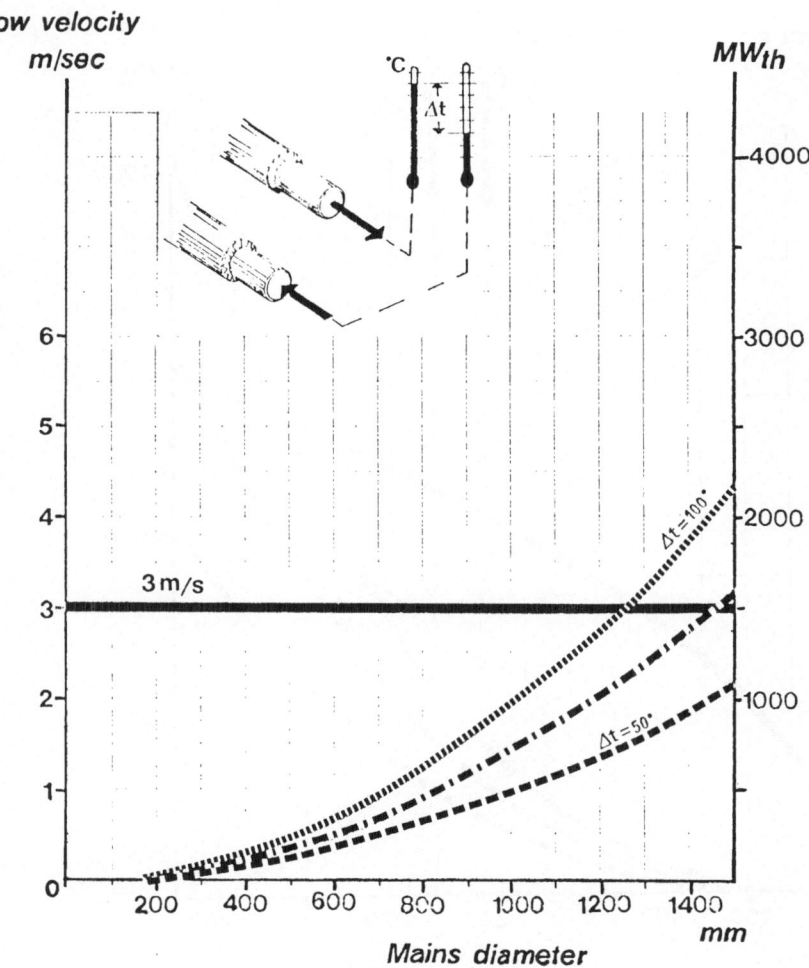

Fig. 32.   Thermal capacity of district heating mains.

330

Water velocity = w m/s
Thermal capacity = $\dot{Q}$ kW$_{th}$

$$w_2 = w_1 \cdot \sqrt{\frac{\lambda_1}{\lambda_2} \cdot \frac{d_2}{d_1}} \quad , \quad \frac{\lambda_1}{\lambda_2} = 1 \quad , \quad w_1 = 3 \ m/s \quad , \quad d_1 = 0.4m$$

$$\dot{Q} = (\varrho \cdot w \cdot \frac{\pi}{4} \cdot d_2) \cdot c_p \cdot \Delta t = 3124.6 \quad \cdot w \cdot d^2 \cdot \Delta t \quad (kW_{th})$$

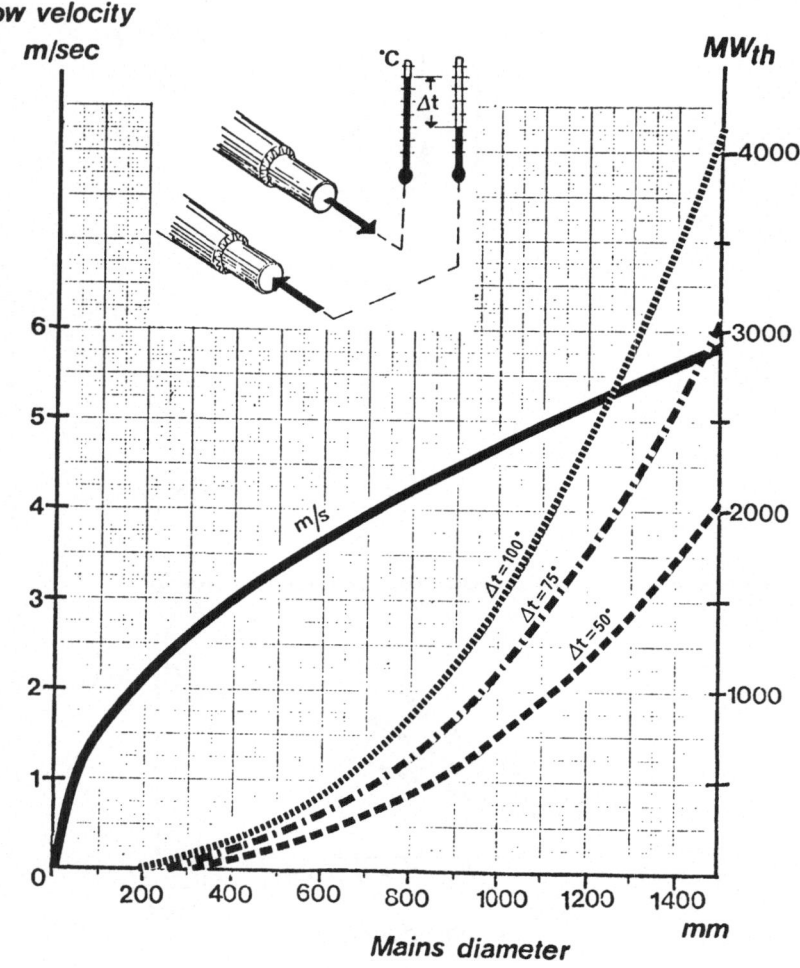

Fig. 33.   Thermal capacity of district heating mains.

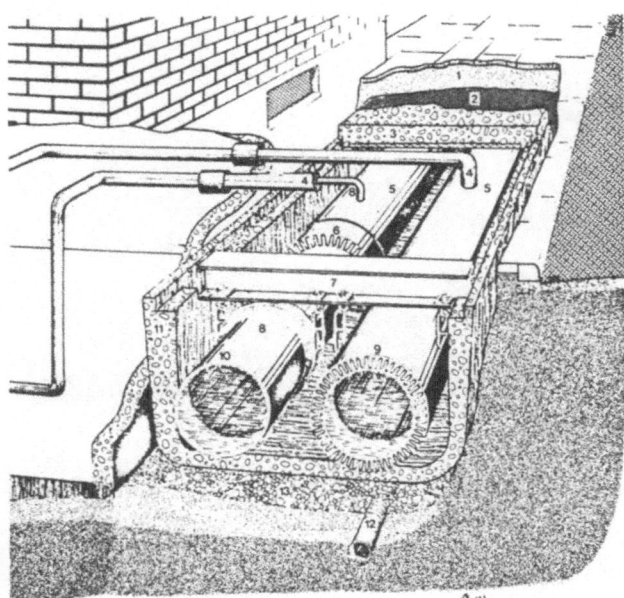

1. Top layer of gravel
2. Asphalt cover. Bitumenmat
3. Concrete slab
4. Feeder line. Plastic pipe culvert with cover of PEH or PVC and insulation of mineral wool, polyurethane foam or similar
5. Main- or distribution pipeline
6. Mineral wool insulation covered with plastic sheet
7. Pipe support including steel beam, movable pendelum and support ring
8. Steel tube
9. Outgoing flow pipe. 120—80 C
10. Return flow pipe. 70—50 C
11. Prefabricated U-section from reinforced concrete
12. Drainage pipe of plastic
13. Crushed stone

The picture shows a concrete culvert with feeder lines.

Fig. 34.   A design of a modern district heating culvert.

Fig. 35.   Distribution pipes - diameter 800 mm.

Fig. 36. A load of distribution pipes.

Fig. 37. District heating pipes from Meier-Schenk AG, Zürich, Schweiz.

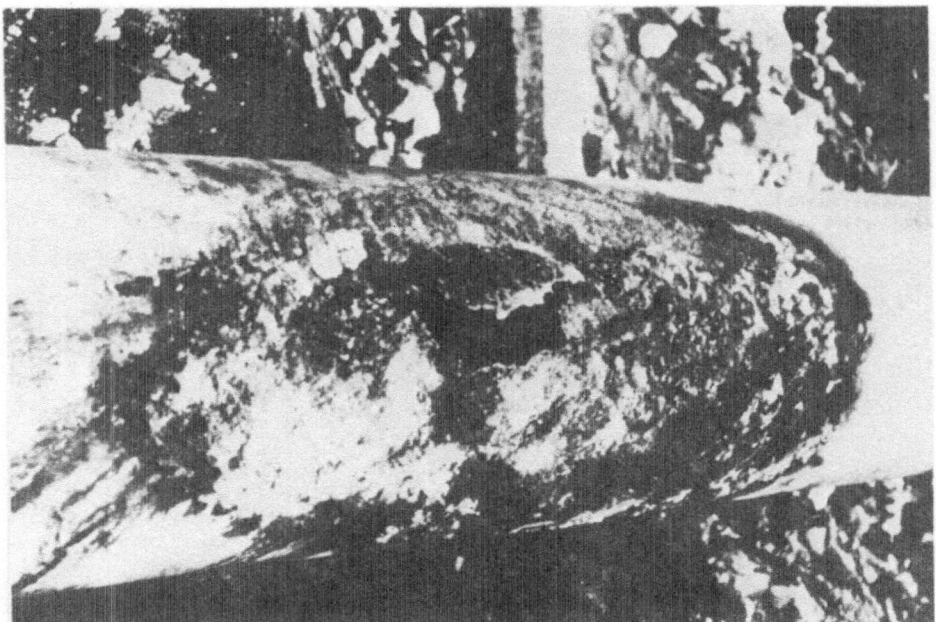

Fig. 38.   Corrosion damage in a 400 mm pipe.

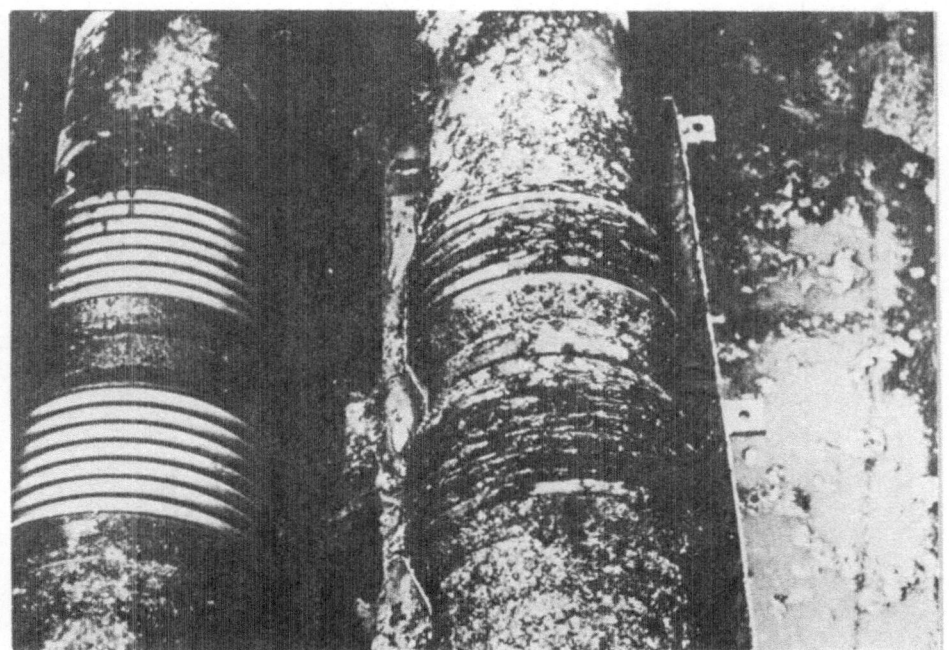

Fig. 39.   Corrosion damage in a 400 mm compensator.

334

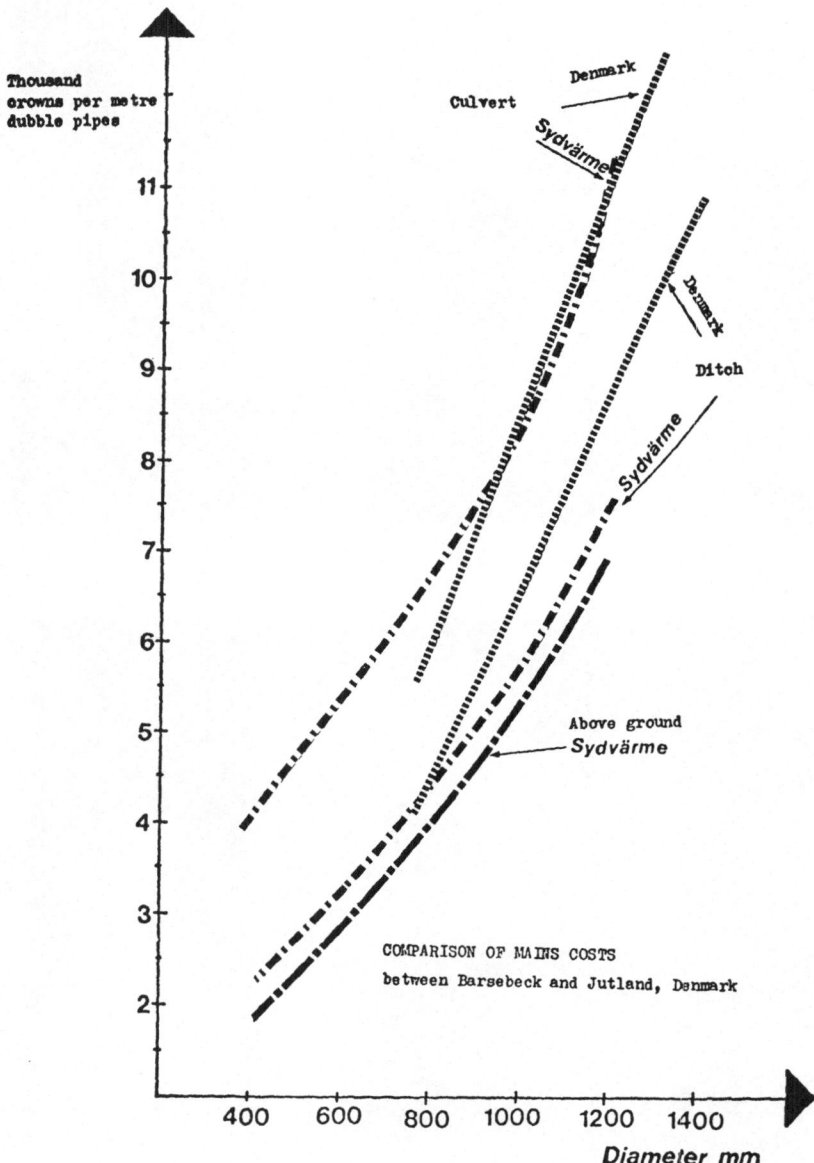

Thousand
crowns per metre
dubble pipes

Culvert

Denmark

Sydvärme

Denmark

Ditch

Sydvärme

Above ground
Sydvärme

COMPARISON OF MAINS COSTS
between Barsebeck and Jutland, Denmark

Diameter mm

Fig. 40. Comparison of mains costs, between Barsebäck and
Jutland, Denmark.

Nominal bores, mm 50, 100, 150

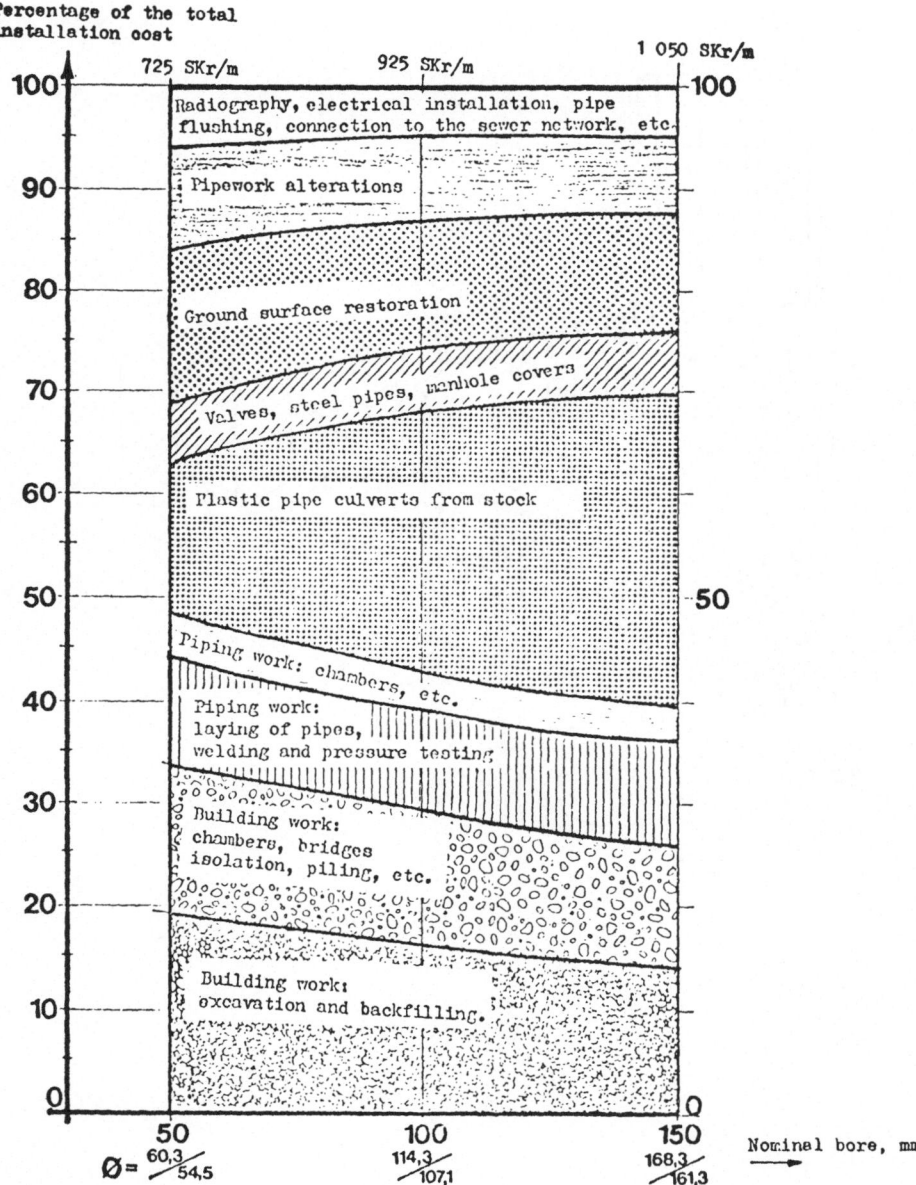

Fig. 41.   Cost distribution for plastic pipe culverts.

Fig. 42. Heleneholmsverket.

Fig. 43.   The Heleneholm district heating and power plant.

Fig. 44.   The Heleneholm district heating power plant at night.

338

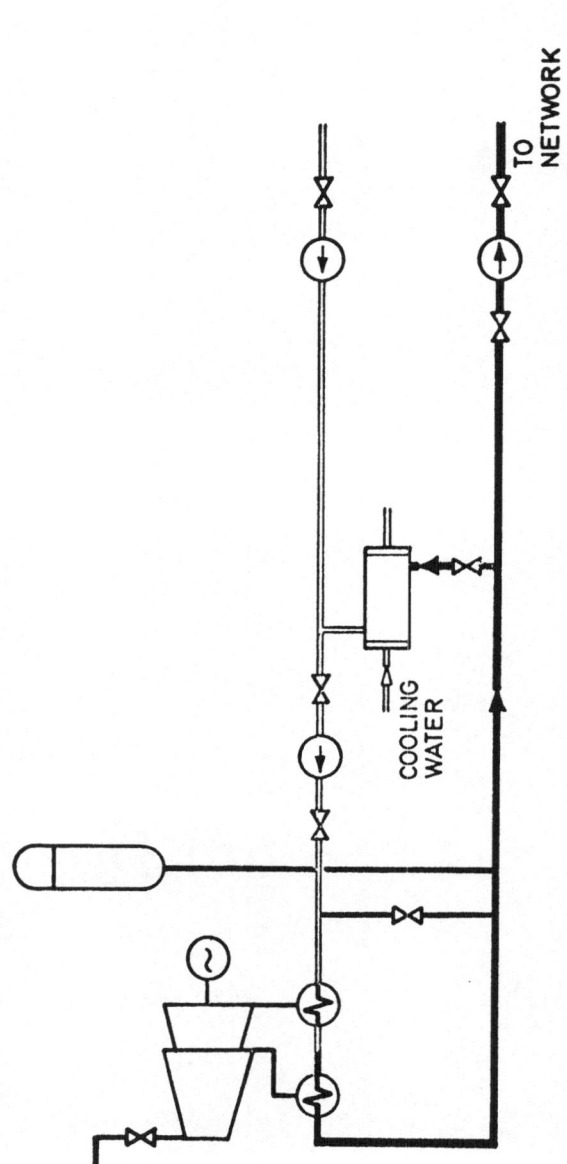

COOLING
WATER

TO
NETWORK

Fig. 45.   District heating turbine with re-cooler.

Fig. 46.   From the left: Feed water storage tank, re-coolers, expansion tanks and accumulators.

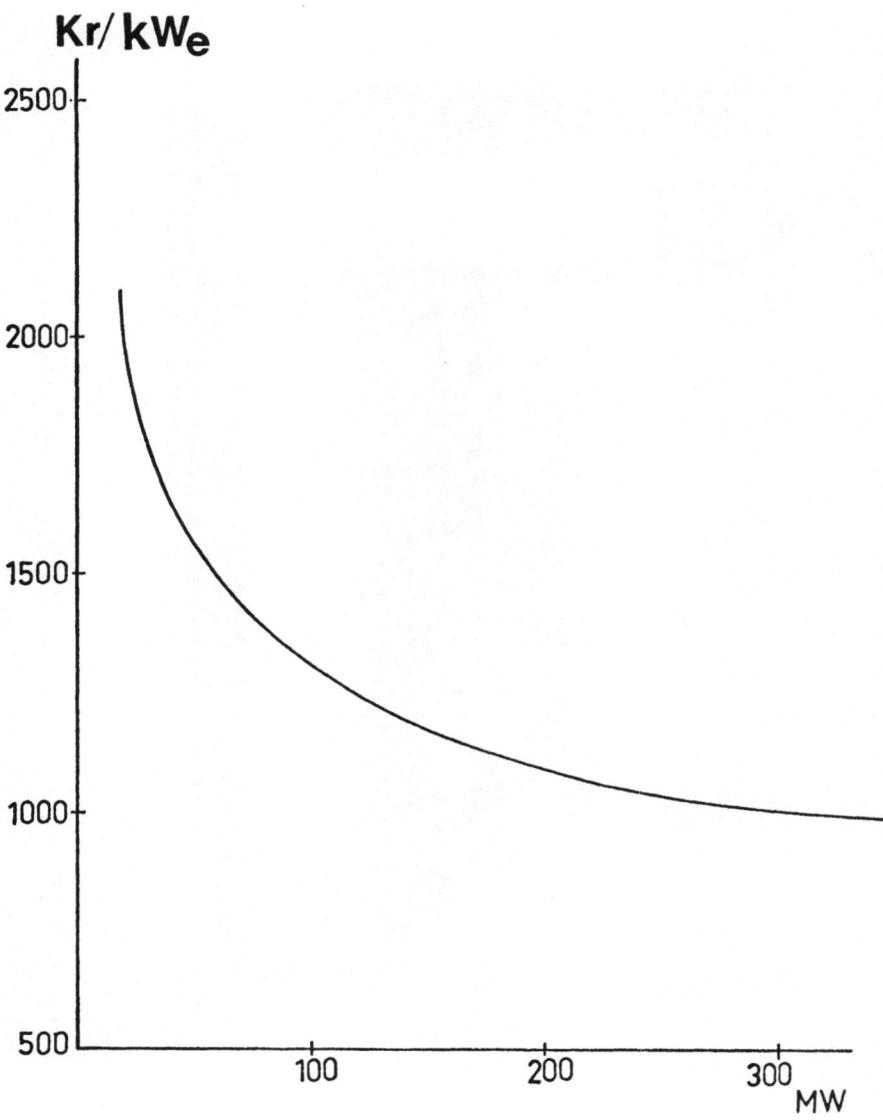

Fig. 47. Specific plant costs.

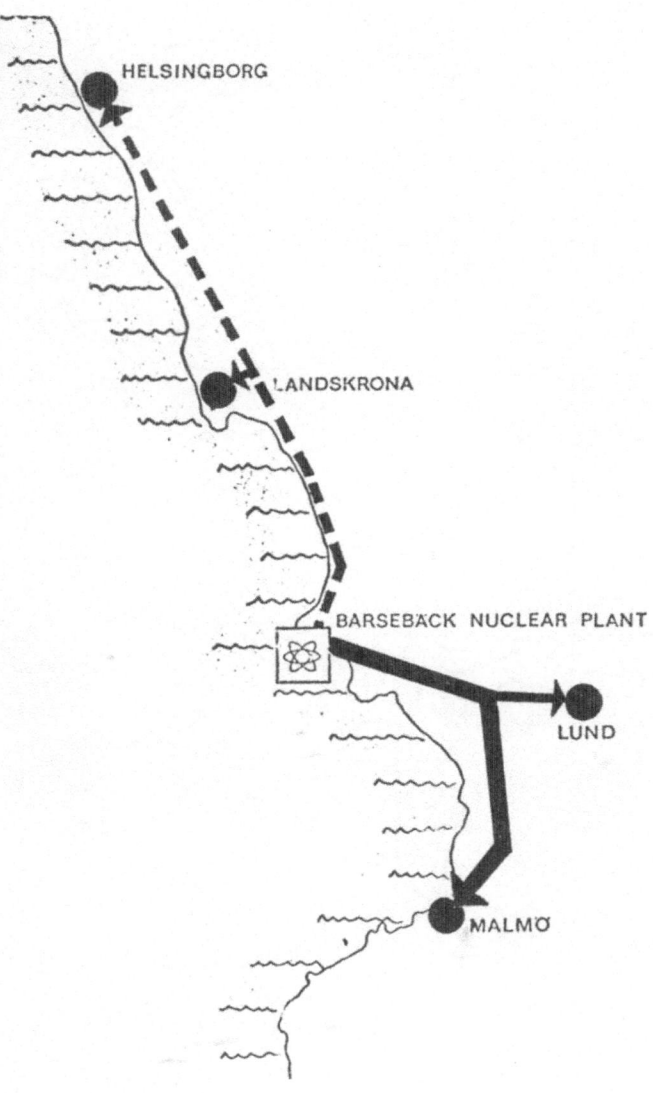

Fig. 48.   District heating from the Barsebäck Nuclear
Power Plant.

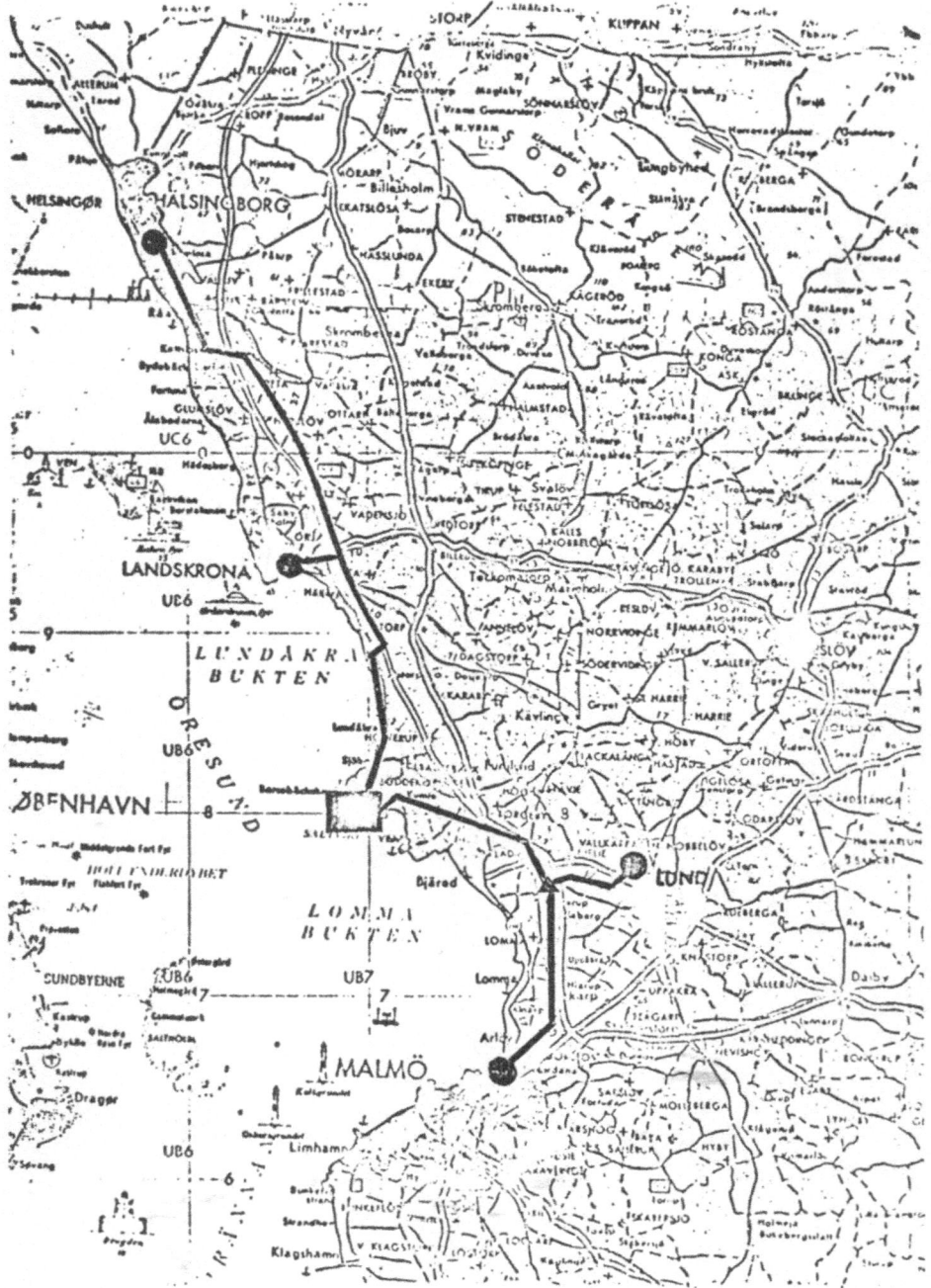

Fig. 49. District heating mains from the Barsebäck Nuclear
Power Plant.

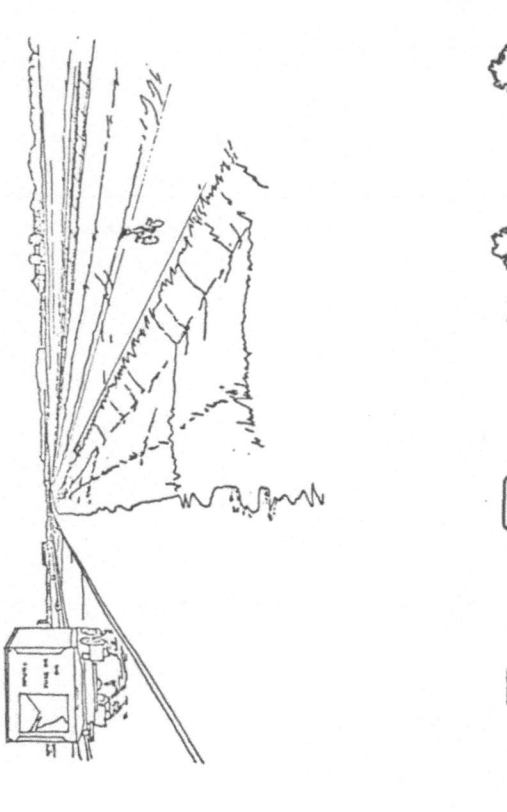

Pipes run beside motorway in ditch with protective banks and cycle track

Fig. 50.   Pipes run beside motorway in ditch with protective banks and cycle track.

344

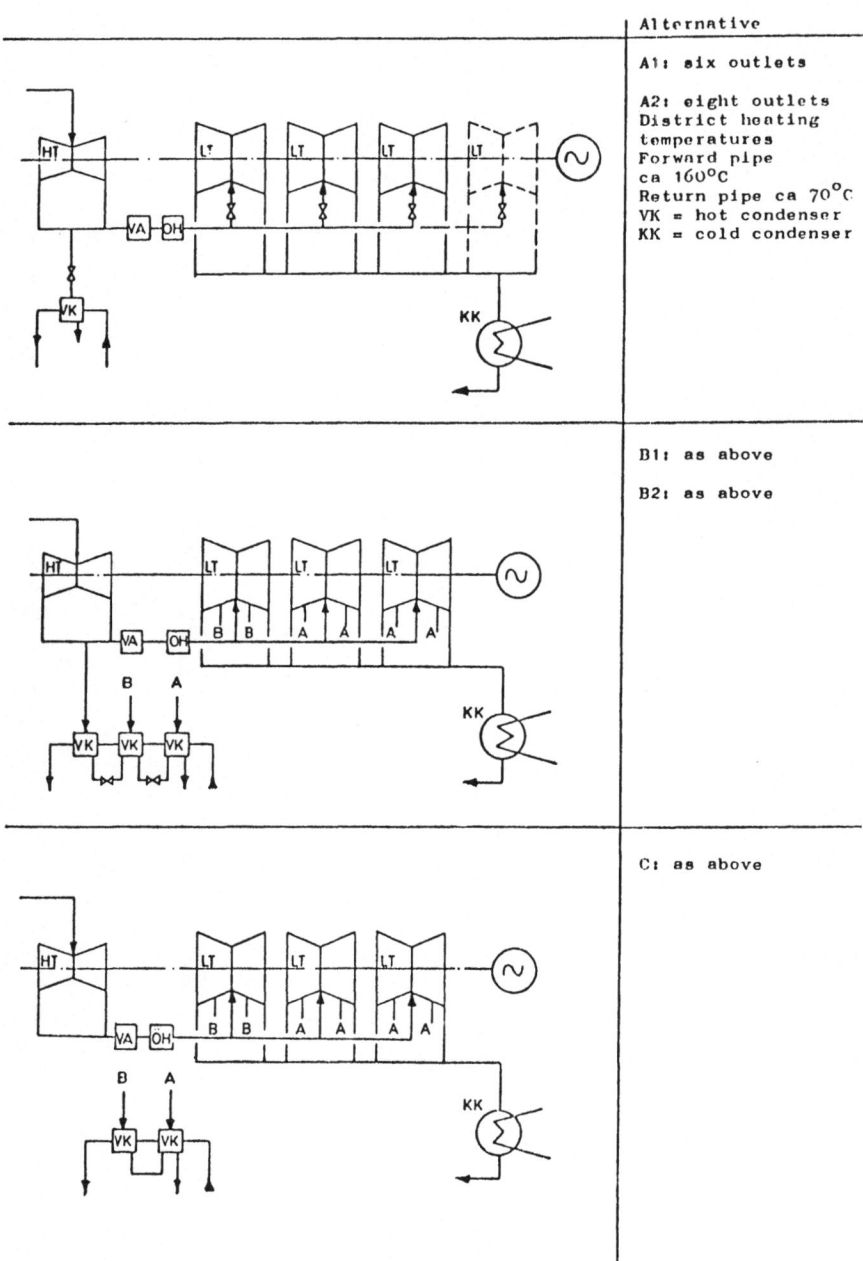

Alternative

A1: six outlets

A2: eight outlets
District heating
temperatures
Forward pipe
ca 160°C
Return pipe ca 70°C
VK = hot condenser
KK = cold condenser

B1: as above

B2: as above

C: as above

Fig. 51. Turbine arrangements in principle.

345

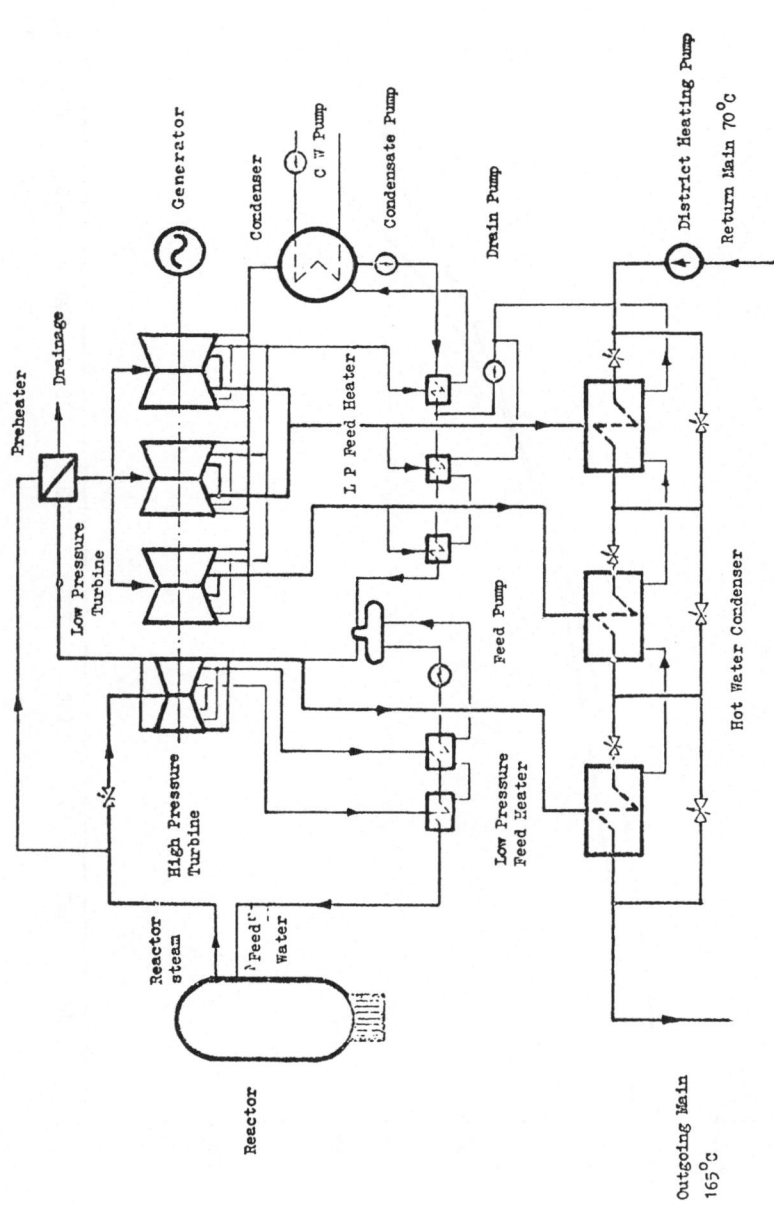

Fig. 52.   Schematic diagram of turbine extraction as alternative B1.

346

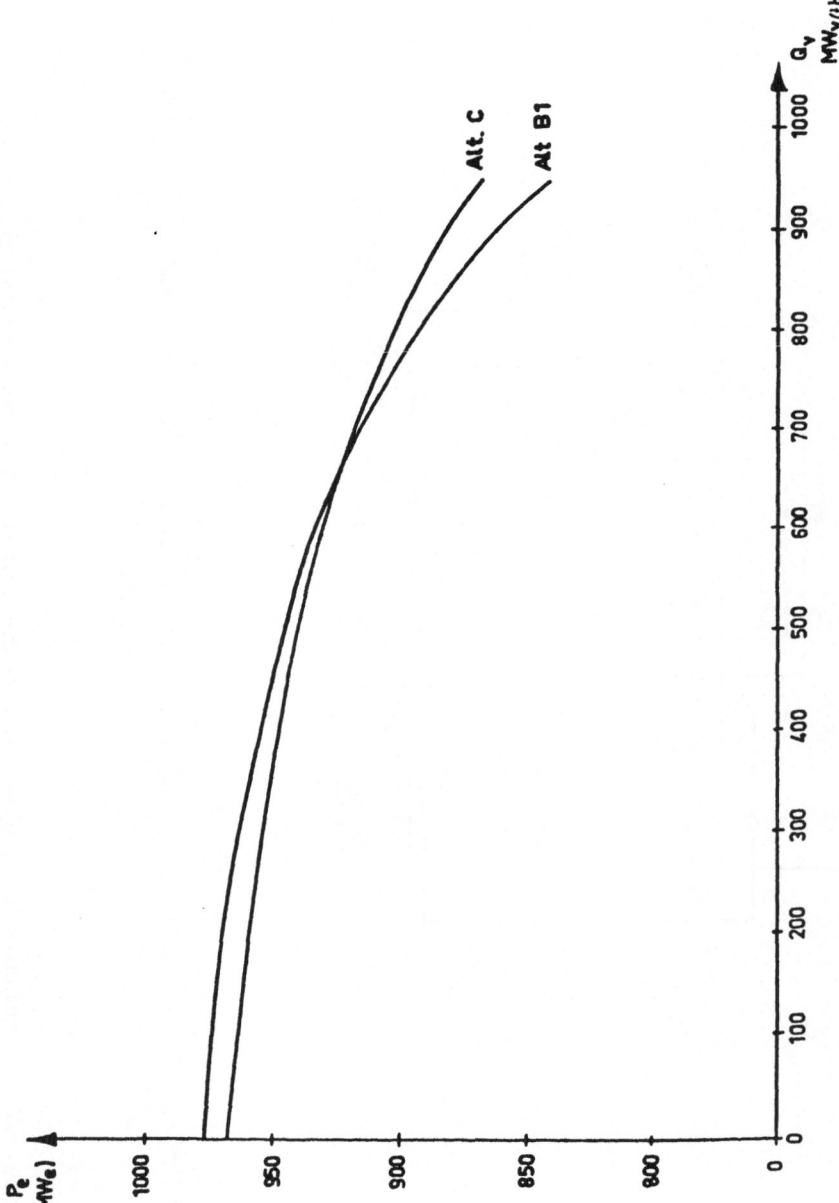

Fig. 53.   Electrical output vs heat load.

347

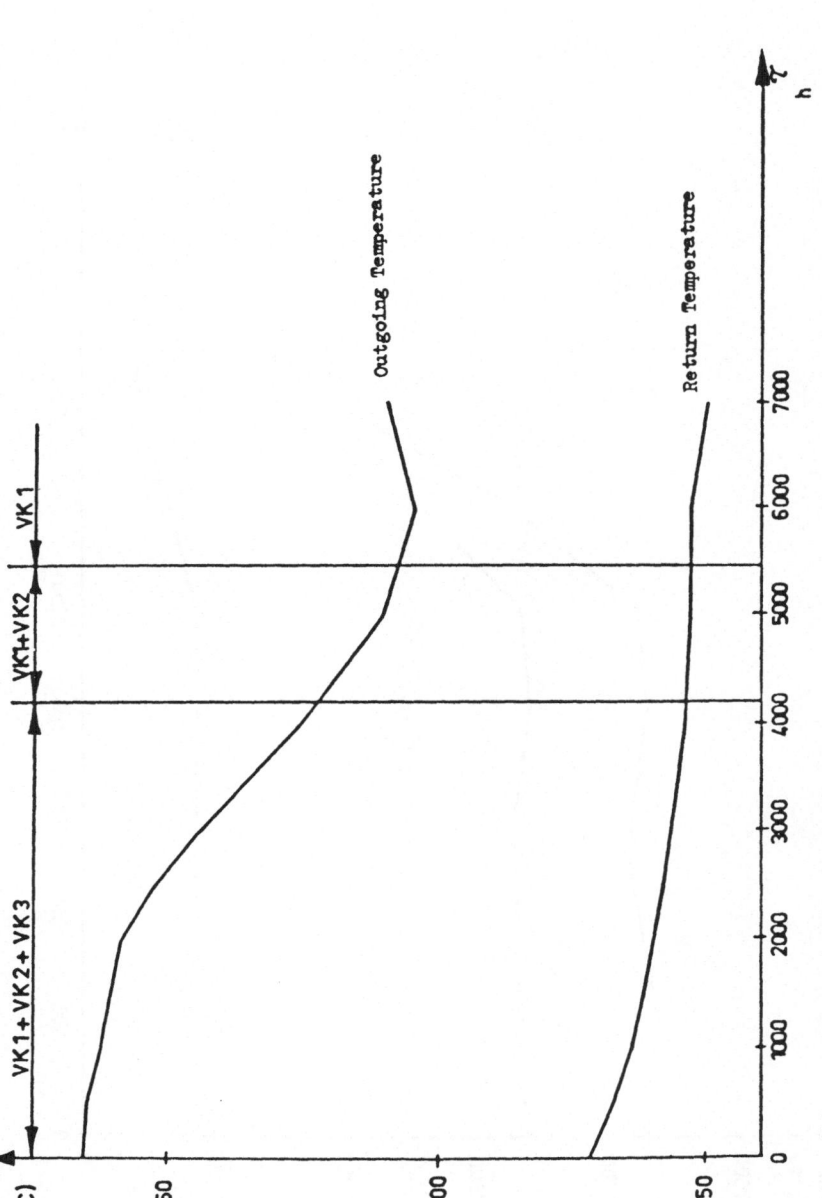

Fig. 54. Temperature duration curves and utilization times for the heat condensers (VK1, VK2 and VK3).

348

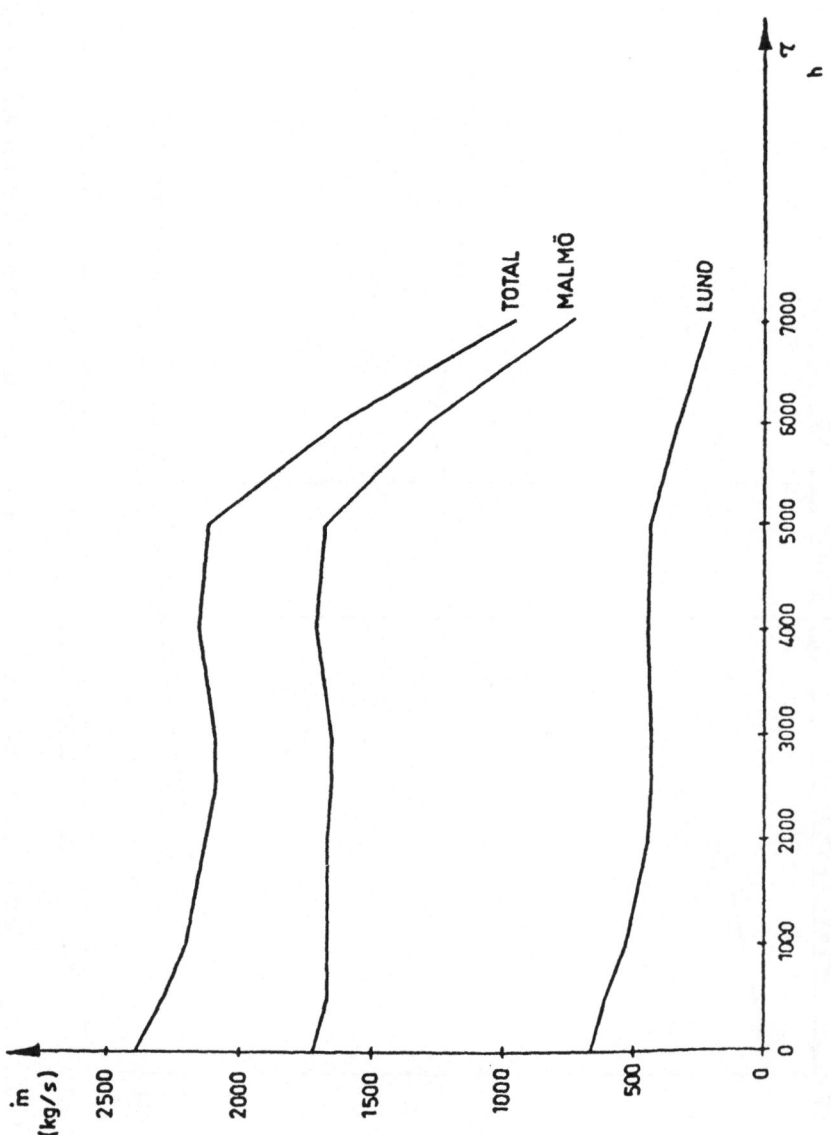

Fig. 55.   Mass flow duration curves.

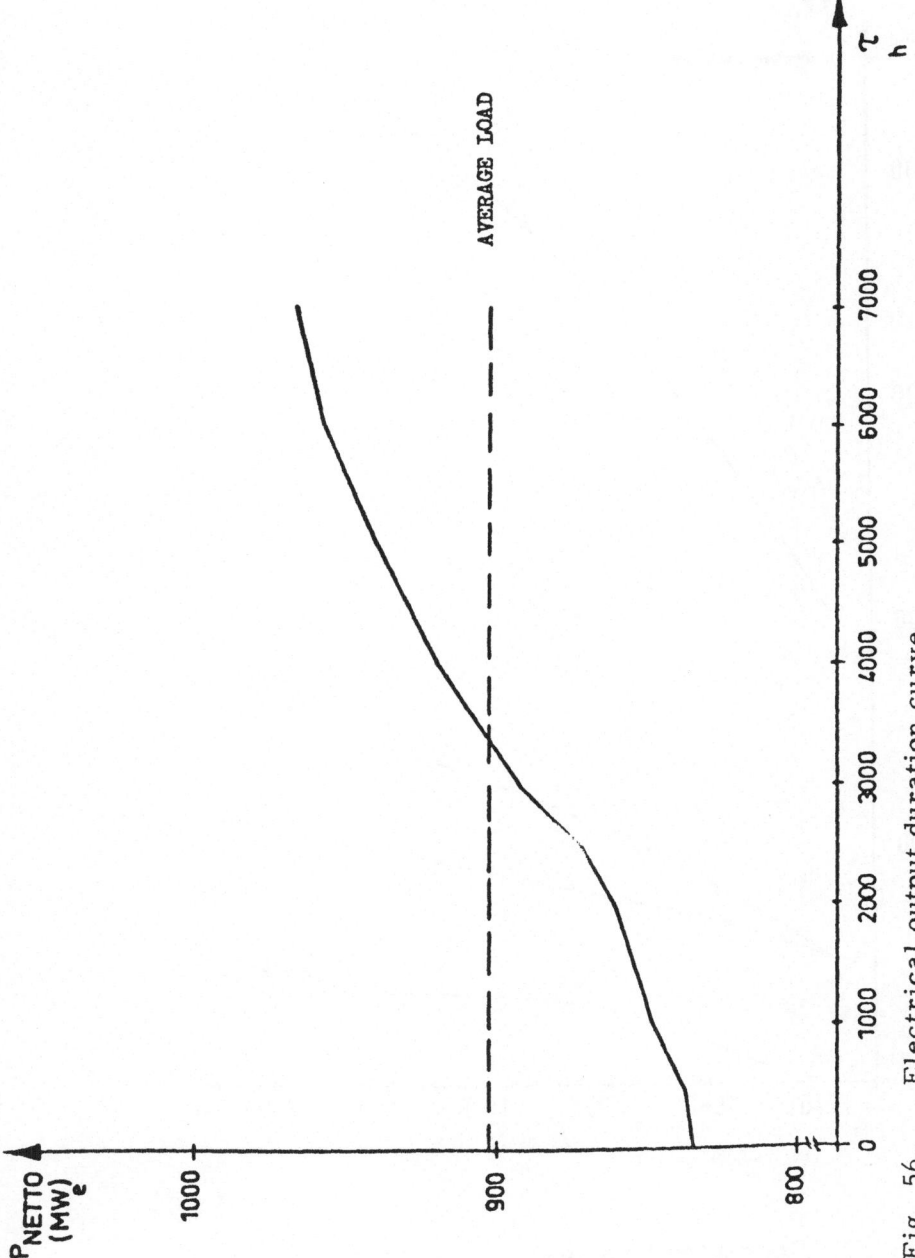

Fig. 56. Electrical output duration curve.

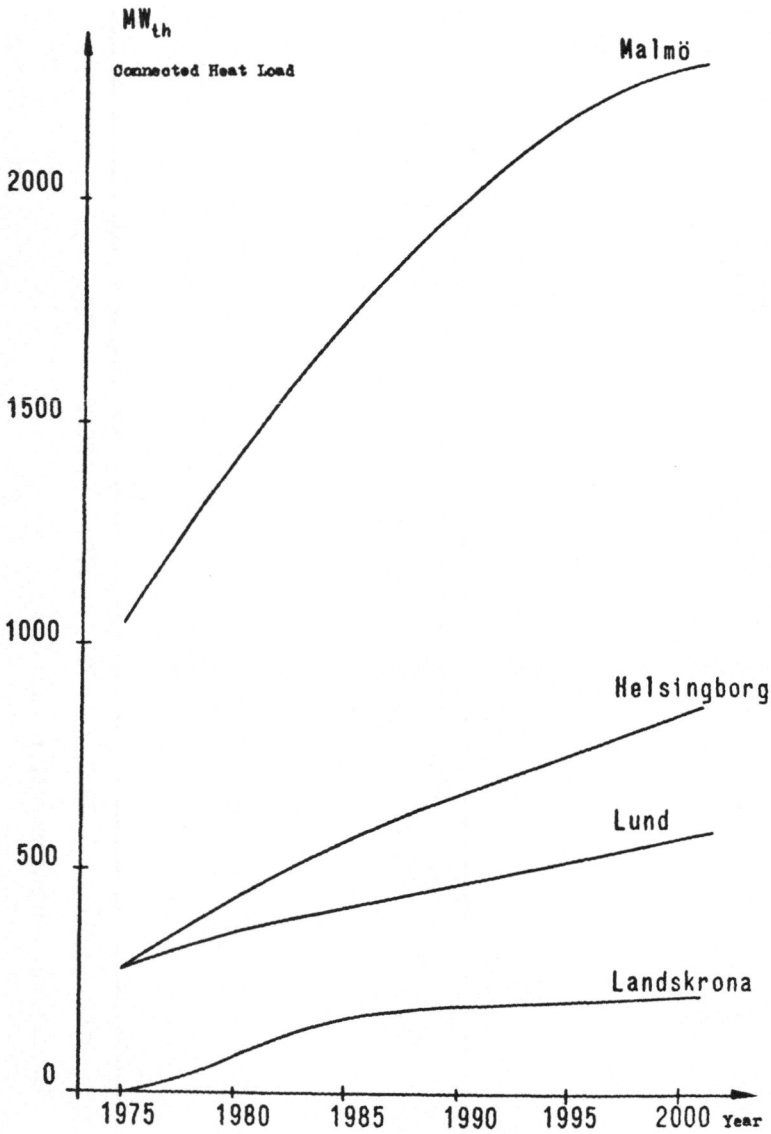

Fig. 57. Connected heat load.

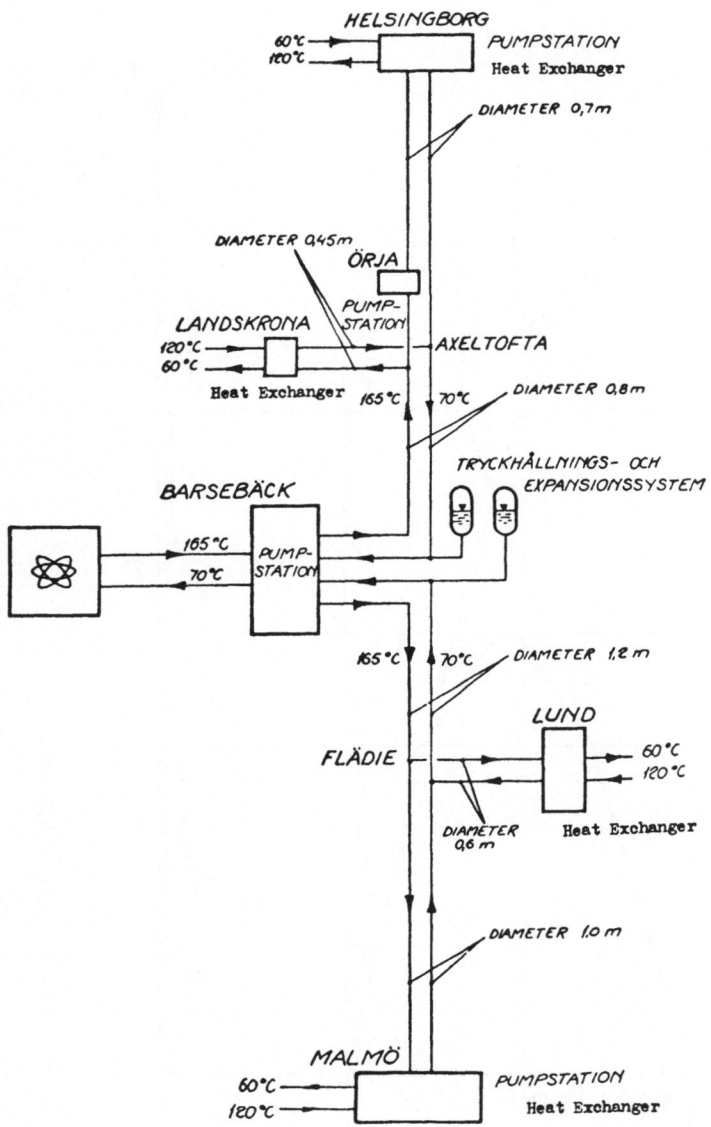

Fig. 58. Heat exchangers and pumping stations.

352

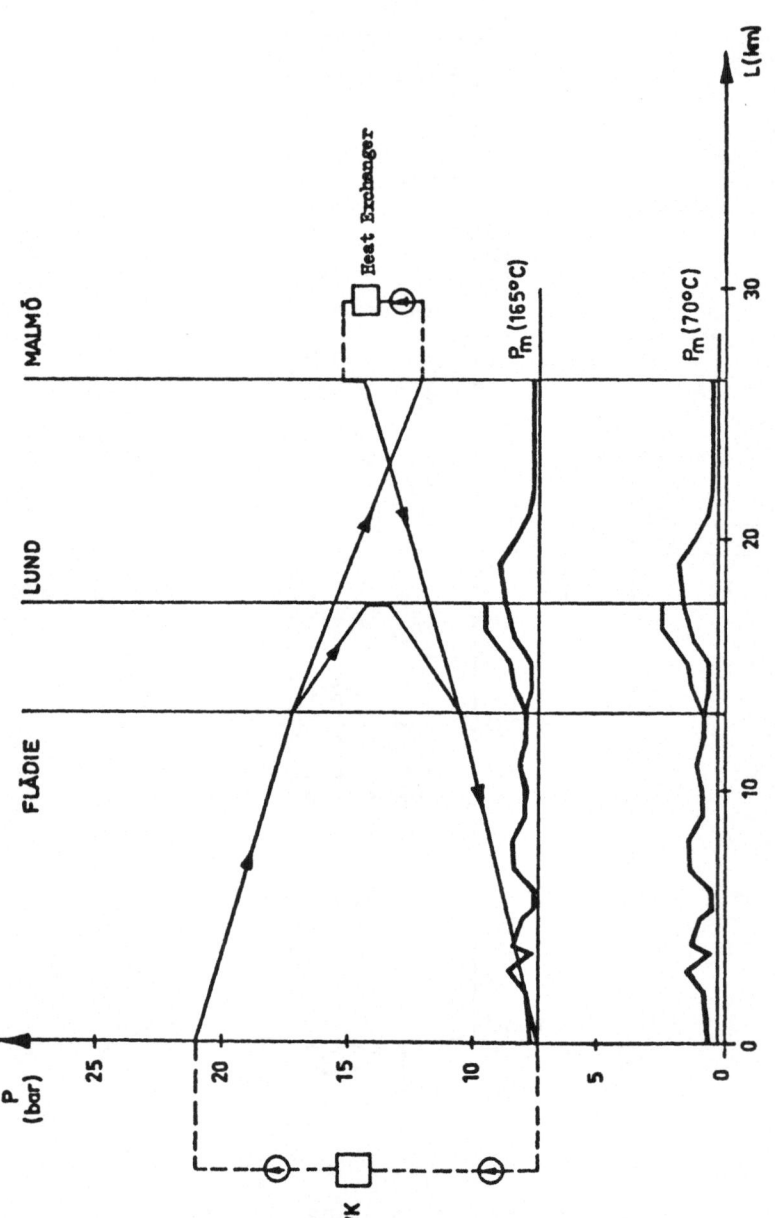

Fig. 59.   Pressure drop diagram.

353

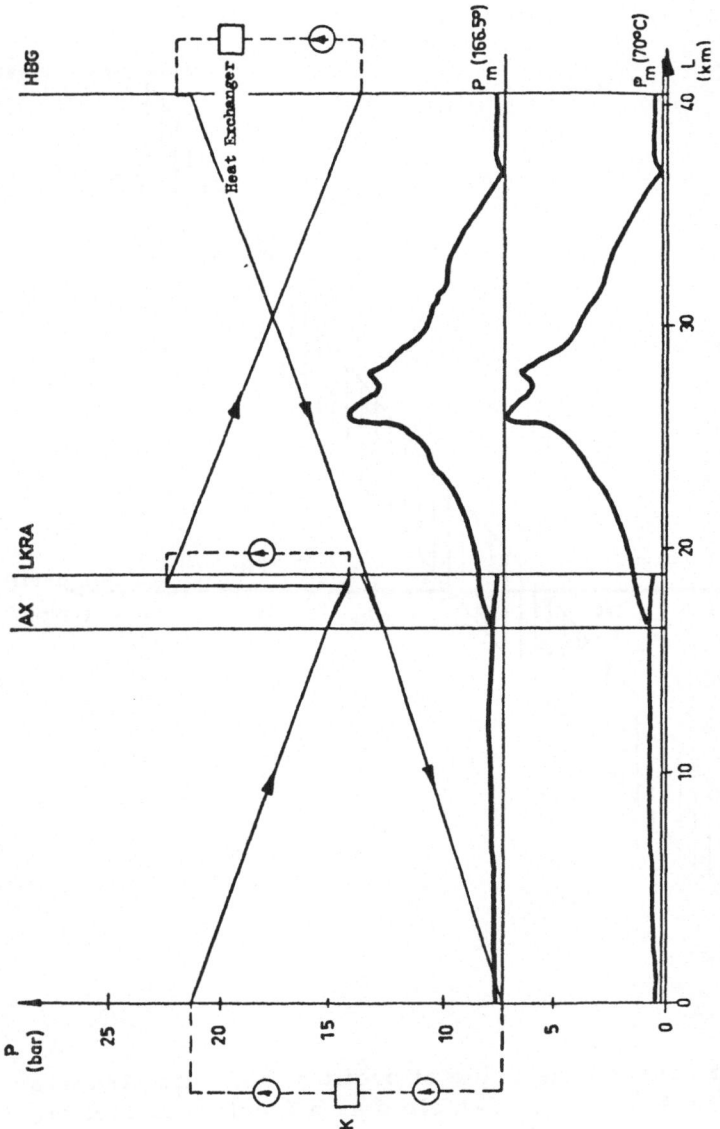

Fig. 60.   Pressure drop diagram.

354

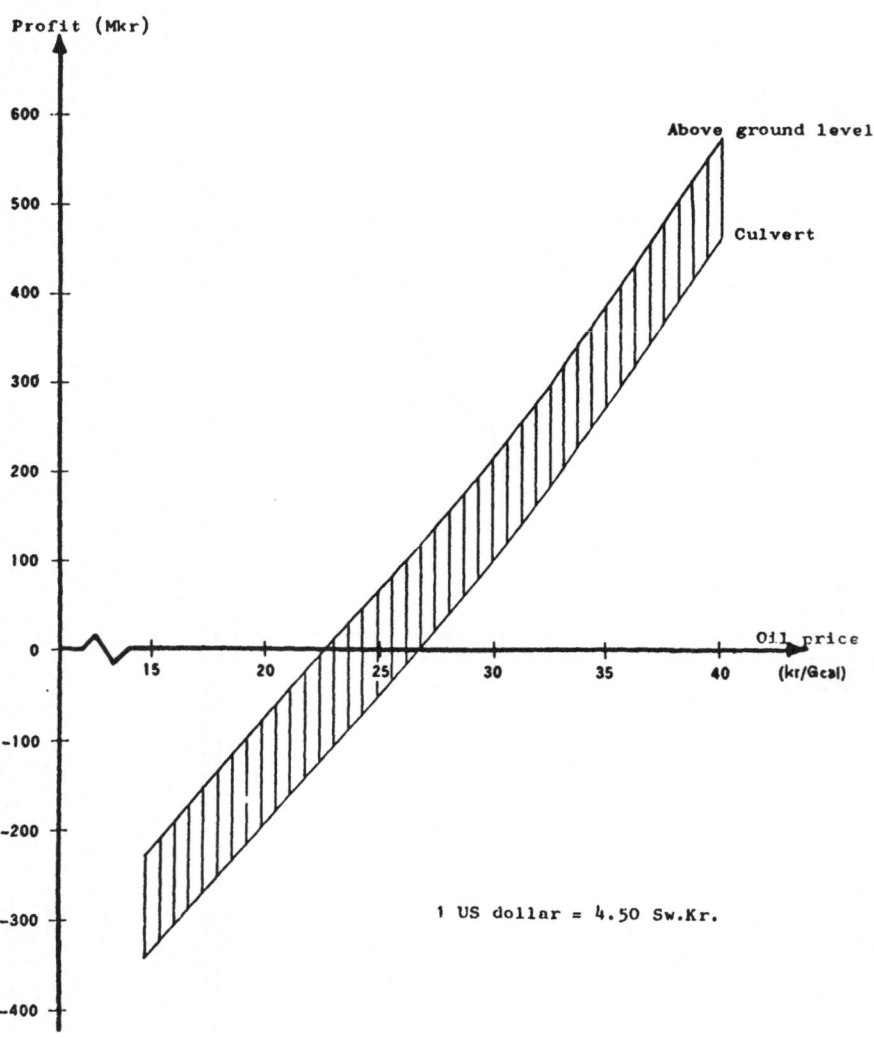

Fig. 61.   Present time value of profit of hot water transfer
Barsebäck-Lund-Malmö as a function of fuel oil
price

INITIAL ALTERNATIVE Starting year: 1983
Rate of interest: 10 %
Depreciation period: 25 years
Erection costs of nuclear power: 2,000 Sw.Kr./kW
Availability of nuclear power: 80 %

DIFFERENCES IN RELATION TO THE INITIAL ALTERNATIVE

Preliminary

Alternative	Break-even (oil price) Sw.Kr./Gcal	Profit 1990 Million Sw.Kr./year	Capitalized profit Million Sw.Kr.	Mean profit yield %/year
Initial alternative	22,75	25	210	8.0
Locating in trench	+ 0.25	- 1	- 10	- 0.5
Locating in culvert	+ 4.00	- 12	- 110	- 5.0
5 years shorter depreciation period (20 years)	+ 0.25	- 2	- 20	- 0.5
1 % lower interest rate (9 %)	- 1.25	+ 4	+ 70	+ 1.5
10 % higher costs of pipeline (330 million Sw.Kr.)	+ 1.00	- 3	- 30	- 1.6
10 % higher nuclear power costs (2,200 Sw.Kr./kW)	+ 1.35	- 5	- 40	- 1.5
5 % higher availability of nuclear power plants (85 %)	- 2.00	+ 7	+ 55	+ 2.0
Commission one year later (1983)	- 0.25	0	+ 10	+ 0.3
Without HVK and RVK in Malmö	- 2.50	+ 8	+ 70	+ 2.7
25 % higher steel prices	+ 2.00	- 7	- 60	- 2.8

Fig. 62.   Dependence of the result on changed conditions.

HEATING OF GREENHOUSES BY WASTE HEAT OF POWER STATIONS

Dr. L. Forgó and J. Bódás

Institute for Energetics, Hőterv, Budapest, Hungary.

ABSTRACT. In the near future a great number of greenhouses having large heat demand will be built in many countries and the heating of them is solved by means of standard heating installations. For this purpose, in co-operation with Soviet experts a solution was found, where the conditioning of the greenhouses is done by low temperature heat, rejected from condensers of power stations. The heating, conditioning and control of the whole system is solved according to agrotechnical requirements by circulated hot air, utilizing the waste heat of a thermal power station. The attractiveness of this suggestion lies in the fact, that with the combined operation of a power station and a greenhouse, both components of this system present economic advantages. The greenhouse is supplied solely by waste heat and the power station is using the air heaters of the greenhouse as air coolers for the condensing system. In addition to this, by means of the abundant heat exchanger surfaces, a much better vacuum is produced for the turbine, resulting in a higher output of the power station, even during the hottest days of the summer period.

1. INTRODUCTION

There are vast territories on the populated area of the earth, which are not able to supply the food demand of the inhabitants - especially vegetables and fruits - because of their climatic conditions. This situation will become worse in the future, as - owing to industrialization - just these territories are rapidly increasing. At the same time greater attention is paid to up-to-date nourishment all over the world and, thus, on those territories, where the adequate growing of vegetables cannot be ensured by ancient agricultural methods, the heated and conditioned greenhouses are of greatest importance. Naturally, these have to be built and operated according to up-to-date agrotechnics and economical large scale production.

Calculating the economy of modern greenhouses numerous factors have to be taken into consideration, e. g. the location, the climatic conditions, the possibilities of transport, the potential consumers, the costs of labour and fuel, etc. All these have, naturally, a decisive influence on the economy of the whole project. With regard to the increasing interest in greenhouses planned in many countries, studies have been elaborated during the last years to clear up the problems involved. Among them especially those have to be mentioned which were prepared in co-operation by the Soviet Teploelektroprojekt (Moscow) and the Hungarian Institute for Energetics (Budapest), because in them - beside the agrotechnical problems - great stress was laid upon the possible most economic energy supply. Detailed review of these studies does not belong to the scope of this conference, therefore only those details will be dealt with, which are of thermodynamical and energetical interest and represent a contribution to conventional solutions.

The design described later on may be of special interest, because it makes possible the utilization of very large waste heat quantities of very low temperature which otherwise would be rejected into the atmosphere.

2. COMBINED OPERATION OF THERMAL POWER STATIONS AND GREENHOUSES

Based on the beforementioned studies it can be stated that greenhouses for vegetables have to be built with an area of 6-30 hectares - in the Soviet Union up to 100 hectares - to assure economy i. e. to offer suitably low prices on the consumer's markets.

For the growing of vegetables usually 15-25°C air temperature and 80-90 per cent relative humidity should be maintained. Assuming such conditions the heating demand of a greenhouse in Central Europe equals 200-250 kcal/m^2, h in the coldest winter period, while in the Soviet Union it is 600-700 kcal/m^2, h. This means, that in case of a greenhouse of e. g. 34 hectares the heat demand in the coldest winter period is 68-85 Gcal/h in Central

Europe and 204-240 Gcal/h in the Soviet Union. The latter means
that the yearly oil consumption of such a greenhouse amounts to
approximately 66,000 tons.

At present this heat demand is generally supplied by oil or
gas fired boilers. In principle this solution is disadvantageous
because of the very low temperature of utilization. It is reasonable therefore, to supply this heat quantity in another, more
attractive way. In case of proper technical solutions this can be
realized by utilizing for heating purposes the heat rejected to the
condenser of a thermal power station. The temperature level of
this rejected heat is - according to present technology - in the
range of 35-40°C. The concept is to utilize the warm water of
35-40°C temperature taken from the condenser of a conventional
thermal power station for the heating of the greenhouse at a
15-25°C temperature level. If the heat demand of a greenhouse
of 34 hectares is taken into consideration under Soviet climatic
conditions, its max. value of 240 Gcal/hr comes up to the heat
rejected from a conventional thermal power station of 200 MW
capacity.

Obviously, in case of the combined operation of a power
station and a greenhouse, the most economic way has to be ensured to both of them so that - if the waste heat of the thermal
power station gets utilized in the greenhouse - no kind of restric-
tion shall become necessary in neither of the plants, due to
momentary climatic conditions, or any other reasons. Thus, it
has to be ensured to the thermal power station that - even if the
heat demand of the greenhouse is minimum, or occasionally zero
- heat rejection from the condenser shall be effectuated undis-
turbedly, while at the same time the greenhouse shall operate in
accordance with prevailing climatic conditions and dispose of
adequate safety with regard to the continuity of heating.

The above outlined principle requires, naturally, the solution of a series of further conditions not mentioned here. In the
following, only those are referred to, which are necessary for
understanding the system described below.

The economic advantages of the proposed system are as
follows:

(1) No fuel consumption arises when heating the greenhouse, as
    this heat would otherwise be rejected to the environment.

(2) The cooling water consumption of the power station can be
    eliminated, thus, in places where water is scarce, the existing
    water sources can be used for irrigation.

360

(3)   The heat-exchanger surface ensures the re-cooling of the
      circulating cooling water of the power station, and, simul-
      taneously, the heating of the greenhouse. Otherwise in both
      plants separate heat-exchanger surfaces would be necessary.

(4)   During the summer period the ventilation within the greenhouse
      can be assured without any further equipment.

The connection diagram of the system is shown on Fig. 1.

Steam produced in the boiler flows to the steam turbine.
The exhaust steam enters the jet condenser proved good in conventional air-cooled power plants. From the condenser warm
water is delivered by means of circulation pumps to the heating
system of the greenhouse. The amount of water corresponding
to the exhaust steam quantity gets delivered by the condensate
pump through the low-pressure preheaters to the feed pumps,
and through the high-pressure preheaters to the boiler.

The above outlined connection is identical to the air-cooled
plants of the power stations e. g. in Hungary (Gagarin Power
Station) or Soviet Union (Rasdan Power Station), and many other
countries.

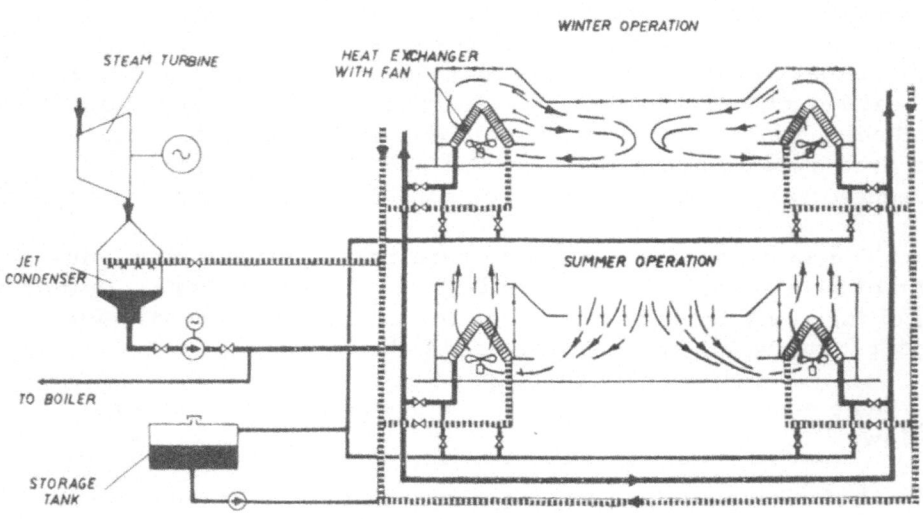

Fig. 1.   Connection of the greenhouse heating system.

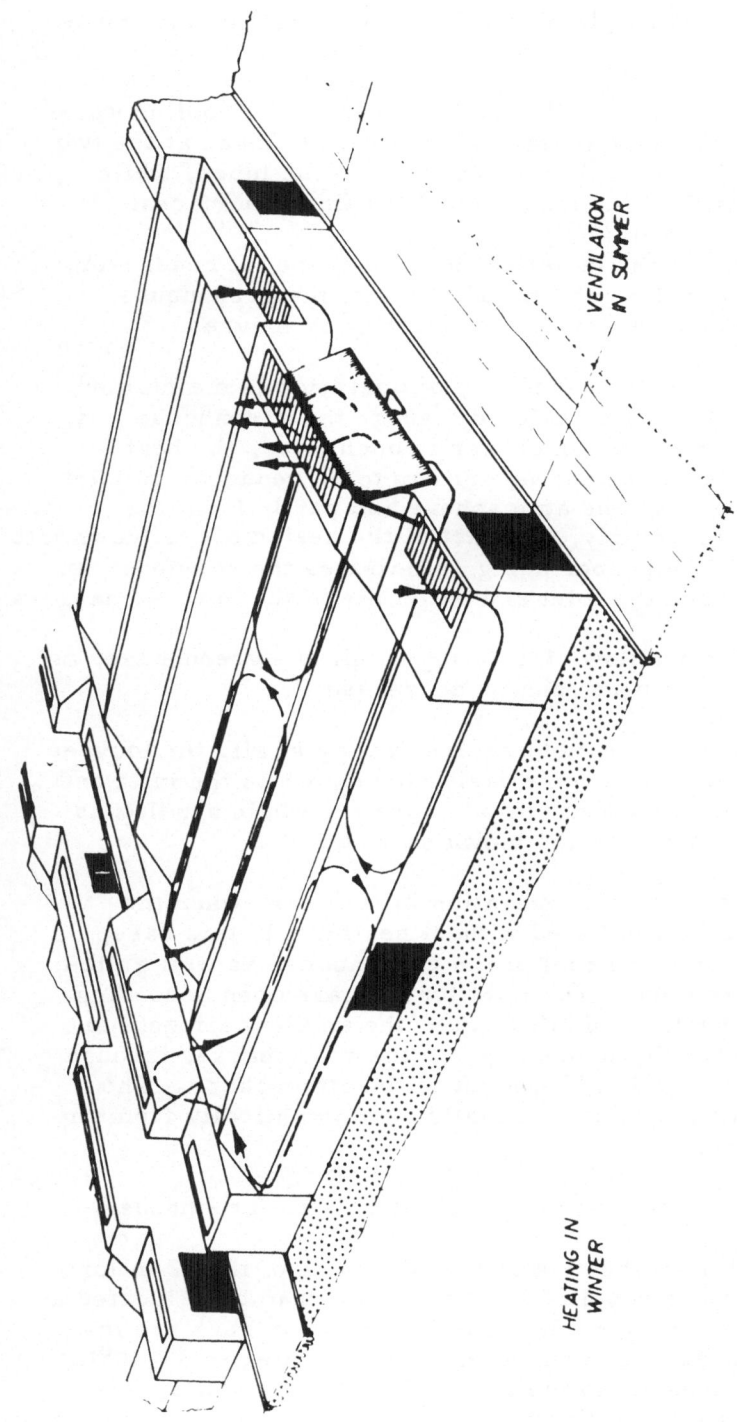

VENTILATION IN SUMMER

HEATING IN WINTER

Fig. 2. Heating and ventilation of the greenhouse.

The operating principle of the greenhouse heating system is as follows:

The water of about 35-40°C is delivered by the cooling water pumps to the finned surface heat-exchangers arranged at the two ends of the greenhouses. The water flows in the tubes, while the air is circulated by the fans around the finned surfaces.

The re-cooled water is delivered back to the jet condenser, while the warmed up air gets introduced into the greenhouse. The air flow can be controlled automatically by louvres.

In wintertime the described way of operation is maintained, and the whole air flow is circulated through the greenhouse. If the heat rejected from the condenser is unchanged, the heat demand of the greenhouse, however, has to be reduced, the inlet air is taken partly from the atmosphere and partly from the heated air and, accordingly, one part of the heated air is blown into the atmosphere. The proper design eliminates the possibility of recirculation between the inlet and outlet air of the heat-exchangers.

In this way the air flow circulating within the greenhouse, as well as the introduced heat, can be decreased.

If the greenhouse does not require heating at all, the louvres for the air circulation are completely closed, while the inlet and outlet louvres are fully open. In this case the whole air flow is taken from the atmosphere and blown back to it.

There is another way of operation in summer, when the greenhouse has to be ventilated without heating. In this case, the openings built into the roof of the greenhouse, as well as the outlet louvres, are open. The inlet louvres are open according to the rate of ventilation. In this case, the ambient air gets to the greenhouse through the openings on the roof, passes through the heat-exchangers and is blown out to the atmosphere. The louvres are controlled either manually or automatically from the remote control room.

Fig. 2 shows the heating and ventilation of the greenhouse.

Air serving for the heating and conditioning of the 72 m long greenhouse sectors is supplied by the heating chambers located at the two ends of the sectors. In these heating chambers are installed the heat-exchangers supplied by warm water of 35-40°C temperature which warm up the air to 15-25°C. In winter this warm air streams in an entirely closed circuit - according to the fingers represented on the left side of the figure - and performs the heating of the greenhouse. In this case only a relatively small amount of fresh air has to be introduced for the sake of ventilation. The fingers on the right side of the figure show the operation of the greenhouse in summer, when it does not need any heating. In this case, the fans of the heat-exchangers take the air from the atmosphere, circulate it through the inner space of the greenhouse and reject it - warmed up by the heat-exchangers - to the atmosphere. By means of the louvres placed in the heating chambers, the most favourable air flow can be set up between the two extreme operational states in accordance with the prevailing climatic conditions, as it may be seen on the left, resp. right side of the figure. In these cases the fans take the air partly from the atmosphere, partly from the air warmed up by the heat-exchangers and blow this mixed air through the greenhouse. Thus, the air-conditioning of the greenhouse can be ensured under all circumstances. At the same time - in case of whichever mentioned operational way - the amount flowing through the heat-exchangers is always sufficient for rejecting the necessary heat quantity from the cooling water of the condenser. Thus, the power station is also independent of the operation of the greenhouse, which is conformed to weather conditions.

Fig. 3 shows the main dimensions of a module of the greenhouse, which contains the entire heating and control system

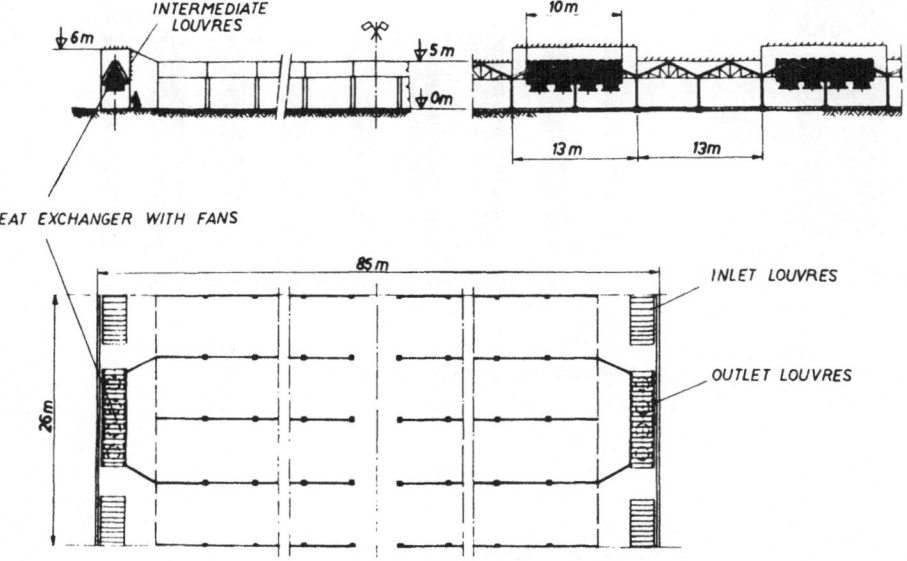

Fig. 3. Main dimensions of the greenhouse unit.

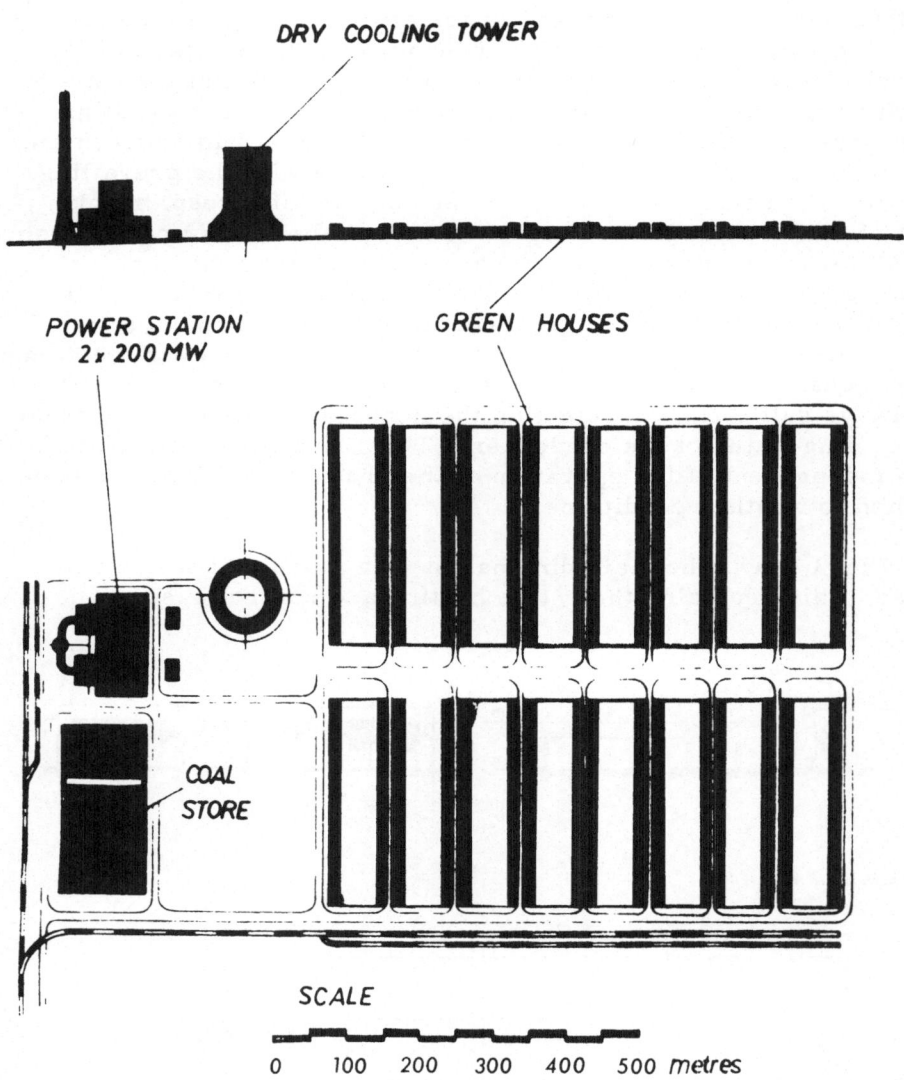

Fig. 4. General layout of a power station combined with greenhouse.

necessary to operation. The productive area of a module for the growing plants is 0.184 ha. From these units a greenhouse of optional area can be built up, with suitable passages for transportations.

Fig. 4 shows the general layout of a power station combined wi'h greenhouse. The power plant is equipped with two steam turbines, each of 200 MW capacity and a coal storage area; the rejected heat of one unit gets transferred to the dry cooling tower, while the heat of the other one passes through the heat-exchangers of the greenhouse. With this solution the heating of the greenhouse can be ensured even if one of the two turbine units is out of service. In this case it is only necessary to connect the warm water pipes to each of the condensers. In this way the waste heat of the power station is rejected by the heat exchangers of the greenhouse and the dry cooling tower can be shut off. A power station of this type, consisting of four 200 MW turbines, each with dry cooling towers, is running in the Soviet Union, in Rasdan. The installation of a fifth unit in this power station has already been taken into consideration, but instead of the installation of a dry cooling tower, combined with a greenhouse.

In order to study flow conditions in the greenhouse, model tests have been carried out in the laboratory of the "Institute for Energetics", Budapest. Fig. 5 shows the photo of the model used for this purpose. The aim of the model tests was to determine how flow conditions can be brought about in such a way that air velocity shall be, possibly uniform, without exceeding certain upper limits and that temperature differences within the greenhouse shall also remain little.

Apart from the direct technological task, that power station and greenhouse heating shall be independent of each other, the installation, described above, has to meet several safety and operational demands. From among the most important ones are the following:

(a) Protection of the greenhouse against unexpected breakdown of the heating.

As the heating of the greenhouse has to be ensured by all means, even in case of unexpected breakdown of the whole power station, suitable standby heating has to be ensured. As power stations usually comprise several units, this requirement can be met with the suitable connection of the pipelines as mentioned before. In these cases it can be taken into consideration, that

Fig. 5. Model test for the greenhouse unit.

under emergency conditions - even if the turbine is shut off - steam can be supplied by the boiler to the greenhouse.

(b) Frost protection of the heat-exchangers installed in the greenhouse.

Though in case of greenhouse heating the water-air heat-exchangers are located in an entirely closed room, water circulation may stop if the whole power station falls out of service. In this case the heat-exchangers have to be drained off into the tank marked on the scheme of Fig. 1.

(c) Putting into service and shutting down of the power station combined with a greenhouse.

From the viewpoint of the power station, the above-described combined system does not differ fundamentally from a power plant operating with a conventional air-cooled installation. As to the dimensions, however, the heating system of the greenhouse, which is at the same time the condensing system of the power station, is relatively widespread, much larger than in the case of a simple air-cooled installation. Therefore, when putting into operation the power station, many heat-exchangers must be put into service and a great amount of water has to be set in circulation. This can, however, be effectuated by starting the water circulation only in the mains, bypassing the heat-exchangers and, at due time, the heat-exchangers can be connected one after the other to this closed circuit by means of remote control.

Apart from these main safety, resp. operational requirements - naturally - many other factors have also to be taken into consideration, this, however, will not be dealt with in this paper.

The above described solution is justified - as mentioned in the introduction - by savings of energy. This, however, has to be ensured in such a way that the operation of the power station combined with a greenhouse shall be economic as a whole. In order to prove this, detailed economy calculations have been carried out, which included the thermal investigation of the operation, as well as studies with regard to the agrotechnical part of the greenhouse. The results of this investigation are summed up in the following table which gives only the main characteristics, referring to the thermodynamical economy of the combination. The table referring to a greenhouse area of

TABLE

Investment costs 10^6 Ft*

		A	B
		Separate power station and greenhouse	Power station and greenhouse combined
1.	Wet cooling tower for the power station 200 MW	120	-
2.	Greenhouse 34 ha, without heating	358	358
3.	Heating plant + pipeline to the greenhouse	338	-
4.	Heating installation inside of the greenhouse	119	-
5.	Combined condensing and heating installation	-	473
	Total investment costs:	935	831

Operational costs
10^6 Ft/year

		A	B
1.	13. 5% of the investment costs	126. 00	110. 20
2.	Fuel consumption in the greenhouse	80. 00	-
3.	Self-consumption of electric energy	3. 15	23. 9
4.	Make-up water for the power station	4. 45	-
5.	Maintenance (surplus to A.)	-	3. 97
	Total annual costs:	213. 60	138. 07

*Ft = Forint = unit of Hungarian currency.

34 hectars (one hectare = 100 x 100 m*) operating under Soviet climatic conditions shows in the first column the separately operated power station and separately operated greenhouse, and in the second column the operational data of power station combined with a greenhouse.

As the table shows, against the separate operation of the power station and the greenhouse, their combined operation offers considerable economic advantages and important savings in fuel oil equalling 66,000 tons per year. Referring the economy to the production costs of the goods produced in the greenhouse, it was found that - setting out only the results of the performed calculations - in case of combined operation, the prime costs of greenhouse products are about 25% lower than in case of conventional greenhouses operated with separate heating.

---

* 1 hectar = 2.17 acres.

SOLAR ENERGY - RESEARCH AND APPLICATION

David Wolf

Weizmann Institute of Science and Ben-Gurion
University, Israel.

INTRODUCTION

The sun is the source of energy of our planet. It supplies us every year with an amount of energy which, if accumulated and saved in a useful form, would be sufficient for the use of the whole world at a high standard of living for thousands of years. And yet, we have great difficulties in using solar energy in our daily life, since it comes to us in a way which is inconvenient for industrial use. The energy reaching the earth is very diffuse; it comes only at daytime and even then, changes constantly according to the day and the season of the year; it disappears for various periods of time on cloudy and rainy days, depending on the geographical location and the season. Solar radiation energy cannot be used as it is, and must be converted into other energies (heat or electricity) by the use of solar collectors or solar cells. The energy collected also depends on the orientation to the sun of the collectors.

Conversion of solar energy into organic material by photosynthesis is another possibility and is actually the way by which life is maintained on our planet. All these factors affect its possible application for domestic and industrial use. Most research and development work is oriented towards trying to overcome all these problems, and in the most economical way. The effort to utilize solar energy has a long history and one can find a vast literature on this subject[1,2,3]. A comprehensive review of the utilization of the sun was recently written by Yellot[4].

GENERAL

As a result of the recent energy crisis, a renewed interest arose in solar energy. Due to the low cost of fossil fuel before the energy crisis, solar energy could not compete with it in most parts of the world, if calculated on the basis of cost per Kcal produced. Moreover, since solar energy produces mainly low-level energy through heat conversion, this source of energy became even less attractive. Not only was solar energy out-priced by fossil fuel, but most of the other available energy sources were also outpriced, including coal, which exists in the world in huge quantities and could supply world energy needs for hundreds of years.

The reason for the high cost of energy obtained from solar radiation is due to the large and expensive equipment installations needed for solar energy collection. Large installations are needed because the intensity of solar radiation is low and large collection areas are required in order to harness a reasonable amount. Moreover, as previously mentioned, the sun is not present at night nor during cloudy days and therefore a storage system is normally necessary which renders the system even more expensive, first, because of the storage system itself and second, because of the additional collectors needed for the accu-mulation and storage of heat for use at times when the sun is not available. The option of using back-up heating systems for nights or cloudy days is also expensive and increases the costs of the energy obtained.

Even under the difficult economic conditions which prevailed before the energy crisis, there were cases where solar energy was still economical and relatively widely used. One of those places is Israel, where solar energy is used in domestic and industrial applications. About one-fifth of the domestic water heating is produced by solar collectors.

Figure 1 shows a quite common apartment building in Israel with solar collectors on the roof. Two collectors and one tank are a typical heating unit for a family.

Also Morse[5] reports that in Australia in 1973 there were about 20,000 square metres of solar collectors and Nogouchi[6] reports that in Japan in the same year there were 2.6 million domestic solar water heating units in operation.

The best known industrial use of solar energy is in the pro-duction of potash at the Dead Sea Works in Israel. Salt water

Fig. 1.   Solar collectors on the roof of an apartment building in Israel.

Fig. 2.   Solar ponds at the Dead Sea Works in Israel.

from the Dead Sea is pumped into huge solar ponds covering an area of some 130 square kilometers where the water is evaporated by solar radiation and the sodium chloride and afterwards, carnalite, crystallize. The latter salt is then decomposed and potash is obtained. A general view of the solar ponds is shown in figure 2. It is interesting to note that the amount of heat used by the ponds for evaporation is equivalent to the total amount of energy presently used in Israel. In other words, Israel is the only country in which approximately half of its energy comes from the sun.

Another industrial application which was used for some time in Israel was in evaporation of the water from the reject brine of an experimental reverse-osmosis desalination plant, operating on brackish water and placed on land in the desert area. This evaporation process was instituted in order to prevent return of the salt into the ground-water of the aquifer and thus avoid increasing its salinity above the present level. The solid salt can easily be discharged in places where it can do no damage.

As a result of the energy crises and the increased cost of fossil fuel, solar energy again becomes attractive even if the increase in cost of the materials normally needed for solar heating systems has worked against it. Therefore, a significant amount of research and general studies are being done now in developing more efficient but less expensive collectors and solar cells, new materials and better surface properties, and computer programmes for system analysis, economic evaluations and optimization studies. We shall now discuss in more detail the various aspects of solar energy, the problems involved in its application and various application possibilities.

SOLAR RADIATION

Solar radiation intensity is a function of location, time of the day and weather conditions. The maximum radiation is outside the atmosphere and on a surface perpendicular to the sun; it is 1162 Kcal/hr. m^2 [3]. Due to the atmosphere, especially water vapour and carbon dioxide which absorb part of the solar energy, there is an attenuation in the radiation and thus the solar radiation reaching the surface of the earth is reduced considerably. The amount of radiation differs in each location, due to the various weather conditions and angles of the solar rays at various locations on the earth and at different hours of the day. The average monthly global insolation at the Israel Meteorological Station in Beit Dagon for the years 1962-1968 is given in Table I with an average annual value of 4630 Kcal/m^2 day [7].

TABLE I

Daily Average Global Insolation of Solar Energy on a Horizontal
Surface Measured in K-Cal/m^2 day.

Month	Beit Dagon Israel	Zurich Switzerland	West Virginia USA	Colorado USA
Jan	2,580	370	1,970	1,900
Feb	3,330	950	2,720	2,710
Mar	4,340	1,830	3,370	3,900
Apr	5,320	2,630	3,870	4,950
May	6,300	3,220	5,220	5,500
Jun	6,760	3,750	5,410	6,400
July	6,600	3,690	5,280	6,350
Aug	6,070	3,180	4,570	5,720
Sept	5,100	2,210	3,890	4,450
Oct	3,910	1,080	3,330	2,150
Nov	2,960	410	2,110	2,100
Dec	2,290	230	1,610	1,600
Yearly Average	4,630	1,960	3,610	4,060

TABLE II

Hourly Average of Global Radiation for the months of January
and July at Beit Dagon, Israel.

Appearance of the sun hr.	January Kcal/hr. m^2	July Kcal/hr. m^2
0	0	0
1	75	100
2	220	250
3	320	440
4	345	620
5	370	740
6	350	800
7	300	825
8	200	790
9	80	715
10	0	620
11		480
12		260
13		100
14		0

For comparison, data is also given in Table I of the insolation in Zurich, Switzerland[8], West Virginia, USA[9] and Colorado, USA[10]. The average hourly global radiation at Beit Dagon, Israel for the coldest winter month of January and the hottest summer month of July are given in Table II[7]. The global radiation was measured using an Eppley Pyranometer, a commonly used instrument for solar radiation intensity.

The electromagnetic radiation of the sun is in the range of $0.1-2.0 \mu$ wave lengths, with half of its radiation in the range of $0.38-0.76 \mu$ which is the visible range and with a maximum intensity around 0.5μ. The solar radiation is normally changed into heat if absorbed on a black surface and is directly transformed into electricity by means of solar cells. At this stage, we will be more concerned about the transformation into heat since it is more economical than the transformation to electricity (which is extremely expensive due to very high cost of solar cells). The application of solar cells is therefore limited so far to power satellites and space ships or to some military installations in remote locations.

The solar energy that we obtain on earth is composed of direct solar radiation and diffused radiation. The diffused radiation exists even on cloudy days and is in the range of 15-20% of the total radiation. In places where the albido is high, like areas with snow or sand, this radiation has a significant positive effect on the amount of radiation collection.

SOLAR COLLECTORS

Most of the solar energy in use today is based on solar collectors which convert the electromagnetic radiation into heat. The heat is then transferred to a fluid (mainly water or air) and is transported where needed or to a heat storage unit to be used at a later stage. One can find a great variety of commercial solar collectors at various costs and with various degrees of efficiency. The regular solar collectors are also called "flat-plate collectors" and are all made of an absorption plate, pipes or channels for the fluid which transfers the heat for use, a transparent cover between the plate and the sun, a thermal insulation in the back of the collector and a case in which all the elements are assembled and enclosed.

The main requirements from a solar collector are:

(1) High temperature output.

(2) High efficiency of solar energy collection.

(3) Low weight.

(4) Low cost.

A diagrammatical representation of the collector and its various parts is given in figure 3.

THE ABSORBER

The major part of the collector is the absorber. The flat-plate absorber of the solar collector is either painted black for increased absorption or is specially treated so as to have a so-called "selective surface" property. A selective surface has a high absorptivity for the wave length of the solar rays which are in the visible range and up to 2μ; and has a low absorptivity (thus a low emissivity) in the far infra-red region in the vicinity of 9-10 μ, which is the wave length region of the thermal radiation at a temperature of around 50°C. A diagrammatical representation of solar radiation and thermal radiation is shown in figure 4. The thermal radiation spectral emittance can be obtained from Planck's Equation[11] while the wave length with maximum emittance can be obtained from Wien's displacement Law[11].

The selective surface was developed in Israel by Tabor[12] and is being used successfully by the Miromit Co. in Israel in the production of their solar collectors. They report absorptivities and emissivities in the ranges of 0.9 and 0.15 respectively[13]. The selective surface property can be obtained by oxidation, electroplating or other surface treatments. The plate is heated up by the solar radiation and in the case of the flat-plate collector, it absorbs both direct and diffuse radiation and the heat is then transferred to the fluid which is in contact with the plate, either directly or through intermediate pipes as shown in figure 3. The direct contact type of collector is of the Roll-bond type, as found in the freezer of the domestic refrigerator unit. The solar collectors also differ from each other in their construction materials. One can find collectors made of steel, copper, aluminium and now even plastic materials are being considered. The selection among the various materials is based on the cost of the thermal properties of each material. The pipes, usually made of the same material as the plates, are connected to the plate (preferably soldered together) in order to get good thermal transfer conditions and thus high thermal fluxes. We found[14] temperature differences between the water in the pipes of the

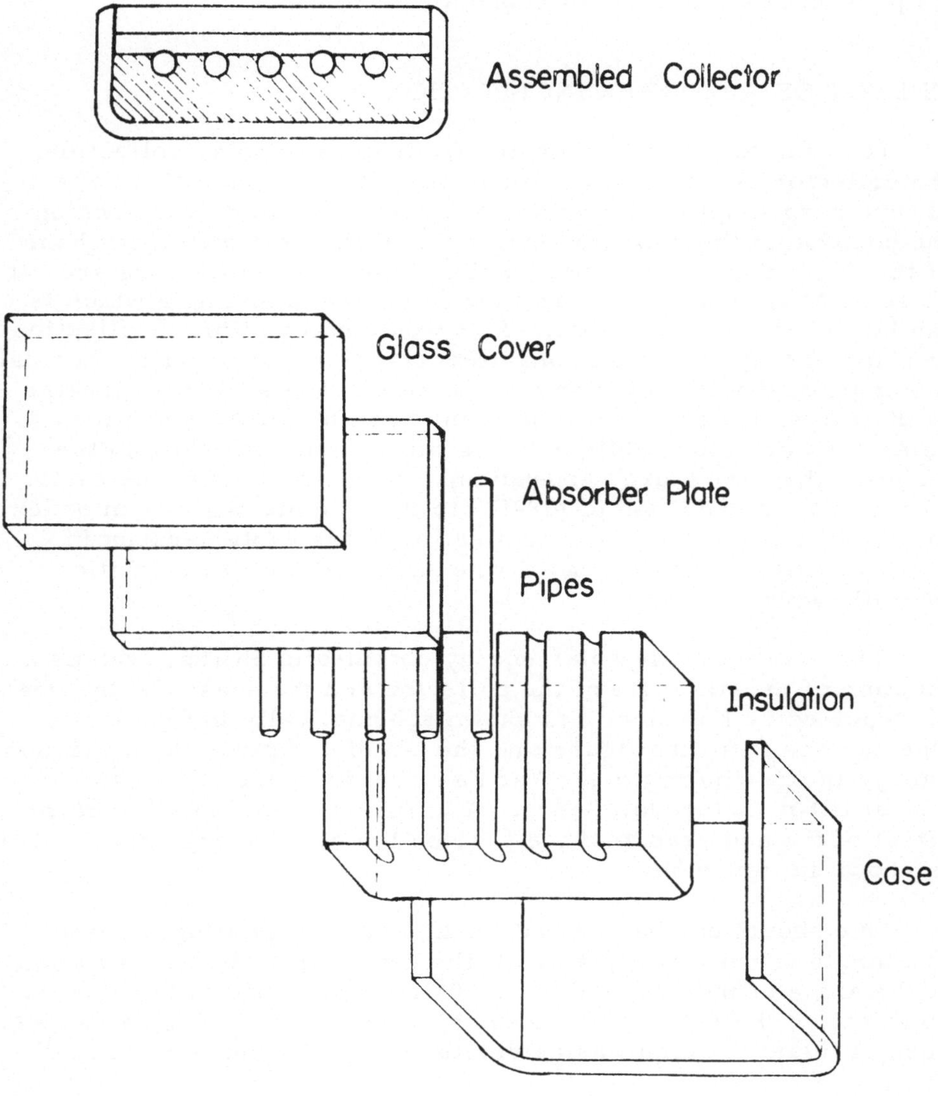

Assembled Collector

Glass Cover

Absorber Plate

Pipes

Insulation

Case

Fig. 3.   A solar collector and its various parts.

collector and the absorber plate of 2°-12°C for various types of
collectors and under similar experimental conditions. The better
the binding between the plate and the pipes, the smaller is the
difference found in the temperatures. In order to increase the
heat flux from the absorber to the fluid, we have recently considered utilizing the device known as a heat pipe. However, at
its present cost, this is not economically feasible.

EFFECT OF ABSORBER INCLINATION

In order to obtain maximum efficiencies of solar collection,
the collector should follow the sun so as to have the solar rays
always perpendicular to the surface. This would mean changing
the position of the collector every day of the year and every hour
of the day. Since this is not feasible, the solar collectors are set
up so as to face the south and they are tilted at an angle which is
the latitude angle, +10°-15°. This would give optimum collection
of solar energy with a reasonable average all year round. Normall
the angle of the collector and the earth surface should be greater
in the winter time and smaller in the summer time, since the
solar rays are more oblique in the winter months, while in the
summer they are more perpendicular to the surface of the earth.
Therefore, the addition of 10°-15° to the latitude angle is in order
to increase the solar energy collection in the winter season in
relation to the summer season when a higher rate of insolation
prevails anyway.

The average daily solar energy obtained in Shiraz, Iran as a
function of the tilt angle of the collector and the season of the year
as reported by Bahadori[15] is shown schematically in figure 5a.
One can see from the figure that the tilt of 45° gives the maximum
energy in the winter time. Ward and Löf also use 45° as an
annual tilt in Colorado, USA[10]. A different representation of the
effect of tilt and season of the year on the solar energy received
is shown in figure 5b.

One should emphasize that an angle approximating the latitude angle is quite acceptable all the year round since a deviation
of the angle of incidence of up to 30° from the normal has almost
no effect on the reflectance in the air-glass system. This can be
seen in figure 6 which was calculated using Fresnel's Equation[11].

Even with an angle of incidence between 30° and 50° of the
collector, reflectance does not increase by more than an additional 3%. However, from a 50° angle of incidence and over, the
reflectance increases in an exponential manner and therefore, a

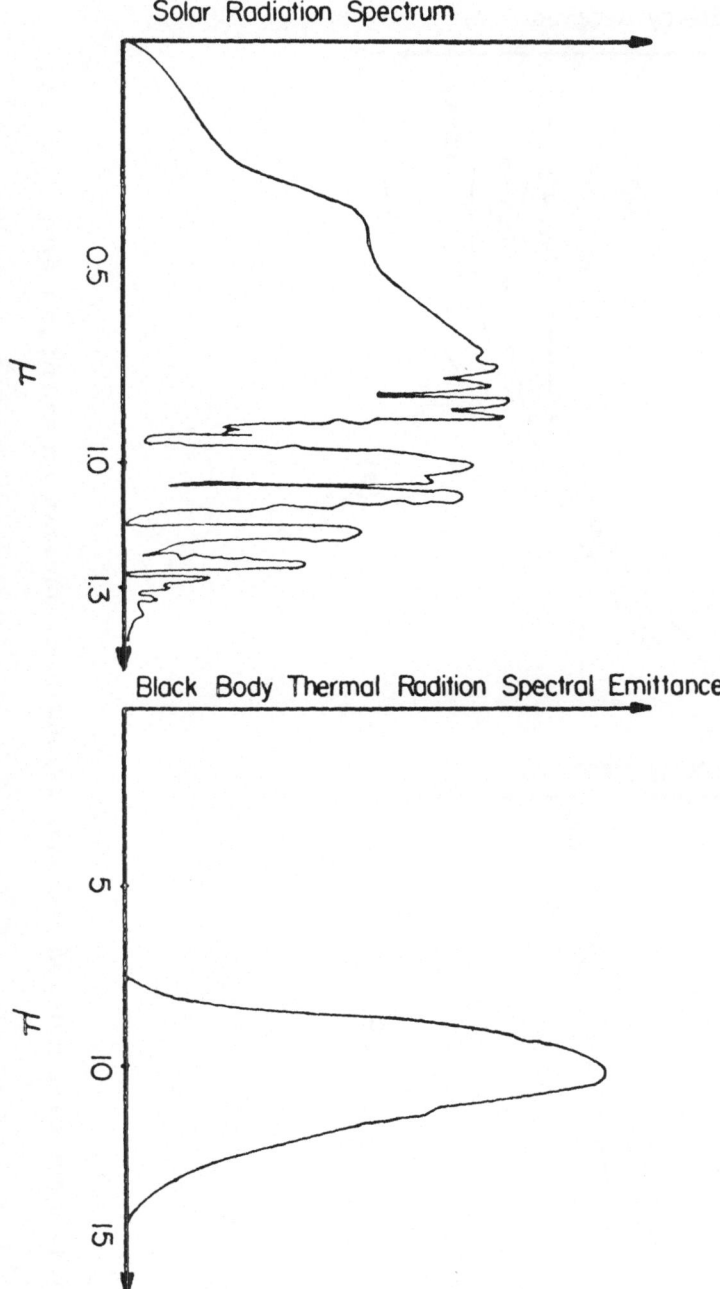

Fig. 4. Diagrammatic representation of solar and thermal radiation.

382

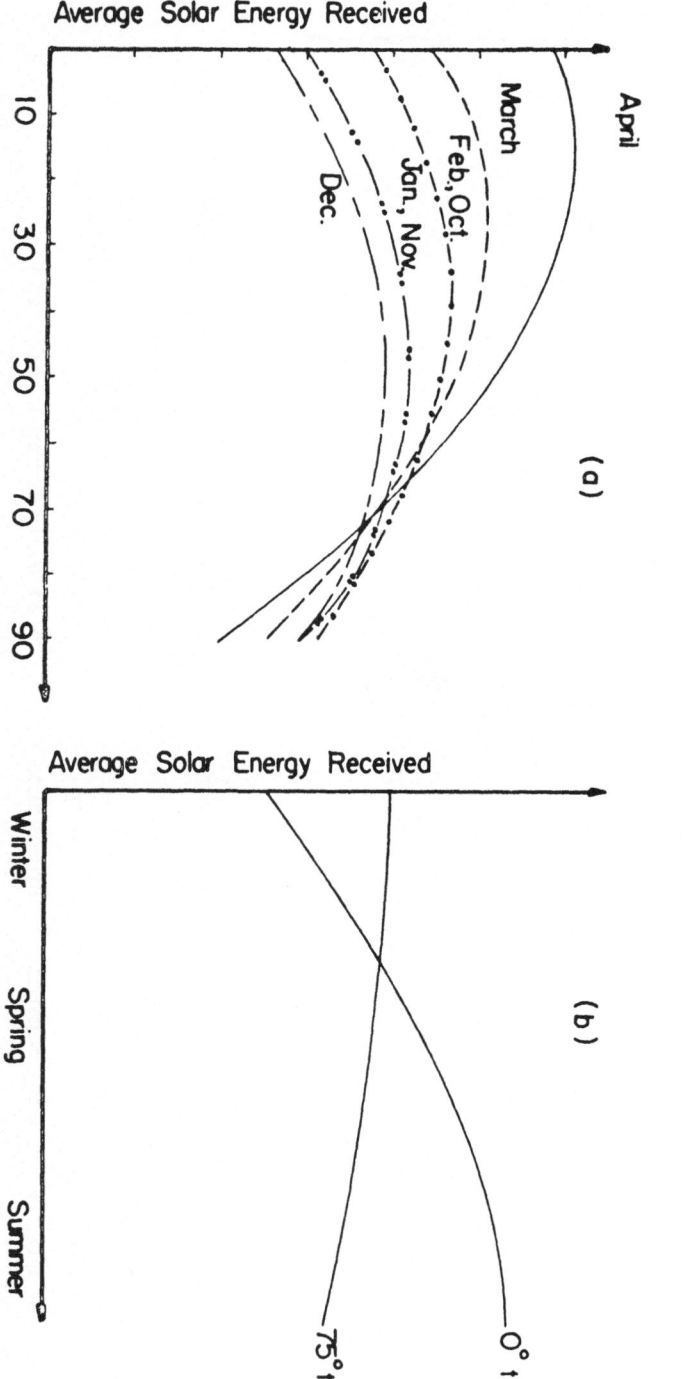

Fig. 5. Effect of tilt on the average solar energy received per unit area and unit time.

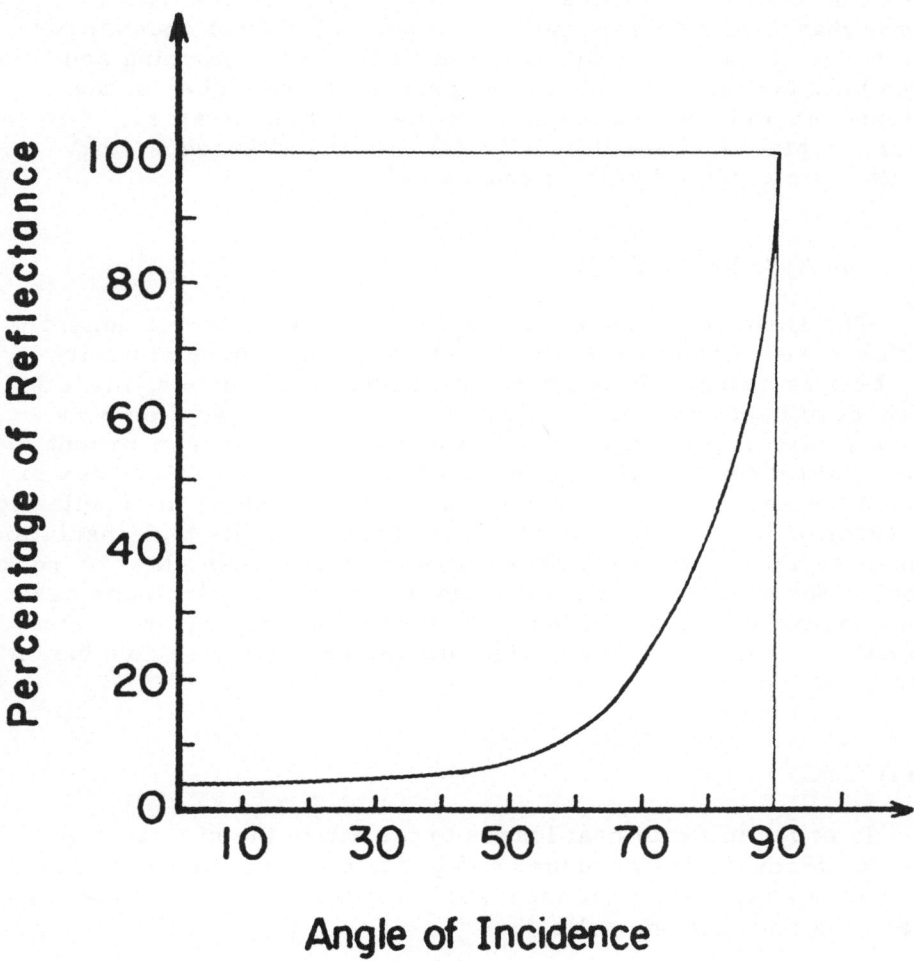

Fig. 6.   Variation of reflectance with angle of incidence for
          natural light.

horizontal collector is unacceptable in the winter time, while a vertical one is not recommended in the summer time. The other way around is of course recommended, except that the technical problems in changing the tilt of the collector with the season of the year makes the installation too expensive.

With regard to the East-West orientation, there are two different cases. In places where hazy mornings prevail, the somewhat South-West orientation is preferred first because of the reduced radiation, due to the haziness in the morning and also because the surrounding temperatures are higher in the afternoons and thus higher efficiencies are then obtained. However, in places where it usually rains in the afternoons, the South-East orientation is of course better.

CASING AND ISOLATION

The absorber is normally kept in an enclosed environment with a case made of galvanized steel which has low emissivity for heat radiation. Between the case and the absorber, there is a sheet of heat-insulating material such as glass wool or rock wool, while in the front of the absorber a cover of transparent material (normally glass) is used which allows the solar rays to reach the absorber, but avoids heat losses. Among the insulating materials, polyurethane is also used because of its high insulation characteristics, even so, it is more expensive than glass or rock wool. However, recently some claim that the polyurethane has some aging effects and it loses its insulation properties. This question needs serious investigation before a decisive conclusion can be reached.

GLAZING

In order to avoid heat losses to the air in the direction of the sun from the hot absorber and yet to allow the sun rays to reach the absorber, a transparent cover which in most cases is glass, is added to the collector.

Ironless glass is preferable in order to avoid excessive heat absorption by the glass. The glass has the required properties such as transparency for visible light and low wavelength of IR; it is opaque to the long wavelength of IR in the range of 9-10 μ , (which is the heating radiation region); it is resistant to UV radiation and other weather conditions except that it is breakable and could be damaged by heavy hail. The glass reduces convective heat losses from the absorber and also prevents radiation losses from the absorber, thus trapping the heat in the collector even if the collector plate does not have a selective surface. However, the glass does heat up by absorbing thermal radiation and therefore loses considerable heat to the atmosphere. Consequently, the application of a selective surface to the absorber plate and the lowering of the emissivity of the glass are both very important in reducing the heat loss from the solar collector.

In cases where selective surfaces cannot be obtained, double glazing can be helpful in reducing the thermal losses. However, in this case, the solar radiation to the plate is also reduced, since each sheet of glass reflects about 8% of the perpendicular solar insolation[11] and therefore the gain in the heat loss by adding a second glass cover is sometimes negligible in comparison with the loss of 8% of the insolation. In fact, the use of double glazing along with a selective surface for getting higher temperatures would probably be unsatisfactory, since the 8% loss in isolation by adding an additional glass cover would be more than the gain in heat loss. Obviously, a surface treatment such as covering the glass with an anti-reflecting coating is recommended if the cost of such a treatment is not prohibitively expensive. The reflectivity for any angle of incidence for the non-polarized radiation can be obtained from Frenel's and Shell's Equations[11] as shown in figure 6 for the air-glass system.

PLASTIC COVERS

Transparent plastic sheets have always been considered for replacement of the glass covers, due to the fact that glass is breakable, quite expensive and quite heavy. Although most of the plastic materials compete favourably with glass from these points of view, they are not so good in other needed properties. Thus they do not withstand well most weather conditions, are chemically degrading under the effect of the UV radiation, have lower transmittance to visible light[16] and they are not always opaque enough in the range of the thermal IR radiation. Among these materials are polyethylene, polyester, polyvinyl chloride and others. In the last few years a good plastic material sug gested was polyvinyl fluoride, also known by its trade name as Tedlar. Tedlar has good physical and chemical properties, withstands UV radiation and is nearly opaque in the range of 8-10 μ. The spectral transmittance of infra-red radiation for several plastic films and glass[16] is shown in figure 7. One can see that polyethylene has a high transmittance for thermal infra-red radiation in the range of 9-10 μ and thus is not suitable for glazing, while polyester and polyvinyl fluoride are both opaque in that region, practically similar to glass.

If the extinction coefficient is known, the transmittance as a function of the thickness of the film can be obtained, using Lambert's Equation. However, if this coefficient is not known, it can easily be obtained experimentally[17].

The Tedlar also has lower reflectance than glass. In fact two Tedlar sheets, each one 0.1 mm thick would reflect no more solar rays than one glass sheet. Tests were made in order to check the lifetime of the Tedlar and the other plastic materials. Edlin and Willaver report[18] that the projected life of Tedlar films is many years, compared to all other plastic films, which under the same experimental conditions have less than one year of lifetime.

EFFECT OF DIRT

The effect of dirt can be significant in reducing the transparency of the glazing cover. Although Hottel and Woertz[17] report very small effects of dirt, Garg[19] in India has made systematic experiments and has found that the dirt reduced the transmittance considerably. In the dirty areas in Roorkee, he found after 30 days of exposure, losses in transmittance from 90% for vertical conditions to 30% for horizontal conditions. As the inclination of the transparent material was more vertical, the losses in transmittance were less. Garg also found that the dirt correction factor (which is the ratio of transmittance of uncleaned plates to clean plates) is lower for plastic material than for glass for any inclination, meaning that dirt settles more on plastic material than on glass.

STORAGE

Two different storage systems have to be considered for solar energy accumulations; heat storage and electrical storage. We will refer mainly here to heat storage although electrical storage is a very important subject in itself. We will only mention that electricity can be stored in accumulators or batteries or it can be used for pumping water to higher levels and then allowing it to fall down and produce electricity when needed. Electricity can also be used for the decomposition of water, and then when energy is needed, the hydrogen can be utilized either by direct combustion or by fuel cells or even by producing from it other organic materials (such as in the coal purification process).

387

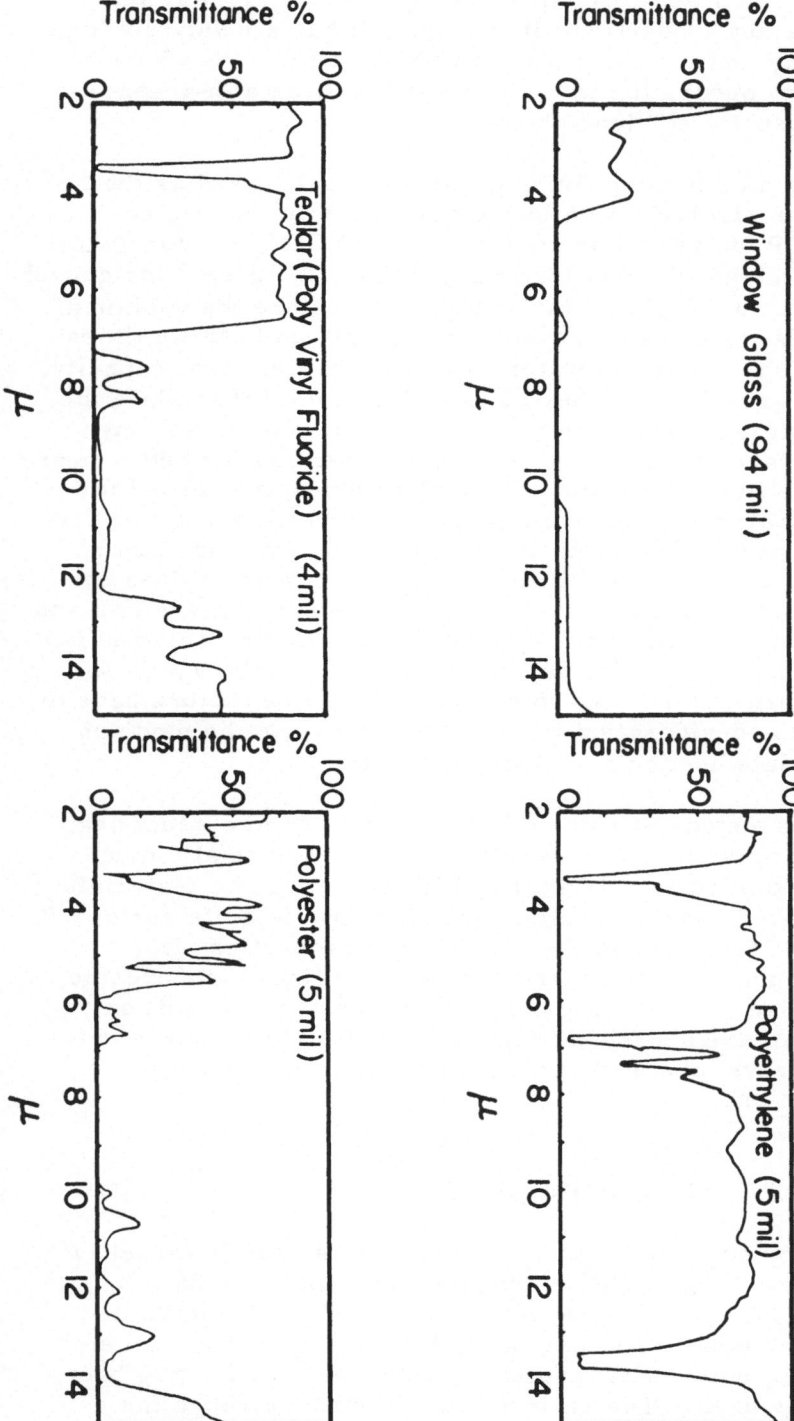

Fig. 7. Spectral IR transmittance for several plastic films and glass.

The heat storage medium is usually water, since it has some important properties: it is cheap, it has a relatively high heat capacity and is easily transportable. For areas where freezing may occur, the water is replaced by an anti-freeze material, usually ethylene glycol.

For air as a heating fluid, gravel is usually used as the heat storage. Gravel has a lower heat capacity than water (0.2 Cal/gC^o for gravel as compared to 1.0 Cal/g.C^o for water) but it has a higher density than water (about 2.2 g/cm^3 for gravel as compared to 1.0 g/cm^3 for water). Therefore the volume of gravel needed is quite reasonable and is only up to three times larger than the volume of water with an equivalent heat capacity. Moreover, gravel can be heated to high temperatures while the maximum storage temperature is limited for water. Among other advantages in using water, one can mention the better heat transfer coefficient than for air; smaller pipes are needed for the flow of water as compared to larger ducts for air and water has somewhat broader applications. However, air has other advantages, such as no limit in temperature range; it does not freeze in cold weather; there are no serious problems with leaks or corrosion as in the water system; hot air is directly used for space heating; and finally, air solar collectors are usually less expensive than water solar collectors. All these factors have to be taken into consideration when decisions have to be made on the appropriate heating and storage system.

Another method of storing heat would be to use latent heat rather than sensible heat, since the former is normally much larger than the latter. Several hydrates and organic materials were suggested but the problem with their use is quite serious[36]. The hydrates tend to lose their nucleation properties after a certain amount of cycles of freezing and melting. The organic materials also have similar characteristics and, in addition, some of them are dangerous and some, expensive. Therefore water and gravel are still the most important heat storage materials in use.

DOMESTIC WATER HEATING UNIT

The domestic solar water heating unit as used in Israel is seen in figure 1 and is shown schematically in figure 8a. It consists normally of two collectors of 1.5 m^2 each and an insulated water tank of approximately 120 litre capacity. An electrical heating element is inserted into the tank for heating back-up for cloudy days. The recirculation of water through the

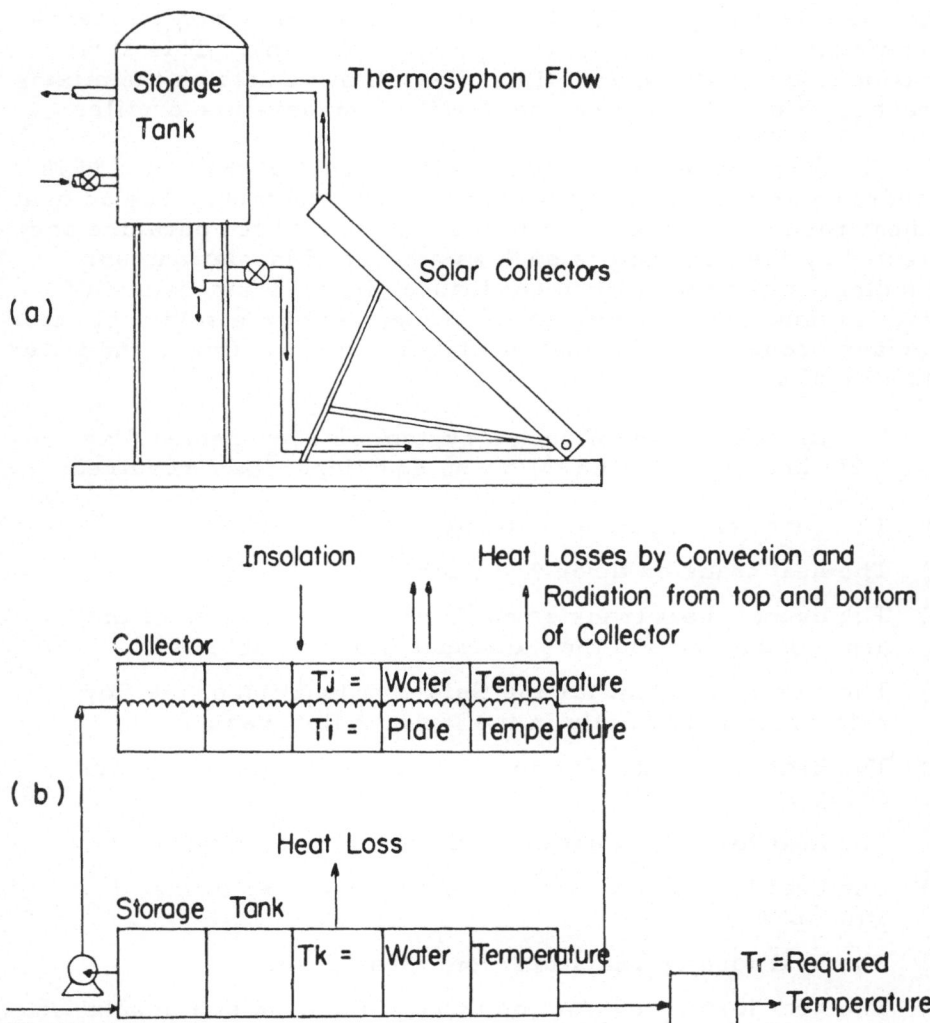

Fig. 8.   Schematic and diagrammatic representation of a
domestic solar water heater.

collector is obtained by the thermosyphon effect. The water
heated up in the collectors rises to the top and enters the upper
part of the tank. The colder water in the lower part of the tank
flows to the bottom of the collector, thus closing the flow cycle.
This flow system provides the water in the tank with a temperature profile from the top to the bottom of the tank. This strati-
fication is quite stable with time. Quantitative data on the mixing
effect is being collected now for various temperature profiles.

The fresh water is fed into the lower part of the tank and is
withdrawn as hot water in the upper part of the tank. The amount
of heat received by the water in the tank and its temperature are
affected by the parameters such as amount of insolation, surrounding temperature, wind conditions, initial temperature of
water in tank and the engineering properties of the solar collector
(surface properties, thermal insulation, thermal flux to the water
and type of cover).

For the simulation of such a system as represented diagrammatically in figure 8b, the following equations[20] are required:

(1)   The heat balance in the collector.

(2)   The heat transfer through the pipes.

(3)   The overall heat transfer coefficient as a function of the
      heat conduction and the individual film coefficients.

(4)   The film coefficient for the fluid as a function of the flow
      rate which is normally in the laminar flow region.

(5)   The heat loss by convection from collector plate to glass
      cover.

(6)   The heat loss by radiation from the collector plate.

(7)   The heat loss by convection from glass cover to the
      atmosphere.

(8)   The heat loss by radiation from glass cover.

(9)   The heat loss by conduction through the back side insulation.

(10) The heat loss by convection from the back side to the
      atmosphere.

(11) The heat loss by radiation from the back side of the
      collectors.

(12) The heat balance for the collector plate.

(13) The heat balance for the glass cover.

(14) The heat balance for the back side of the collector.

(15) The heat loss by conduction and convection from the tank storage.

(16) The heat loss by conduction and convection from the various pipes.

(17) The heat balance for the water in the storage tank.

(18) The solar insolation as a function of time.

(19) The surrounding temperature as a function of time.

(20) The wind velocity as a function of time.

If double glazing is used, then instead of equations 5 to 8 we have the following equations:

(21) The heat loss by convection from collector plate to first glass cover.

(22) The heat loss by radiation from collector plate.

(23) The heat loss by convection from first glass cover to the second glass cover.

(24) The heat loss by radiation from first glass cover.

(25) The heat loss by convection from second glass cover to atmosphere.

(26) The heat loss by radiation from second glass cover.

In addition, equation 13 should be applied for both glass covers.

If a high output temperature of water is needed and auxiliary heating is used for elevating the temperature, then an additional equation is used. This is the heat balance of the auxiliary heating unit. The possibility also exists of having a variable flow through the pipes or variable input to the storage tank as a function of the output temperature from the collector, thus increasing the efficiency of the heating system.

The above equations, including the differential equations, have to be solved simultaneously in order to find the dynamic and static behaviour of the system and the performance of the collectors. The calculations were done by using an IBM digital computer of the 370 series and using the digital simulation language called CSMP. A typical result of temperature profile along the tank as a function of the hour of the day is shown in figure 9.

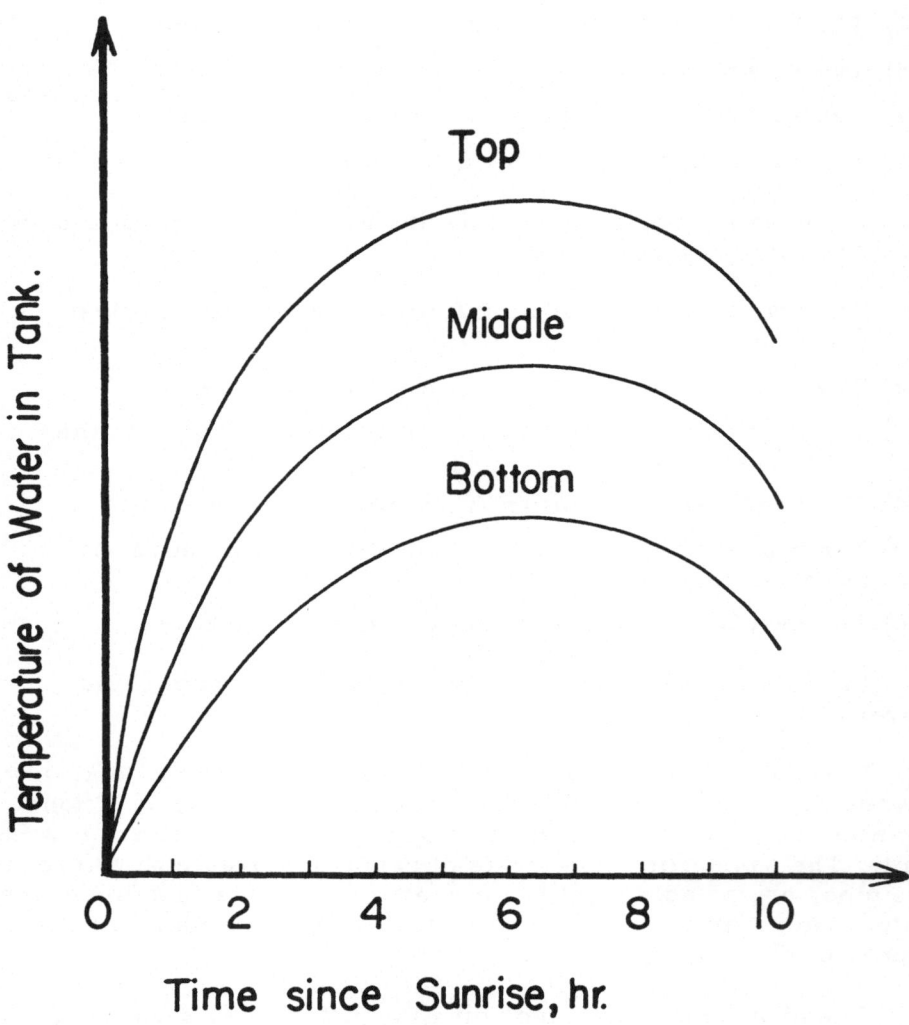

Fig. 9.   Water temperature profiles in the storage tank as a
function of location in tank and time.

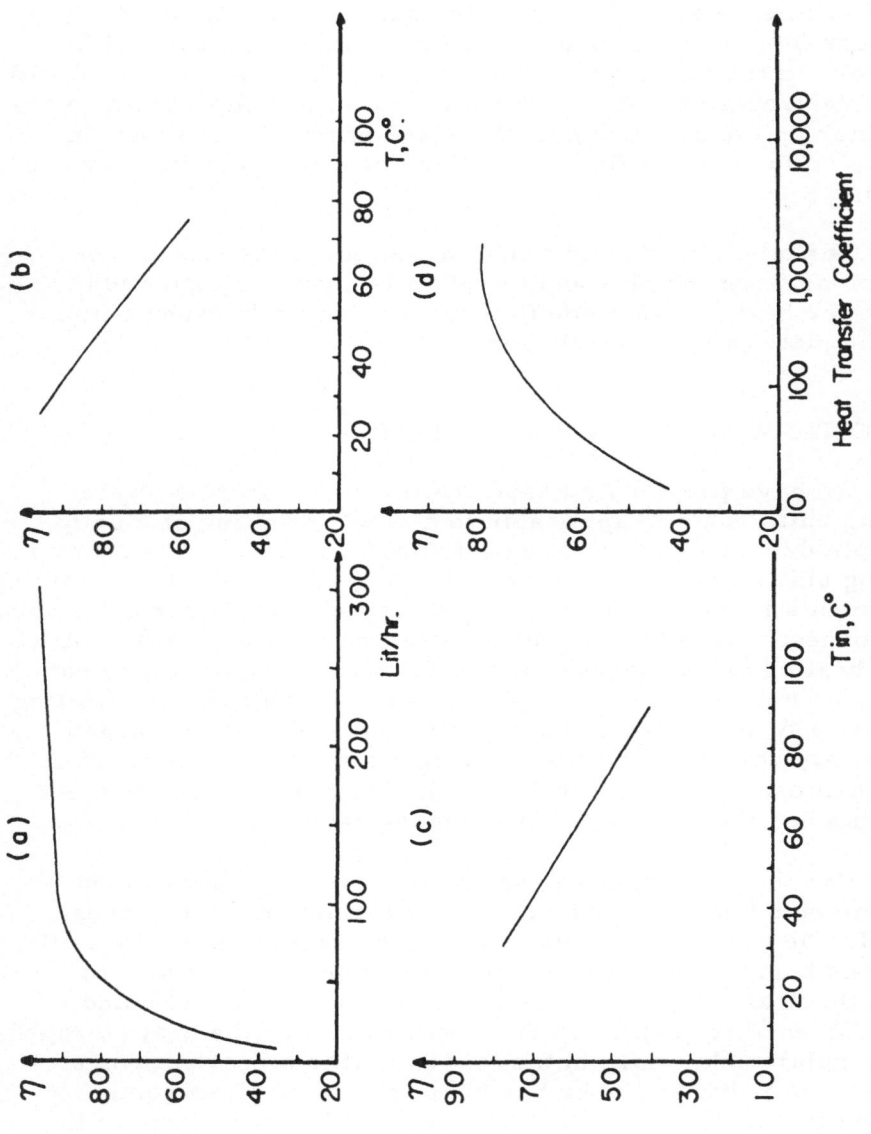

Fig. 10.   Effect of some parameters on the efficiency of a solar collector.

394

The effect of water recirculation rate on the efficiency of heat collection is shown in figure 10a for certain fixed operating conditions and weather conditions. After a certain recirculation rate, the efficiency does not increase any further. The behaviour of the efficiency as a function of the outlet temperature is seen in figure 10b. One can see that the efficiency increases with decrease in temperature. Higher temperatures are obtained with lower recirculation rates. The effect of inlet temperature to the collector on the outlet temperature and efficiency is shown in figure 10c.for certain fixed operating conditions and weather conditions.

For other fixed conditions, the effect of the type of connection between the pipe and the plate is shown in figure 10d where the heat transfer coefficient for the bond between the pipe and the plate is the variable parameter.

DOMESTIC WATER HEATING SYSTEMS

We have previously described the domestic solar water heating unit. This is applicable to a single dwelling unit or for multiple dwelling units where each individual family has its own heating unit. However, a central heating system for such building apartments is now being considered, in which the solar collectors are connected together and supply heat for all apartments. Auxiliary heating is considered for cloudy days. The arrangements should be made so that each family pays for the auxiliary heating proportional to its use and obviously it should be so arranged that maximum use of the solar energy is made. Some possible arrangements are shown in figure 11. Each one of the different systems has its advantages and disadvantages.

For the cases d and e, in figure 11 we have closed loop systems and thus no problem of scale formation. Anti-freeze can also be used as the fluid, making the system more versatile. In cases b and c, water softeners should be used in order to avoid the scale formation. Between cases b and c and d and e, the difference is small. In the first two cases, the heat storage is a regular tank, while in the last two, it is a heat exchanger where each individual user has his own pipe or pipes going through the shell of the heat exchanger. This arrangement is needed if a water flow meter for hot water is not available, if a closed loop system with anti-freeze is required through the collector, when the pressure vessel can be made cheaper for smaller pressures or when the collectors cannot withstand high pressures. Between b and c or between d and e, the difference is only in the

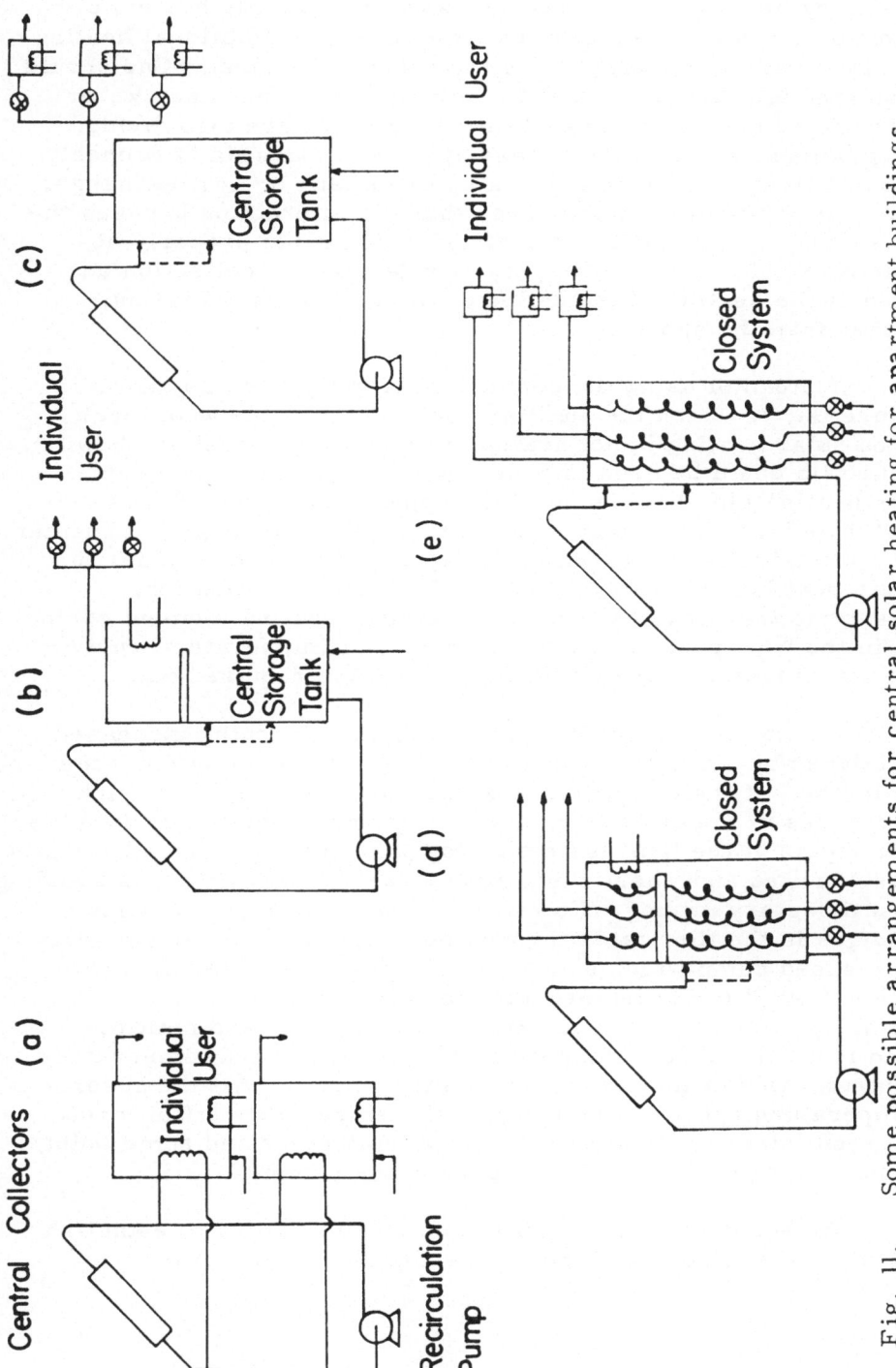

Fig. 11.   Some possible arrangements for central solar heating for apartment buildings.

auxiliary heating. In the second case, each family has his own additional heating, while in the first case, the additional heating is also common for all (as are the solar collectors). One should also mention that cases d and e actually do not need any water flow meter unless the solar heating expenses are also divided proportionally to the use of the hot water. Case lla is probably the simplest one, since it has a common tank or heat exchanger. But unlike the other cases here, there is interaction between the solar heating and auxiliary heating, which if not properly designed, might reduce efficiency of solar energy collection and even cause heating of the collector by the auxiliary heating, rather than the opposite.

Instead of using the commonly existing tanks as shown in figure lla, in which the auxiliary heating heats the whole tank when solar heating is not available, a storage vessel as shown in figure llb could be used in order to overcome these problems. The heating element, being in the upper region, would heat only that water which is ready to use and not the entire tank. Whether or not the baffle (as shown in the figure) is needed in order to avoid heat losses to the collector is under investigation. Assuming the baffle exists, the electrical heating element could be in the "on" position all the time and would operate whenever the water temperature would drop below the required one.

All cases, except for case a, permit obtaining increased efficiency by suitable control of the hot water coming from the collector. If the temperature of the water coming to the tank decreases in the afternoon, one can introduce this water into the lower part of the tank and not on top (where the previous hot water was supplied and where the temperature was higher), thus avoiding unsatisfactory mixing. However, all cases of figure ll allow changes in the flow rates through the collectors by using a variable speed pump which should also improve collector operation at least by allowing temperature to coincide with the need. Actually, a combination of variable speed pump and multiple feed locations should improve both efficiencies and temperature control. One of the control strategies could be that the absorber temperature operates the pump while temperature differences between inlet of collector or outlet of collectors and some point in the storage tank affect the speed of the pump.

An economical study should be made in order to establish the type of system which should be applied in each case.

AUXILIARY EQUIPMENT

Unlike the small domestic unit, these larger systems require more auxiliary equipment for better control, improved efficiencies and higher temperature. Thus variable speed pumps, thermostats, control units and calorimeters are desirable. The variable speed motor should enable changing the flow rate through the collector as a function of the insolation rate; the thermostat should be able to operate the pump or the inlet system to the tank as a function of the absorber and water temperature; and the calorimeter should enable measuring the use of calories by individual users. This is even more important in space heating where every user of heat would like to pay according to his demands. Such meters are now available and they include a flow meter, a temperature difference measurement and a multiplier of both in order to obtain calories.

CONNECTION OF COLLECTORS

For large installations where many solar collectors are needed, the collectors have to be connected so as to form one operating unit. There are several possibilities of connecting these so that they form cascades in parallel or in series. Garg[21] has studied this problem and found that a true parallel connection is the most efficient arrangement. This is to say that each collector is fed from the same common manifold and has its output in a common manifold.

OPTIMIZATION OF SOLAR COLLECTOR AREA

For central domestic heating and in any other case where hot water is needed all year round, solar heating can be economical if properly designed. This would involve the consideration of back-up heating in cold seasons or cloudy days, since if designed for the worst weather conditions, the collector area would be over-designed for the summer time (and also large heat storages would be needed). However, if designed for any other conditions, the area of collectors would be too small for the winter time and back-up heating must be included. Therefore, an optimum area of collectors should exist as shown in figure 12a. This picture is obtained if the fuel or electricity is expensive enough or the solar collectors are of reasonable cost. The savings are higher the more these two conditions are satisfied. We refer here only to actual savings in money. Obviously the saving is also important, as is the ecological advantage of using solar energy.

398

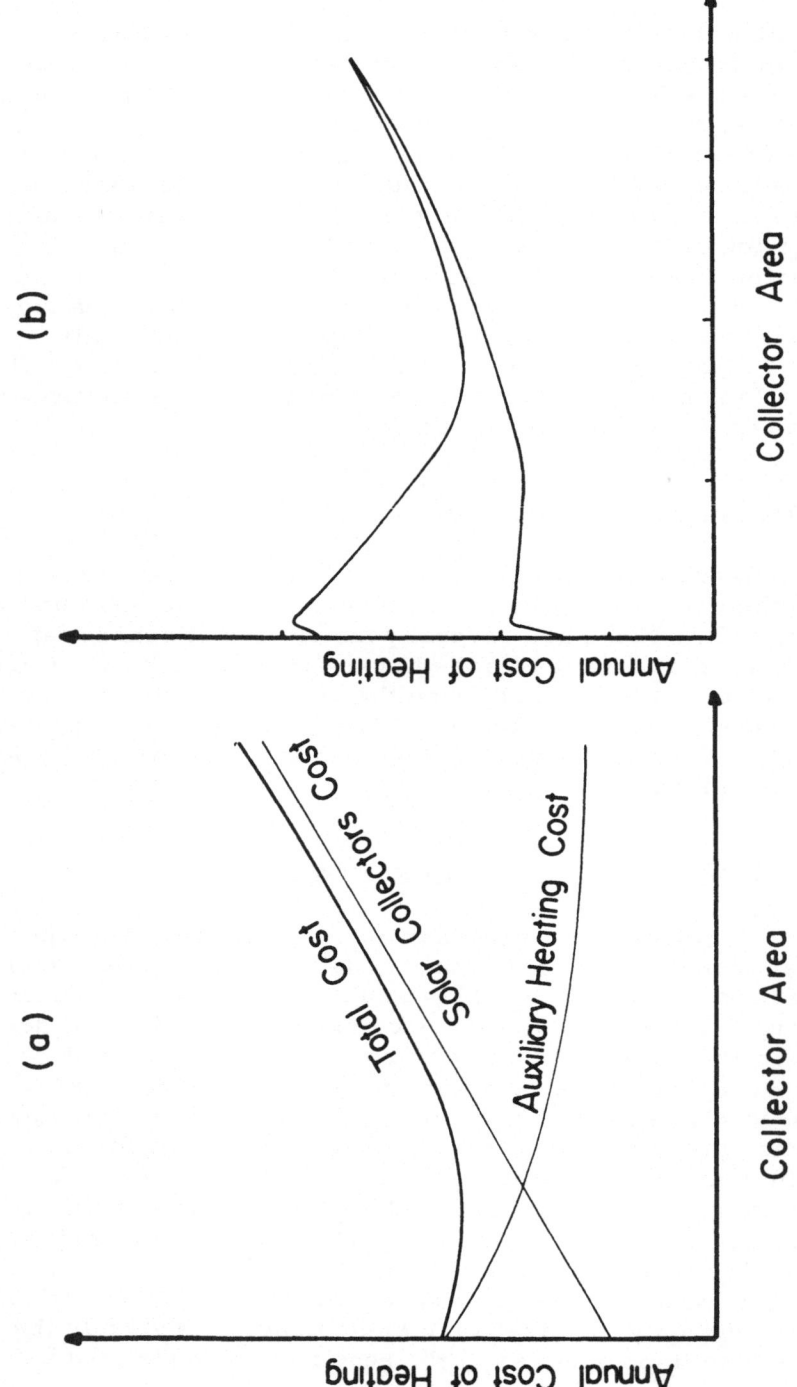

Fig. 12.   Effect of cost of collectors on annual cost of heating.

For the optimization study of such a combined system, we need the following data:

(1)   Amount of solar insolation and efficiency of collection.

(2)   Amount of water needed and inlet and outlet temperatures.

(3)   The cost of the heat supplied by the back-up systems.

(4)   The cost of the solar energy collection system including collectors.

The costs include investment maintenance and operational costs and are based on certain assumptions of amortization time and the interest rate on capital.

The calculations were made[22] for various surface areas of collectors, starting with no solar collectors (in which case the heating is all done by the back-up system), and ending with the area of collector which would be sufficient for heating water even on the coldest winter month without using any back-up heating.

The data on the monthly solar insolation was used as given in Table I. As the insolation rate is different in each month, the cost of heating is also different in each month of the year and the optimal cost was therefore calculated on an annual basis.

Two typical results of the cost calculations as a function of the collector area are shown in figure 12b. One can see that in one case there is an optimum condition and solar collectors are desirable, while in the second case, the cost is always higher when solar collectors are incorporated. The second case is obtained either when the back-up heating is very cheap or the solar collectors are very expensive. The increase in costs at the beginning when a small area of collectors is added is due to the fixed investment in installation needed and most of it is not a function of the collector area. As a general remark one could say that if the addition of solar collectors is economical at their present costs, the optimum collector area as a first approximation would be about 50% of the maximum collector area needed in the coldest months of the year with solar collectors as the sole heating system. However, this is only a rough estimate and for exact data, the complete optimization calculation is needed.

SPACE HEATING AND COOLING

The need for space heating is much greater than that for domestic water heating and it is approximately in the proportion of 5:1. Therefore, the use of solar space heating should be seriously considered and investigated.

Except for places where the energy is very cheap, domestic water heating by solar energy is generally economical. This is not the case with space heating for two reasons. First, space heating is required only for a part of the year, namely in the winter months and secondly, space heating is needed when solar radiation availability and intensities of radiation are low. Therefore, care must be taken when decisions are to be made on using solar energy for heating. In areas with long winter months and relatively long periods of sunshine, it may prove economical to use solar energy for heating. One might expect that the economics would improve significantly if the same solar systems would also be used in the summer time for air conditioning and cooling or for heating swimming pools in cooler seasons.

Since the present economics in using solar energy are not always favourable, due to the high investment cost, one should expect a reduction in costs if large scale production will start with the increased use of solar energy. The economics will also improve due to advanced technology and improved performance and efficiency of solar collector systems. Moreover, if solar systems will be incorporated in buildings during the construction period, a part of the cost of the solar collectors will be on the account of the construction cost itself. Since the use of solar energy saves oil as well as money, governments should encourage the use of it by lowering or eliminating any tax on solar units and in some cases even by giving loans for the installation of solar energy systems for heating, cooling and other uses as well.

The cooling system is obviously one which can use solar energy, since when cooling is needed, solar energy is available and at high insolation rates. Moreover, the collectors on the roof reduce the heat load of the house, thus reducing the cooling rate needed.

The use of air conditioning for making the solar system economical is justified only where air conditioning would be used anyway, otherwise it would be unreasonable to add air conditioning just to make the solar system economical on an over-all basis. Presently the amount of air conditioning is much less than heating. Another problem with solar air conditioning is the fact that with present technology and knowledge, one needs heating temperatures for an absorption cooling unit of around 95°, which is hard to achieve with present flat plate collectors at reasonable efficiencies. The systems commonly considered are H_2O-NH_3 and $LiCl$-H_2O. The $LiCl$-H_2O system, although it requires lower temperatures, tends to crystallize, which is of course a problem. Therefore, in order to promote the use of solar energy in space conditioning, one should strive to reach higher temperature outputs of the collectors and lower temperature needs by the absorption cooling systems.

So far the heating and cooling systems are in the experimental and the feasibility study stages and many demonstration houses have been built[23] for testing the feasibility and economics of heating and cooling by solar energy. A comprehensive overview on the heating and cooling of buildings by solar energy with special emphasis on the architectural points of view was recently written by Givoni[24].

COOLING AT NIGHT

Solar collectors in principle could also be used for cooling at night by radiation to the clear sky. The cool fluid obtained then cools gravel or water, which can be used the next day for cooling the home.

In principle, radiation to the sky should be considered to be to zero temperature, but in practice there is an effective temperature to which the collector is radiating. This is the temperature which when used in the Stefan Boltzman equation gives the heat loss as obtained in practice.

The reason for the low cooling effect is the fact that the radiation is absorbed by the atmosphere and then re-radiated back to the earth. The effective temperature in the Beer Sheva, Israel area is about 15°C below the atmosphere temperature. One square metre of collector can dissipate 5.4×10^6 joule of heat per night[25] from water at an initial temperature of 37° and final temperature of 17°C.

For an initial temperature of 29°C, the heat loss was $3.8 \cdot 10^6$ joule/m^2 night. This cooling can also be used for cooling ponds where algae are grown. This cooling method is also important where water is scarce and expensive and regular cooling towers cannot be used. The size of the area for cooling water with a cover is 3.5 times higher than the area which would be needed for cooling water without any cover and where evaporation would take place.

SOLAR-ASSISTED HEAT PUMPS FOR HEATING AND COOLING

A heat pump in principle takes heat from one area and delivers it to another area; thus one place is cooled while the other is heated. Therefore, if the cooled place is of interest (i. e. cooling is needed) the heat pump is a refrigerator, while if the heated place is of interest (i. e. heating is needed) the heat pump is a heater. Normally such heat pumps either take the heat from the surrounding area or transfer heat into the surroundings. In each case, work is added to the system and the principle of operation of a heat pump is the reverse of that of a heat engine. The H&T diagrams for a heat engine, a heat pump for refrigeration and a heat pump for heating are shown in figure 13.

Analysing the diagrams in figure 13, one can see that in order to obtain high values of COP for refrigeration, it is desirable that T_2 be as high as possible; for heating T_1 should be as high as possible and in both cases ΔT should be as small as possible.

This suggests that for the refrigeration cycle for a given requirement of T_2 in the cooler, T_1 (which is normally the surrounding temperature) should be as low as possible in order to keep ΔT small. For heating to a temperature of T_1, the surrounding temperature T_2 should be as high as possible for the same reason. In other words, the heat pump for cooling is very efficient on a cold day while a heat pump for heating is very efficient on a warm day. However, in practice, the opposite case prevails. Refrigeration is needed on hot days, while heating is needed on cold days and therefore, the COP of heat pumps are not always high, especially in extreme weather conditions. Therefore, river waters, which have less fluctuating temperatures with the season of the year and during both seasons than the air, were used[26] as the heat source in the haating mode and as the heat sink in the cooling mode. Solar-assisted heat pumps taking heat from the heat storage tank should be even more efficient for heating than river-assisted heat pumps and for refrigeration, the COP should improve if the same storage tank would be used for storing cold water during the night and serve as a heat sink during the day.

If the same operating system could be used for assisting the heat pump in both cycles (i. e. refrigeration in summer and

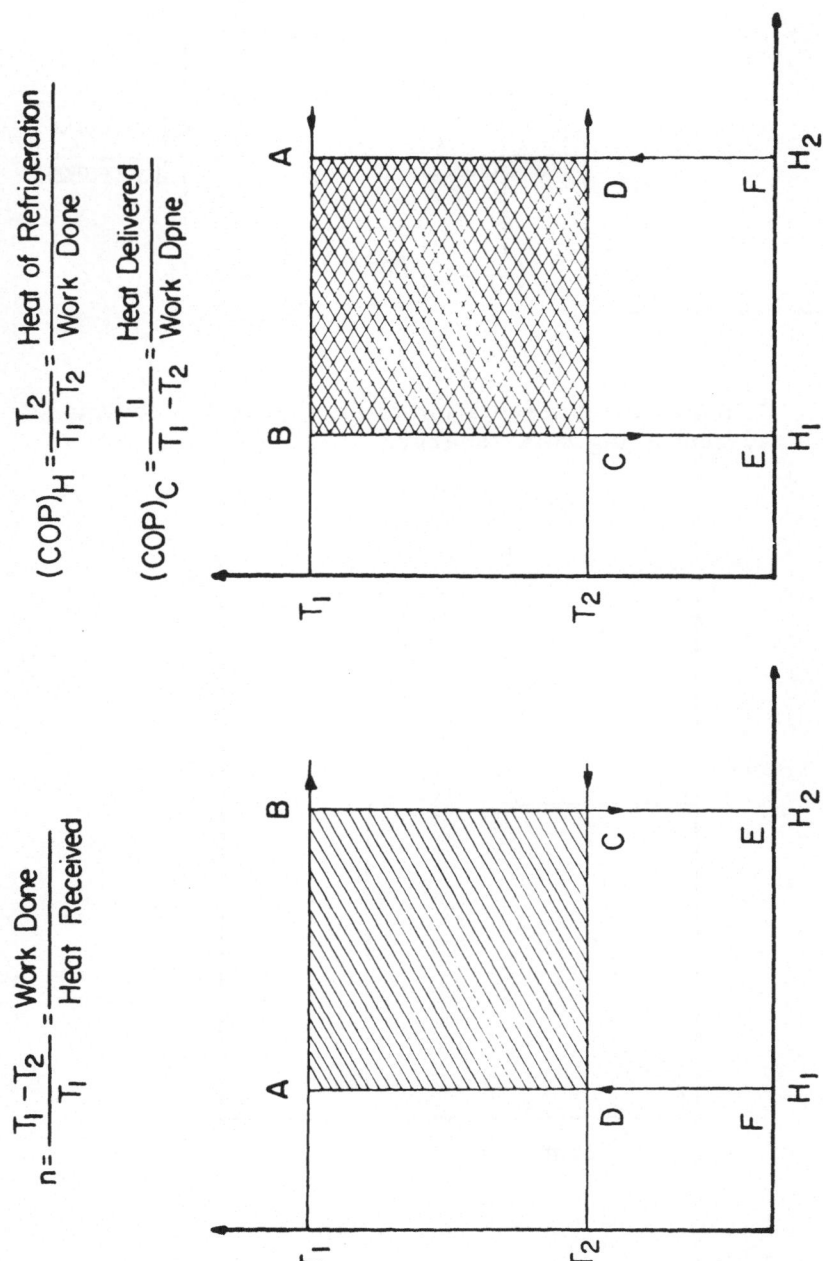

$$n = \frac{T_1 - T_2}{T_1} = \frac{\text{Work Done}}{\text{Heat Received}}$$

$$(COP)_H = \frac{T_2}{T_1 - T_2} = \frac{\text{Heat of Refrigeration}}{\text{Work Done}}$$

$$(COP)_C = \frac{T_1}{T_1 - T_2} = \frac{\text{Heat Delivered}}{\text{Work Dpne}}$$

Fig. 13. H & T diagrams for a heat engine and heat pumps.

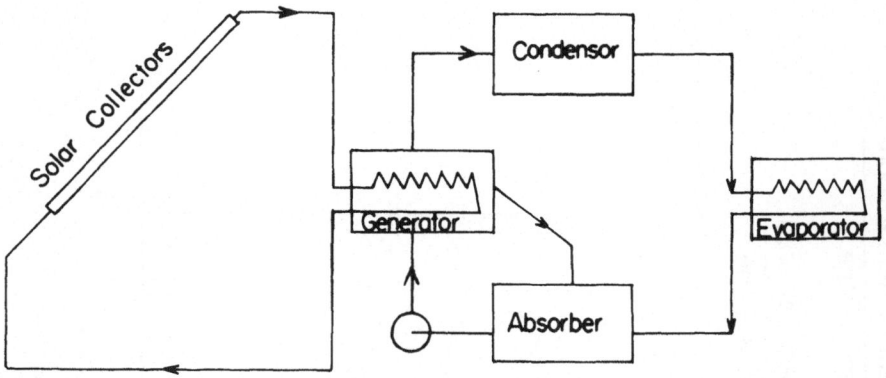

Fig. 14.   Schematic representation of an absorption cooling
system using solar energy.

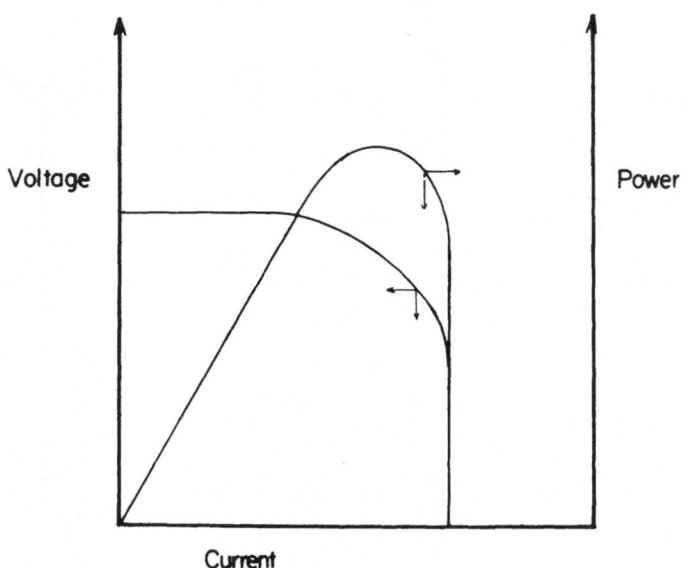

Fig. 15.   Voltage, current and power relationship for a
solar cell.

heating in winter) the whole system should prove economically justified. The solar collectors would raise the temperature of the storage tank in the winter, thus improving the COP for heating, while the same collectors could operate in the summer during the clear nights, for cooling the water in the storage tank, thus improving the COP for cooling in the day time. In places with low off-peak night power, one could even use the heat pump itself in reverse during the night time. The savings would be twofold. First, the amount of electrical energy needed would be less and secondly, the size of the heat pump would be smaller for a certain load, since the design is normally for the worst operation conditions which are much improved in the solar-assisted heat pumps.

The economics of such a system also depend strongly on the total annual heating and cooling demand in any particular region under consideration and in the ratio between heating and cooling demands. Presently operated heat pumps have supplementary resistance heating in severe weather conditions. COP as low as 1.5 can be expected when the temperature drops to -20° while at the temperature of 15°C the COP is around 3.0[27]. The COP includes, of course, the resistance heating and on a seasonable basis it is sometimes called SPF (Season Performance Factor) as defined by Dunning[28]. The above numbers show that if solar heating could be used to raise the temperature T_2 in the cooler part of the heat pump, the COP would also rise. In fact it would increase steadily as T_2 increases, as we have seen from the H&T diagram. Theoretically COP of 10 and more could be obtained. If heating and cooling are required at the same time, the advantage of using the heat pump is obvious.

Since solar collectors are more efficient as heating units than as cooling heat exchangers, space heating and cooling using only the heating system of solar collectors is more desirable. This can be achieved by using absorption units rather than compression units as considered above.

Moreover, the sun radiation intensity in the summer time is very high and that is exactly the time when air-conditioning is required. A solar-operated refrigeration unit was successfully operated by Farber[29] and the system is shown schematically on figure 14. However, as already mentioned, the flat plate collectors now commercially available do not supply temperatures of around 95°C at high efficiencies and thus the COP of an absorption unit operating at the lower temperatures received from flat plate collectors (e.g. 75°C) are quite low and make the system uneconomical. Therefore, an effort should be made to increase the efficiency of collectors by raising their output temperature, since this also immediately increases the COP. This effect has been expected, especially from the concentrating collectors.

HUMIDITY CONTROL

In arid zones, air conditioning can be obtained by desert coolers in which the dry and hot air is passed through water, thus humidifying the air and cooling it to the wet bulb temperature. This of course is not possible in humid areas, but solar energy could be used indirectly in such areas for cooling, as follows: The hot and humid air could be dried by some chemical absorber which would absorb the water and the dry air obtained can then be passed through desert cooler. The chemical, after absorbing the water, can be regenerated by hot air obtained from solar air heaters. Silica Gel as the absorbent chemical has probably most of the needed properties for such a drying material. The same collectors which are used during the day for drying the absorber could also be used during the cool months for cooling gravel or water to be used the following day for cooling the home.

The hot dry air could also be applied to an industrial use such as drying, especially in the food industry.

CONCENTRATING COLLECTORS

The amount of insolation received by a flat plate collector is related to its collection area. However, concentrating collectors are available[30] where insolation from areas larger than the collector itself are concentrated on the collector, thus obtaining higher temperatures and higher efficiencies. These collectors have reflecting surfaces so arranged that the solar rays are all reflected to a common centre or focus. The main drawbacks of such a system are the facts that it does not work on diffuse radiation and that in most cases either the collector or the reflecting surface have to change positions with regard to the sun during the day. The higher temperatures received by the concentrating collectors should enable better use of solar energy in the absorption cooling units and even power generation. However, the cost of such collectors is still too high to make them economical and many technical problems still exist.

TESTING SOLAR COLLECTORS

Although there is a great variety of solar collector units, yet there is no accepted standard testing for the collector or the storage tanks. Therefore, it is quite difficult sometimes to see which collector is better and under which conditions. Recently methods for testing solar collector and storage systems have been suggested by the USA National Bureau of Standards[31, 32].

SOLAR PONDS

For the production of power or large quantities of heat, solar ponds were proposed[33] instead of huge numbers of solar collectors. The solar pond principle of operation is as follows: At the bottom of the pond there is solid salt (for example $MgCl_2$ which dissolves into fresh water poured into the pond) then a concentration gradient is obtained with heavier concentrated solution being at the bottom of the pond and the lighter one at the top. By diffusion, the salt would reach the top of the pond and the concentration would become uniform, unless we withdraw the concentrated solution continuously from the bottom of the pond and add fresh water continuously to the top of the pond, thus keeping a constant concentration gradient. This concentration gradient, and consequently, the density gradient are the essential part of the solar pond. If solar insolation heats up the bottom of the pond, the hot solution will not rise to the top, since the effect of heating on the density gradient in the pond is less than the effect of concentration on the density gradient. Thus a non-convection condition is obtained, where the hot solution is at the bottom of the pond while the cold water is at the top.

There are several advantages in using solar ponds. First it is more economical than vast numbers of collectors. Second, relatively high temperatures can be obtained since the heat is trapped at the bottom of the pond. For the same reason, high efficiencies are obtained. Third, the solar pond is the collector and the heat storage at the same time, the storage being the water itself and also the ground on which the pond is built. However, the mixing-by-wave motion due to winds should be avoided and this is quite a problem. The hot brine can be used for power production by flash evaporation and the vapour for driving a turbine, or it can be used for other purposes, among which water desalination suggest itself immediately. This is especially true now when desalination units have been developed and operated in Israel with relatively low input temperatures of brine, even lower than the temperatures expected from the solar pond. The solar ponds project investigated in Israel was terminated in the mid-sixties, since at that time it proved uneconomical due to the cheap fossil fuel then available. Now with the increase in cost of the fossil fuel, this project has received renewed interest and an economical feasibility study has been started. At the same time, research is being done on the technical problems related to the mixing-by-wave motion.

Research on solar ponds has also been done in India and the USA and the latter has considered smaller ponds for house heating. Also being studied is the use of transparent plastic sheets for separating the layers of water of various temperatures rather than counting on the layers obtained by using salt solutions. The use of such sheets introduces some technical problems in addition to some additional radiation losses. However, it avoids the problems involved in using highly concentrated salt solutions. It also prevents accumulation of dirt, which is a very important factor. Even for the large solar ponds described previously, the use of cover sheets is now being considered, since the accumulation of dirt in a solar pond can be found anywhere in the pond and not on the top or the bottom only, all depending on the density of the trapped dirt. One can expect floating, sinking and suspended dirt as well as biological growth material. The optimum depth of such a solar pond is expected to be about one meter. The idea of using a part of the Dead Sea as a natural solar pond has also been suggested[34], although many problems have to be cleared up before such an idea can be realized.

SOLAR CELLS

Solar radiation can be converted directly to electricity by solar cells. These cells are extensively used in space ships but they are yet too expensive to be used on earth except for some exceptional cases like remote military installations. One kilowatt power installed of solar cells is about 2 orders of magnitude more expensive than one kilowatt installed at a conventional power plant. Moreover, a storage system such as accumulators is needed for accumulating electrical energy during sunshine periods for nights and cloudy periods and this makes the system even more expensive.

The most common solar cells are made of silicon, cadmium sulfide, selenium, and a few others, but the silicon will probably dominate the future market. The maximum voltage obtained from one silicon cell is 0.54-0.61 volt. and the maximum current is around 30-35 milliamps/cm^2. At maximum power the voltage and current are obviously less than their maximum values. A typical relationship between the voltage of a cell and its current and between the power and the current is shown in figure 15. The silicon cells react to radiation rays of 0.3-1.1 μ. The maximum current is 34 milliamps/cm^2. Efficiencies of silicon cells are 10% and more of the insolation.

By connecting solar cells in series and in parallel, one can get higher voltages and higher currents. Units as big as several hundred watts have already been obtained. A large scale installation is now in an experimental stage in Israel and the arrangement of the solar cells can be seen in figure 16. In order to increase the efficiency of solar cells, Glaser[35] proposed an idea of putting several square miles of solar cells into orbit and then by microwaves, transfer the electricity produced to a terrestial station. The amount of insolation in orbit is about 15 times more than that on earth. This figure is obtained from the following data. Solar radiation density is 5/4 times more in orbit than on earth due to the absorption by the atmosphere. The percentage of clear sky in orbit is about twice that on earth. Another factor of two can be obtained due to the angle of incidence. Finally a factor of three is obtained since insolation in orbit is 24 hours a day instead of about 8 hours a day on earth. The product of all these factors is 15 as mentioned above. Since such a project is a very expensive one and quite ambitious, the implementation of such a satellite power station is not expected soon. However, if the cost of the solar cells will go down significantly, this project may prove feasible,as will be some terrestial aspects.

SUMMARY

Our main emphasis is on the direct use of solar energy to heating. Although the wind, waterfalls and ocean waves are obtained as a result of the solar energy effect, yet we did not deal with these aspects. Neither did we deal with photochemistry or photosynthesis, which also imply the use of solar energy by its conversion to chemical products. The direct conversion of solar energy to electricity is of great interest and was therefore briefly explained, although so far it is not economical for normal use. The combination of solar cells and flat plate collectors, where the solar radiation is used in part for electricity and in part for heat, have great advantages and should be studied.

The major research and development efforts are being made in the most efficient way of harnessing solar energy, in con-

Fig. 16.   Large scale experimental installation of solar cells
in Israel.

verting it into useful energy, in developing the appropriate
storage system for times when the sun is not available and in
developing technical means and systems to overcome the tran-
sients in the supply of solar radiation.   More specifically, the
effort in the application of solar energy is made in the improvement of performance, and reduction in weight and in cost of the
presently available solar collectors which are normally of the
flat plate type, in further development of new medium temperature types of collectors in which new designs, new materials of
construction and new surface treatments are being tested, in
developing efficient high temperature collectors like the concentrating collectors for steam production and in the utilization of
solar ponds also for steam production and power generation.

In the direct conversion of solar energy to electricity, the
cost reduction of both the photo-electric and the thermo-electric
units is the main objective.   The photosynthesis conversion has
great potential.   The growing of forests or algae in ponds is
being suggested.

Great effort is being put into the development of systems
which could be used for space heating, heating swimming pools,

domestic water heating, refrigeration, air conditioning and industrial application. The use of a supplemental supply of energy to solar energy and the incorporation of heat pumps in solar energy systems are being considered as alternatives for large heat storage capacities in most of the system analysis and economic studies. The heat storage materials are being extensively studied so as to get high heat capacities at maximum storage capacity.

Auxiliary equipment and components like heat exchanges, measuring devices, valves, pumps, motors and control units are being developed for the application in solar energy systems.

More demonstration units and houses are being built and operated in order to prove the feasibility of using solar energy for large-scale heating and cooling purposes and for the economic evaluations of such projects.

Although the econogical advantages in utilizing solar energy are obvious, there are still ecological aspects which have to be considered.

Unusual ideas like power plants in the sky are being suggested.

Mathematical models and simulation programmes are being developed for numerical evaluation, design criteria studies, analysis of systems and for economic evaluations. Other calculations like the determination of optimum orientation of collectors during the day and the season of the year at various locations on earth are also being performed.

It is expected that these research and development efforts will bring solar energy to a high level of utilization in the foreseeable future.

REFERENCES

1. Zaren, A. M. , and Erway, D. D. , "Introduction to the Utilization of Solar Energy", McGraw Hill, New York, (1963).

2. Daniels, F. , "Direct Use of the Sun Energy", Yale University Press, New Haven, (1964).

3. Williams, J. R. , "Solar Energy - Technology and Applications", Ann Arbour Science, Ann Arbor, Michigan, (1974).

4. Yellot, J. I. , "Utilization of Sun and Sky Radiation for Heating and Cooling of Buildings", ASHARE - p. 31, (Dec. 1973).

5. Morse, R. N. , "Solar Power in the Australian Energy Scene", Nature, 246, 271, (1973).

6. Noguchi, T. , "Recent Developments in Solar Energy Research and Application in Japan", Solar Energy, 15, 179, (1973).

7. Manes, A. , Teiteliman, A. and Frueling, I. , "Solar Radiation and Radiation Balance at the Central Meteorological Institute", Beit Dagon, Israel Meteorological Service, Research Division, Beit Dagon, (1970).

8. Ginsburg, T. , Scheinder, B. and Woodman, T. , "Sonnenenergie - Herausforderung an die Technik", Neue Zuricher Forschung und Technik, II, No. 117, (Marz 1974).

9. Rittelman, P. R. , "Using Solar Energy in Residential Housing", C. S., p. 20, (July 1974).

10. Ward, D. S. and Lof, G. O. G. , "Design and Construction of a Residential Solar Heating and Cooling System", Solar Energy 17, 13, (1975).

11. Sears, F. W. , "Optics", Addison-Wesley Press Inc. , Cambridge, Mass. , (1969)

12. Tabor, H. , "Selective Radiation", Bull. Res. Council of Israel, 5A, 119, (1955).

13. Matza, I. , Director Miromit Ltd. , Israel, private communication.

14. Kudish, A. , Dostrovsky, I. , Epstein, H. , Cohen, H. and Wolf, D. , "Experiments with Flat Plate Collectors", Symposium on Solar Energy Utilization, Rehovot, Israel, (June 1975).

15. Bahadori, M. H. , "A Feasibility Study of Solar Heating in Iran", Solar Energy, 15, 3, (1973).

16. Hanson, K. J. , "The Radiative Effectiveness of Plastic Films for Greenhouses", Journal of Applied Meteorology, 2, 793, (1963).

17. Hottel, H. C. , Weortz, B. B. , "The Performance of Flat-plate Solar Heat Collectors", Transaction of the ASME - p. 91, February, (1942).

18. Edlin, F. E. and Willaver, D. E. , " Plastic Films for Solar
     Energy Applications ", United Nations Conference on New
     Sources of Energy, Rome, Italy, (August 1961).
19. Garg, H. P. , "Effect of Dirt on Transparent Covers in
     Flat-plate Solar Energy Collectors", Solar Energy, 15, 299,
     (1974).
20. Wolf, D. , Dostrovsky I. , Epstein. M. , Kudish, A. and
     Sharaga, J. "Simulation of Solar Collector and the Heating
     System", Symposium on Solar Energy Utilization, Rehovot,
     Israel, (June 1975).
21. Garg, H. P. , "Design and Performance of a large-size
     Solar Water Heater", Solar Energy, 14, 303, (1973).
22. Epstein, M. , Dostrovski, I. , Kulish, A. , Wolf, D. and
     Sharaga, J. , "Economical Aspects in the Design of Water
     Heating Systems which Incorporate Solar Energy and
     Conventional Fuel", Symposium on Solar Energy Utilization,
     Rehovot, Israel, (June 1975).
23. Shurcliff, W. A. , "Solar Heated Buildings and Brief Survey",
     Solar Energy Digest, San Diego, California, 6th ed.
     (October 1974).
24. Givoni, B. , "Heating and Cooling of Buildings by Natural
     Energy. An Overview", "Man, Climate and
     Architecture" Chapter 8, 2nd ed. , Elsevier, London,
     (in press).
25. Bar-Cohen, A. and Rambach, C. , "Heat Dissipation by
     Night Radiation to the Sky". Symposium on Solar Energy
     Utilization, Rehovot, Israel, (June 1975).
26. Davies, S. J. , "Heat Pumps and Thermal Compressors".
     Constable and Comp. Ltd. , London, (1950).
27. Weinstein, A. , "Heat Pumps are Good, Solar Assisted
     Heat Pumps are Better", Workshops on Solar Energy Heat
     Pumps, Pennsylvania State University, (June 1975).
28. Dunning, R. L. , "Relative Efficiencies of Gas Furnaces and
     Electric Heat Pumps". Nuclear Energy Digest, 4, (1973).
29. Furber, E. A. , "Design and Performance of a Compact
     Solar Refrigeration System", Eng. Progress at University
     of Fla. 24, 70, (1970).
30. Winston, R. , "Principles of Solar Concentrators of a Novel
     Design", Solar Energy, 16, 89, (1974).
31. Hill, J. E. and Kusuda, T. "Method of Testing for Rating
     Solar Collectors based on Thermal Performance ", Thermal
     Engineering System Section, NBS, Washington D. C. ,
     Interim Report, (December 1974).
32. Kelly, G. E. and Hill, J. E. , "Method of Testing for Rating
     Thermal Storage Devices Based on Thermal Performance".
     Thermal Engineering Systems Section, NBS, Washington D.C. ,
     Interim Report, (May 1975).

414

33. Tabor, H. and Matz, R., "Solar Pond Project", Solar Energy, 9, 177, (1965).
34. Assaf, G. "The Dead Sea - A Scheme for Solar Lake", Symposium on Solar Energy Utilization, Rehovot, Israel, (June 1975).
35. Glaser, P. E., "A New View of Solar Energy", Proc. IECEL paper No. 719002, (July 1971).
36. Telkes, M., "Solar Energy Storage", ASHRAE, p. 38, (S ptember 1974).

420

LIST OF CONTRIBUTORS

Prof. A. M. Angelini,
Ente Nazionale per l'Energia
  Elettrica,
ROMA, Italy.

Prof. G. Angelino,
Politecnico Di
Milano,
MILAN, Italy.

Prof. L. Borel,
Ecole Polytechnique Federale,
Inst. de Thermodynamique,
33, Av. de Cour,
CH-1007 LAUSANNE,
Switzerland.

Prof. E. Camatini,
Politecnico Di
Milano,
MILAN, Italy.

Prof. M. Duminil,
Ministere de l'Education
  Nationale,
Conservatoire National des
  Arts & Metiers,
Inst. Francais du Froid
  Industriel,
292, Rue Saint-Martin (IIIe),
PARIS, France.

Dr. K. F. Ebersbach,
Forschungsstelle für
  Energiewirtschaft,
Am Blütenanger 71,
8000 MUENCHEN 50,
Germany.

Dr. L. Forgó,
"HOTERV"
Inst. for Energetics (H-1027),
Postafiok 15,
BUDAPEST 11, Hungary.

Dr. B. Geeraert,
LABORELEC,
Section 8 - Electrothermie
  et Electrochimie,
1640 RHODE-ST.-GENESE,
Belgium.

Dr. Allan Haag,
Teknologie Licentiat,
Sorlabacksgatan 46 A,
216 MALMÖ, Sweden.

Dr. G. E. Kelly,
Mechanical Systems Section,
Bldg. Environment
  Division, IAT,
National Bureau of Standards,
Bldg. 226, Room B 126,
US Dept. of Commerce,
WASHINGTON, DC 20234,
USA.

Dr. T. Kester,
Scientific Affairs Division,
NATO,
1110 BRUSSELS, Belgium.

Dr. P. Kolbusz,
The Electricity Council -
  Research Centre,
CAPENHURST,
Chester CH1 6ES, UK.

Prof. D. Wolf,
Ben Gurion University,
Chemical Engineering Dept.,
PO Box 2053,
BEERSHEVA 84 120, Israel.
(most recent address)

INDEX

418